Lecture Notes in Electrical Engineering

Volume 1244

The book series *Lecture Notes in Electrical Engineering* (LNEE) publishes the latest developments in Electrical Engineering—quickly, informally and in high quality. While original research reported in proceedings and monographs has traditionally formed the core of LNEE, we also encourage authors to submit books devoted to supporting student education and professional training in the various fields and applications areas of electrical engineering. The series cover classical and emerging topics concerning:

- Communication Engineering, Information Theory and Networks
- Electronics Engineering and Microelectronics
- Signal, Image and Speech Processing
- Wireless and Mobile Communication
- Circuits and Systems
- Energy Systems, Power Electronics and Electrical Machines
- Electro-optical Engineering
- Instrumentation Engineering
- Avionics Engineering
- Control Systems
- Internet-of-Things and Cybersecurity
- Biomedical Devices, MEMS and NEMS

For general information about this book series, comments or suggestions, please contact leontina.dicecco@springer.com.

To submit a proposal or request further information, please contact the Publishing Editor in your country:

China

Jasmine Dou, Editor (jasmine.dou@springer.com)

India, Japan, Rest of Asia

Swati Meherishi, Editorial Director (Swati.Meherishi@springer.com)

Southeast Asia, Australia, New Zealand

Ramesh Nath Premnath, Editor (ramesh.premnath@springernature.com)

USA, Canada

Michael Luby, Senior Editor (michael.luby@springer.com)

All other Countries

Leontina Di Cecco, Senior Editor (leontina.dicecco@springer.com)

**** This series is indexed by EI Compendex and Scopus databases. ****

Mikhail V. Nesterenko · Victor A. Katrich · Sergey L. Berdnik · Oleksandr M. Dumin · Anatoliy I. Luchaninov · Dmitriy V. Gretskih

Vibrator and Slot Antenna Arrays for Modern Applications

Springer

Mikhail V. Nesterenko
V. N. Karazin Kharkiv National University
Kharkiv, Ukraine

Victor A. Katrich
V. N. Karazin Kharkiv National University
Kharkiv, Ukraine

Sergey L. Berdnik
V. N. Karazin Kharkiv National University
Kharkiv, Ukraine

Oleksandr M. Dumin
V. N. Karazin Kharkiv National University
Kharkiv, Ukraine

Anatoliy I. Luchaninov
Kharkiv National University of Radio Electronics
Kharkiv, Ukraine

Dmitriy V. Gretskih
Kharkiv National University of Radio Electronics
Kharkiv, Ukraine

ISSN 1876-1100 ISSN 1876-1119 (electronic)
Lecture Notes in Electrical Engineering
ISBN 978-3-031-68345-9 ISBN 978-3-031-68346-6 (eBook)
https://doi.org/10.1007/978-3-031-68346-6

This Springer imprint is published by the registered company Springer Nature Switzerland AG
The registered company address is: Gewerbestrasse 11, 6330 Cham, Switzerland

Someone told me that each equation I included in the book would halve the sales.

Stephen Hawking

In memory of our friends and colleagues,
Yuriy M. Penkin and Andrey V. Gomozov

Preface

At present, linear vibrator and slot radiators, which have already become classical, are widely used in multifunctional radio engineering and electronic complexes based on such structures. They may be used as independent transmitting and receiving structures, elements of antenna systems, and devices of antenna-feeder paths. Massive applications of these multifunctional structures including multi-element antenna arrays (AA) have become an objective prerequisite for development of theoretical methods for analyzing electrodynamic characteristics of such systems. A special place is occupied by vibrator, slot, and vibrator-slot AA allowing us to significantly expand possibilities of forming electromagnetic fields with predefined spatial-frequency, spatial-polarization and spatial-energy characteristics. A multitude of literary sources (including publications by the book authors) carried out by researchers over past decades have allowed to create the modern theory of single vibrator and slot radiators as well as their combined structures. This theory combines both the fundamental asymptotic methods for calculating characteristics of single elements, numerical-analytical approaches developing them as well as direct numerical methods of electrodynamic analysis. However, it is not possible to consider the electrodynamics of linear vibrator and slot radiators, and multi-element systems with these elements to be completely developed theory. First of all, this is related to the development of modern technologies of receiving and transmitting antennas, devices for antenna-feeder paths, wireless power transmission systems, and ultra-wideband antenna structures. Also important are issues associated with structural components of devices ensuring minimization of their weight and size parameters, electromagnetic compatibility of radio electronic devices, using composite and metamaterials, and formatting the required spatial distributions of electromagnetic fields in various material media, including dissipative media.

Multiparameter optimization of complex electrodynamic problems requires effective computing and software resources for mathematical modeling to develop multi-element antenna devices. Therefore, despite the rapid growth of the computational potential of modern computer technology, there exists a need to develop new, more efficient methods for electrodynamic analysis of antenna-feeder systems created on the basis of linear vibrator and slot structures with arbitrary geometric and electrophysical parameters which satisfy modern requirements. As is known, the effectiveness of mathematical modeling is determined by rigorous formulations and solutions of boundary value problems. The effectiveness depends on the degree of analytical developmental work and time required to execute computational algorithms under condition that the minimum possible volume of operative memory is used. That is, the weightier the analytical method component, the higher its efficiency. In this regard, *the development of precise numerical-analytical methods of electrodynamic analysis, expanding the possibilities of physically correct mathematical modeling to new classes of boundary value problems, has been and remains a fundamentally important problem for researchers in the field of theoretical and applied radiophysics.*

Known numerical-analytical methods for analyzing single and multi-element structures of linear radiators and antenna-feeder devices developed on their basis can be conditionally divided into two types. The first type includes methods that use analytical solutions allowing to find electric and magnetic currents in individual structural elements. The second type uses varieties of the method of moments allowing to reduce to analytical forms expressions for matrix elements (or their components) in systems of linear algebraic equations (SLAE). Of course, the methods of the first type are more preferable both from the point of view of their efficiency and physical clarity. However, their implementation is limited by the complexity of asymptotic solutions of boundary value problems which concerns finding currents in individual radiators of multi-element systems.

The solutions of problems presented in the monograph aimed at determining the electrodynamic characteristics of impedance vibrator, slotted, and multi-element radiators with arbitrary geometric and electrophysical parameters were obtained within the framework of unified methodological approach to constructing asymptotic solutions of integral equations for currents. This approach has allowed us to study a number of new vibrators, slotted, and combined vibrator-slotted structures and their multi-element systems. The research results reveal the possibilities of using such structures as basic elements for creating modern antenna-waveguide devices operating in the range from meter to millimeter wavelengths which have new technical characteristics and functionalities.

The monograph consists of 9 chapters and 4 appendices. For the reader's convenience, references are presented in each chapter.

Chapter 1 is introductory in nature. Basic equations of macroscopic electrodynamics, exact and approximate boundary conditions, and methods for constructing Green's functions are presented. The information offered in this chapter allows the reader to perceive the information in further sections of the book without searching for special literature.

In Chap. 2, a problem of electromagnetic waves excitation (scattering or radiation) by material bodies of finite dimensions in the presence of coupling holes between two electrodynamic volumes are formulated in most general form. Initial systems of integral equations relative electric surface currents on vibrators and equivalent magnetic currents in coupling holes were derived based on impedance boundary conditions on vibrator surfaces and continuity conditions of magnetic field tangential components at coupling holes. It was shown that the problem solution should be based on integral equations for Hertz vector potentials with Green's function kernels. A physically correct transition from the derived two-dimensional to one-dimensional integral equations for currents is made for following radiators: thin linear impedance vibrators with irregular geometric and electrophysical parameters, narrow slots arbitrary located relative to boundaries of corresponding electrodynamic volumes, and combined vibrator-slot structures. The fundamentals of the method of moments are presented, and the advantages and disadvantages of various approximate analytical methods for solving the integral equations for currents are analyzed. Application of the generalized method of induced electro-magnetic-motive force (EMMF) for studying multi-element vibrator-slot structures is also considered. New theoretical aspects concerning thin impedance vibrators and narrow slots proposed by the authors are considered.

The problems solutions presented in Chap. 3 can be used for multi-parameter optimization of electrodynamic characteristics of radiating multi-element structures based on vibrators with constant and variable distributed surface impedance. A distinctive peculiarity of the method proposed by the authors can be characterized by using approximating functions obtained as solution of the integral equation for the current by the asymptotic averaging method. The correctness of the method is confirmed by the comparative analysis of results obtained by this approach and the method of moments. It should be noted that the proposed concept of the generalized method of induced electro-motive forces (EMF) preserves advantages of numerical-analytical methods as compared with direct numerical methods. It can be used to model vibrators with variable impedance and their systems. Thus, the method allows us to widen the boundaries of numerical-analytical approach to solution practically significant problems related to application of impedance vibrators.

Chapter 4 is devoted to studying characteristics of slotted waveguide antenna arrays, namely calculation results and experimental studies concerning characteristics of electromagnetic fields radiated by multi-element slotted arrays; analysis of radiated fields formation and search for required amplitude-phase distributions on the AA aperture to obtain their frequency-energy characteristics are also offered. In a number of important practical applications, such as, for example, issues related to electromagnetic compatibility of radio electronic systems and medical diagnostics and hyperthermia, distributions of electromagnetic fields in the near and intermediate zones should be known. This chapter is devoted to this problem solution also.

Combined vibrator-slot antenna arrays are considered in Chap. 5. The dual-frequency antennas based on vibrator-slot structures with similar radiation pattern (RP) in main polarization planes are suggested. Such antennas can be realized by diode switching slot and vibrator elements. The traditional combined radiators with passive vibrators are proposed to use at the main frequency. The system of active vibrators operates at an alternative frequency. The formation of radiation fields by the combined structures at the main and alternative frequencies is analyzed. An electrodynamically rigorous mathematical model of combined vibrator-slot structures consisting of narrow radiating slots cut in an end wall of rectangular waveguide and several thin impedance vibrators placed over the infinite screen is presented. Numerical results concerning internal and external electrodynamic characteristics of antennas with optimized structural parameters have confirmed the possibility of constructing the Yagi-Uda combined radiating structures in the microwave and extremely high frequency (EHF) bands.

Chapter 6 is devoted to solving problems associated with impedance synthesis of vibrator antenna arrays. A new method of impedance synthesis of radiation fields for antenna arrays based on a single methodological conception is presented. The approach turns out to be common for radiators with impedance coatings of different configurations and/or different distributions of lumped impedances. The synthesis problem of radiator impedances for the spatial scanning of the RP was solved for the linear vibrator array. The analytical solution of the problem was obtained for the equidistant array of symmetric vibrators excited by equal currents. The generalized methodology of the analytical solution of the impedance synthesis problem is expanded on the two-dimensional planar vibrator antenna arrays.

Aspects related to an impedance synthesis of slot antenna arrays are considered in Chap. 7. The impedance synthesis problem of 2D antenna arrays consisting of narrow linear slots with magneto-dielectric filling is solved. The slots are cut in a perfectly conducting screen whose geometric centers coincide with nodes of a rectangular grid with double periodicity. The slots are excited by voltage δ-generators in the center of slot apertures. The slot cavities are represented by rectangular resonators with magneto-dielectric filling controlling induced magnetic impedance of radiators by varying the material parameters of the slots filling. Also, an impedance synthesis problem of 2D antenna arrays consisting of slotted spherical radiators whose geometric centers are located at nodes of a flat rectangular grid with double periodicity has been presented.

In Chap. 8, generalized scheme of infinite rectennas array is considered and specific features of matrices characterizing its linear subcircuit are analyzed. Equations of state for infinite rectennas array are obtained, and it is proved that calculation of its nonlinear mode is reduced to calculation at fundamental frequency and frequencies of higher harmonics of the parameters of one "equivalent" receiver-rectifier element (RRE). Based on this proof, a technique for analyzing large-aperture rectennas with an equal-amplitude excitable aperture has been developed. Numerical studies of dependences of characteristics of RRE as an element of infinite rectennas array on parameters of the array and degree of nonlinearity of characteristics of rectifier diodes were performed. Technique for analysis of large-aperture rectennas with their nonuniform excitation has been proposed. Possibility of practical application of the proposed methodology is confirmed by comparing results of calculations by it with experimental data.

Antenna arrays consisted of ultrawideband combined vibrator-slot radiators are examined in Chap. 9. The single radiating element of the arrays is the Clavin radiator that contains a ultrawideband slot and two ultrawideband dipoles. The frequency range of the slot is widened by its special form. Metal cones placed on a metal screen near the slot have ultrawideband properties as well. The strong interaction between dipoles and slot additionally improves frequency characteristics of the combined radiator. Slot is excited by the δ-generator with ultrashort impulse of voltage. Cones being short circuited to the screen are the passive elements. Using these elements, time and spatial characteristics of different antenna array radiation are considered.

Expressions for the electric and magnetic Green's functions of vector potentials for various electrodynamic volumes considered in the monograph can be found in Appendix A. Mathematical aspects of the asymptotic averaging method are presented in Appendix B. Formulas for the surface impedances of electrically thin vibrators are represented in Appendix C. The relationships between dimensions of physical quantities in the CGS and SI unit systems are given in Appendix D.

The monograph is written in an academic style. It can be useful for teachers and students of technical universities. The results of numerical modeling presented in the book and asymptotic formulas obtained can be suitable to engineers engaged in developing antenna and waveguide multielement systems.

The authors would like to express sincere gratitude to Anatoliy M. Naboka for editing the English text.

Kharkiv, Ukraine

Mikhail V. Nesterenko
Victor A. Katrich
Sergey L. Berdnik
Oleksandr M. Dumin
Anatoliy I. Luchaninov
Dmitriy V. Gretskih

Contents

Chapter 1
Excitation of Electromagnetic Fields in Electrodynamic Volumes with Coordinate Boundaries

This Chapter briefly presents basic equations, boundary conditions and tensor Green's functions for studying electromagnetic fields, which will be used throughout the book. This chapter with comments on some aspects of electrodynamics theory, will allow a reader to read the book without searching for additional references.

1.1 Helmholtz Equations in Applied Electrodynamics

We will consider only harmonic oscillations, bearing in mind that non-harmonic oscillations can be decomposed into Fourier series or integrals. We also assume that all relevant quantities depend on time as $e^{i\omega t}$, where ω is the circular frequency of the monochromatic process. Then the relationships between vectors of electric and magnetic fields strength, $\vec{E}(t)$ and $\vec{H}(t)$, and their complex amplitudes, $\vec{E}$ and $\vec{H}$, can be written as

$$\vec{E}(t) = \mathrm{Re}\left\{\vec{E}e^{i\omega t}\right\}, \quad \vec{H}(t) = \mathrm{Re}\left\{\vec{H}e^{i\omega t}\right\}, \tag{1.1}$$

where "Re" is the real part of the complex vector standing in the curly bracket, $\vec{E}(t)$ and $\vec{H}(t)$ are the intensity vectors of the electric and magnetic fields, $\vec{E}$ and $\vec{H}$ are the corresponding complex vector amplitudes. The complex amplitude of all time-dependent physical quantities in electrodynamics equations are defined in a similar way.

Let us introduce, along with real extraneous electric current $\vec{j}^E$, virtual extraneous magnetic current with densities $\vec{j}^M$, respectively Then for a stationary, homogeneous, and isotropic medium, the Maxwell equations, symmetric with respect to electric and magnetic quantities, can be written as

M. V. Nesterenko et al., *Vibrator and Slot Antenna Arrays for Modern Applications*, Lecture Notes in Electrical Engineering 1244,
https://doi.org/10.1007/978-3-031-68346-6_1

$$\mathrm{rot}\vec{E} = -ik\mu_1\vec{H} - \frac{4\pi}{c}\vec{j}^M, \quad \mathrm{rot}\vec{H} = ik\varepsilon_1\vec{E} + \frac{4\pi}{c}\vec{j}^E, \tag{1.2}$$

where ε_1 and μ_1 are the complex permittivity and permeability of the medium in which the electromagnetic fields are excited, $k = \omega/c$ is the wave number, and $c \approx 2.998 \cdot 10^{10}$ cm/s is the speed of light in vacuum.

Throughout the book, the absolute Gaussian system of units will be used as the most convenient in theoretical physics as compared with the SI unit system. The Maxwell Eq. (1.2) in the SI system include four vectors: $\vec{E}$, $\vec{D}$, $\vec{H}$ and $\vec{B}$ defining the field at each point in space have different dimensions. Here $\vec{D}$ and $\vec{B}$ are the vector of electric and magnetic inductions. In the vacuum, the four vectors are connected in pairs by the relations: $\vec{D} = \varepsilon_0\vec{E}$ and $\vec{B} = \mu_0\vec{H}$, where the coefficients ε_0 and μ_0 play the role of the permittivity and permeability characterizing the electromagnetic properties of vacuum. These coefficients, devoid of real physical meaning, are included in all relationships for the fields, which is difficult to reconcile. However, if necessary, a mutual transition the CGS and SI systems is always possible (see Appendix D).

The charges are usually not included in the in Eq. (1.2), since, by virtue of the continuity equation the density of electric ρ^E and magnetic ρ^M charges are uniquely determined by currents as.

$$\mathrm{div}\vec{j}^E + i\omega\rho^E = 0, \quad \mathrm{div}\vec{j}^M + i\omega\rho^M = 0. \tag{1.3}$$

Applying to formulas (1.3) the divergence operations and the identity divrot $\equiv 0$ to the Eq. (1.2), we obtain.

$$\mathrm{div}\left(\varepsilon_1\vec{E}\right) = 4\pi\rho^E, \quad \mathrm{div}\left(\mu_1\vec{H}\right) = 4\pi\rho^M. \tag{1.4}$$

Thus, the charges and corresponding field equations may not be considered. Thus, both the charges themselves, and corresponding field equations, have no independent meaning and, therefore, are usually not considered. For the same reason, the wave equations for scalar potentials related to the vector potentials by the Lorentz or gradient invariance conditions will not be considered.

As can be seen, the Eq. (1.2) are coupled with respect to complex vector amplitudes $\vec{E}$ and $\vec{H}$. Therefore, it seems convenient to convert the equations so that they include only one vector

$$\mathrm{rotrot}\,\vec{E} = k^2\varepsilon_1\mu_1\vec{E} - \frac{4\pi}{c}\left(ik\mu_1\vec{j}^E + rot\,\vec{j}^M\right),$$
$$\mathrm{rotrot}\,\vec{H} = k^2\varepsilon_1\mu_1\vec{H} - \frac{4\pi}{c}\left(ik\varepsilon_1\vec{j}^M - rot\,\vec{j}^E\right), \tag{1.5}$$

The Equation (1.5) can be easily transformed to inhomogeneous Helmholtz vector equations relative only $\vec{E}$ or $\vec{H}$ in the form

$$\Delta \vec{E} + k^2 \varepsilon_1 \mu_1 \vec{E} = \frac{4\pi}{c}\left(ik\mu_1 \vec{j}^E - \frac{1}{ik\varepsilon_1}\text{graddiv}\vec{j}^E + rot\vec{j}^M \right),$$
$$\Delta \vec{H} + k^2 \varepsilon_1 \mu_1 \vec{H} = \frac{4\pi}{c}\left(ik\varepsilon_1 \vec{j}^M - \frac{1}{ik\mu_1}\text{graddiv}\vec{j}^M - rot\vec{j}^E \right), \quad (1.6)$$

where $\Delta \equiv$ grad div $-$ rot rot is the Laplaceoperator.

The Equation. (1.6) in many cases can be inconvenient in view of their right-hand side complexities. Therefore, it is useful to introduce electrodynamic vector potentials of the electric $\vec{A}^E$ and magnetic $\vec{A}^M$ types, or the Hertz polarization vector potentials, $\vec{\Pi}^E$ and $\vec{\Pi}^M$ which are interconnected, under the accepted assumptions, by the relations

$$\vec{A}^E = ik\varepsilon_1\mu_1 \vec{\Pi}^E, \quad \vec{A}^M = ik\varepsilon_1\mu_1 \vec{\Pi}^M. \quad (1.7)$$

The formulas (1.7) emphasize the equivalence of both types of vector potentials. Therefore, the choice of one or another presentation, in essence, is a matter of habit or tradition. Further, we will use the equations for the Hertz vector potentials allowing to express the vectors $\vec{E}$ and $\vec{H}$ by the following relation

$$\vec{E} = \text{grad div}\vec{\Pi}^E + k^2\varepsilon_1\mu_1\vec{\Pi}^E - ik\mu_1 rot\vec{\Pi}^M,$$
$$\vec{H} = \text{grad div}\vec{\Pi}^M + k^2\varepsilon_1\mu_1\vec{\Pi}^M + ik\varepsilon_1 rot\vec{\Pi}^E. \quad (1.8)$$

Substituting (1.8) into Eq. (1.6), we obtain the following inhomogeneous Helmholtz equations for Hertz vector potentials:

$$\Delta \vec{\Pi}^E + k^2\varepsilon_1\mu_1\vec{\Pi}^E = -\frac{4\pi}{i\omega\varepsilon_1}\vec{j}^E,$$
$$\Delta \vec{\Pi}^M + k^2\varepsilon_1\mu_1\vec{\Pi}^M = -\frac{4\pi}{i\omega\mu_1}\vec{j}^M. \quad (1.9)$$

As can be seen, the right-hand part of Eq. (1.9) includes the densities of the excitation currents, therefore these equations are more suitable for solving electrodynamic problems rather than Eq. (1.6).

Further we will use the following notations: the vector $\vec{F}^{E(M)}$ denotes the vector either $\vec{F}^E$ or $\vec{F}^M$, and $\varepsilon_1(\mu_1)$ denotes either ε_1 or μ_1. Then, for example, Eq. (1.9) can written as

$$\Delta \vec{\Pi}^{E(M)} + k^2\varepsilon_1\mu_1\vec{\Pi}^{E(M)} = -\frac{4\pi}{i\omega\varepsilon_1(\mu_1)}\vec{j}^{E(M)}. \quad (1.10)$$

The above equations are valid in any orthogonal coordinate system, allowing application of differential operators.

1.2 Exact Boundary Conditions for Electromagnetic Fields

The media parameters ε_1 and μ_1 involved in the equations of the Sect. 1.1 as constant, can vary continuously or discreetly in space. For example, the parameters can have discontinuities at an interface between two media. The vectors of the electromagnetic fields can also have discontinuities at the interface, therefore, Eq. (1.2) must be supplemented with boundary conditions. As is known from classical electrodynamics, the boundary conditions at a smooth interface between two media have the following form

$$\mu_1\left(\vec{H}_1,\vec{n}\right)-\mu_2\left(\vec{H}_2,\vec{n}\right)=4\pi\rho_S^M, \tag{1.11}$$

$$\varepsilon_1\left(\vec{E}_1,\vec{n}\right)-\varepsilon_2\left(\vec{E}_2,\vec{n}\right)=4\pi\rho_S^E, \tag{1.12}$$

$$\left[\vec{n},\vec{E}_1\right]-\left[\vec{n},\vec{E}_2\right]=\frac{4\pi}{c}\vec{j}_S^M, \tag{1.13}$$

$$\left[\vec{n},\vec{H}_1\right]-\left[\vec{n},\vec{H}_2\right]=-\frac{4\pi}{c}\vec{j}_S^E, \tag{1.14}$$

where indices 1 and 2 correspond to the first and second media, $\rho_S^{E(M)}$ are the surface charge densities, $\vec{j}_S^{E(M)}$ are the surface current densities, and $\vec{n}$ is the vector of unit normal to the interface which is directed to the second medium. Equations (1.11 and 1.12) define variation of the normal components of the electromagnetic field at the interface between the two media. Note that conditions (1.13) and (1.14) follow directly from Maxwell Eq. (1.2).

If the extraneous electric currents on the interface surface S are absent, the condition (1.14) simplifies to

$$\left[\vec{n},\vec{H}_1\right]=\left[\vec{n},\vec{H}_2\right], \tag{1.15}$$

which defines the continuity of the tangential components of the magnetic field on the interface surface. Similarly, if $\vec{j}_S^M=0$ the condition (1.13) defines the continuity of the tangential components of the electric fields at the interface.

If the conductivity of the second medium has infinite, i.e., the medium is perfectly conductive), the fields $\vec{E}_2$ and $\vec{H}_2$ are identically zero and the boundary condition can be written as

$$\left[\vec{n},\ \vec{E}_1\right]\Big|_S=0. \tag{1.16}$$

If the body conductivity is great, but finite, the exact boundary conditions (1.13) and (1.14) should be satisfied on the body surface. Hence, to determine electromagnetic fields outside a well-conducting body, the Maxwell equations must be solved both outside and inside the body using the boundary conditions at the interface between the two media. Such a problem is more complicated than that of determining the electromagnetic field only in one media and the boundary condition on the surface. Therefore, it is desirable to replace the boundary conditions (1.13) and (1.14) by the boundary condition connecting the values of the field vectors on the surface only in one media. Shchukin-Leontovich approximate boundary conditions or impedance boundary conditions [1, 2] satisfy this requirement.

1.3 Approximate Boundary Conditions for Electromagnetic Fields and the Limits of Their Correct Application

The one-sided impedance boundary conditions allow to reduce the number of interfacing electrodynamic volumes which should be taken in the problem solution. Eliminating the need to determine fields inside the adjacent metal-dielectric elements at the problem formulation level is the main advantage of the impedance approach. The Shchukin-Leontovich impedance condition on the boundary surface S can be written in the following form [2, 3]

$$[\vec{n}, \vec{E}]|_S = \overline{Z}_S[\vec{n}, [\vec{n}, \vec{H}]]|_S \,, \tag{1.17}$$

where $\vec{E}$ and $\vec{H}$ are the vectors the electric and magnetic harmonic fields, $\vec{n}$ is the impedance surface normal, directed inside the impedance region, $\overline{Z}_S = Z_S/Z_0$ is the normalized surface impedance, and $Z_0 = 120\pi$ Ohm is the resistance of free space.

If the $\overline{Z}_S = 0$, i.e., interface surface is perfectly conducting, the formula (1.17) is reduces to (1.16). It should be noted that the boundary condition (1.17) is approximate, since, in this case, the solution of the electrodynamic problem represents the first term of the asymptotic expansion of the exact solution [4] in powers of the small parameter

$$|\overline{Z}_S| << 1. \tag{1.18}$$

Since only the tangential components are included in the boundary condition (1.17), there exists some restrictions on the surface S geometry. It is evident that condition (1.17) holds if the surface curvature radius is much greater than the length of the incident wave. The conditions which take into account the interface curvature can be read as [2, 5]:

$$\begin{aligned} E_{\tau 1} &= \overline{Z}_S \left(1 + \frac{\chi_1 - \chi_2}{2ik\sqrt{\varepsilon_1 \mu_1}}\right) H_{\tau 2}\bigg|_S , \\ E_{\tau 2} &= -\overline{Z}_S \left(1 + \frac{\chi_2 - \chi_1}{2ik\sqrt{\varepsilon_1 \mu_1}}\right) H_{\tau 1}\bigg|_S , \end{aligned} \tag{1.19}$$

where χ_1 and χ_2 are the main Gaussian curvatures of the surface, E_τ and H_τ are the tangential components of the electromagnetic fields on the interface surface.

The surface impedance of an electromagnetic field is usually interpreted as relationships determining links between the tangential components of the complex amplitudes on the surface S [6]. If the impedance value does not depend on the incidence angle and incident wave polarization, it is known as extraneous impedance [3]. If the impedance value does not depend on the wave incidence angle, but depends on the wave polarization and spatial orientation of the surface S, the surface impedance is a two-dimensional second-rank tensor which components are extraneous impedances. In the general case, a concept of anisotropic surface impedance is introduced as matrix

$$\hat{\overline{Z}}_S = \left\| \begin{matrix} \overline{Z}_{S11} & \overline{Z}_{S12} \\ \overline{Z}_{S21} & \overline{Z}_{S22} \end{matrix} \right\|, \quad \overline{Z}_{Sjk} = \overline{R}_{Sjk} + i\overline{X}_{Sjk}, \ j, k \in \{1, 2\}, \tag{1.20}$$

under conditions that inequalities

$$\overline{R}_{S11} \geq 0, \quad \overline{R}_{S22} \geq 0, \quad 4\overline{R}_{S11}\overline{R}_{S22} \geq \left|\overline{Z}_{S12} + \overline{Z}^*_{S21}\right|^2 \tag{1.21}$$

hold. In (1.20) and (1.21) $\overline{Z}^*_{S21}$ is complex conjugate of $\overline{Z}_{S21}$. The inequalities (1.21) ensure that additional energy sources on the surface S and energy flows through this surface are absent. Of course, the impedance $\overline{Z}_S$ in (1.17) and (1.20) must be replaced by the tensor $\hat{\overline{Z}}_S$. It should be emphasized that the surface S, on which the impedance boundary condition should be satisfied, does not have to coincide with the real impedance boundary surface and can be considered as a conditional boundary surface. A spectral analysis of complex structures and media should require introduction of a partial impedance, the value of which in the general case depends on the frequency and the number of spatial harmonics in the electromagnetic fields representation. Such impedance problems are beyond the scope of this book.

First, let us consider possible formulations of the impedance conditions and the solution accuracies they can provide. According to results obtained in [4], the boundary condition (1.17) is applicable when the following requirements are met: a penetration depth of electric fields into an impedance material and a field wavelength should be small compared to an incident wave wavelength, a distance from a field source, and curvature radii of a boundary surface S. In addition, variations of the material parameters of the impedance layer at distances comparably with the field wavelength or penetration depth should be small. In the general case, the accuracy of the formula (1.17) was estimated to be proportional to $\sim |\overline{Z}_S|^2$, since only the first term of the solution obtained as asymptotic series with respect to the normalized impedance $\overline{Z}_S$ was used. Leontovich obtained a similar estimate by comparing the plane wave reflection obtained in the impedance approximation with that of the exact Fresnel solution [1]. However, for a certain class of propagation models, corrections to (1.17) are proportional to cubic but not quadratic terms in the small parameter $|\overline{Z}_S|$ [4].

Even though the accuracy estimates of the condition (1.17) were obtained on the basis of the skin-effect theory for of conducting body surfaces [1, 4], they can be uniquely extended to the more general case of impedance domains [3]. All the above requirements can be integrated into one of a purely physical nature: the field at the impedance surface must be a plane wave propagating in a direction normal to a boundary S. This condition is always fulfilled for electrically thin impedance structures, including film coatings.

However, the presented requirement cannot be strictly fulfilled if the impedance surface is excited by waves at small incident or Brewster angles. In the first case, the reflected and refracted rays are gliding near the surface, while in the second case they must be mutually perpendicular. In both cases, the directions of the refracted rays do not coincide with the direction of the boundary surface normal. Therefore, it is customary to distinguish between three separate cases of the impedance conditions depending upon the wave incidence angle: (1) normal incidence, when the Shchukin-Leontovich condition are valid, (2) Brewster angle incidence, and (3) tangential incidence. Apparently for the first time, the condition (1.17) was corrected and used in [7] where problem of the reflection of electromagnetic waves from the surface of real soil at angles close to the sliding incidence angle was solved. Subsequently, similar cases were studied for a number of other media including inhomogeneous plasma. The analysis of various variants of the approximate impedance boundary conditions are presented in [8–12].

The accuracy justification of the impedance boundary condition (1.17) cannot give a complete answer to the question: with what accuracy the specific characteristics of the wave fields can be calculated for arbitrary angles of a plane wave incidence on the media interface. General conclusions concerning for the smallest errors relative to exact values for the whole range of incidence angles were formulated in [13] for perpendicular and parallel polarizations relative to the interface surface. For the perpendicular polarization, the reflection coefficients should be calculated based on the Shchukin-Leontovich approximate boundary conditions. For a wave of parallel polarization, the formulas valid for the Brewster angle are preferable. These conclusions were made based on the exact formulas obtained in [14].

The above analysis of the boundary condition (1.17) accuracy was performed under condition that surface impedance $\overline{Z}_S$ presented by a power series only terms proportional to the first degree of a small parameter were taken into account. However, this simplification permits only the small $\overline{Z}_S$, and secondly, and does not provide the necessary accuracy for solving the diffraction problem when the wave is incident at the Brewster angle or tangential to the surface interface. These shortcomings can be eliminated by the method proposed in [15], where a generalized impedance approximation was formulated as:

$$\vec{E}_\tau + \overline{Z}_S\left[\vec{H}_\tau, \vec{n}\right] + \frac{1}{2}\overline{Z}_S^3 n^2\left(N_1 + \sum_{s=1}^{\infty}\left(\overline{Z}_S n\right)^{2s}\frac{(2s-1)!!}{2^s(s+1)!}N_{2s+1}\right)\vec{H}_\tau = 0, \quad (1.22)$$

where the matrices $N_m (m = 2s + 1)$ are defined as

$$N_m = \left\| \begin{matrix} 0 & 1 \\ m & 0 \end{matrix} \right\|, \tag{1.23}$$

$n = k / k_0$ is a dimensionless refraction parameter, k and k_0 are wave numbers in the impedance medium and external space, respectively. In the first approximation, the Eq. (1.22) coincides with the boundary condition (1.17), but the vectors $\vec{E}_\tau$ and $\vec{H}_\tau$ are related through the refraction factor n, which in a reflection problem is uniquely connected to a wave incidence angle.

Thus, the approximate Shchukin-Leontovich condition (1.17) valid for a small surface impedance $\overline{Z}_S$, is generalized for arbitrary $\overline{Z}_S$ as the series expansion (1.22), which expand the applicability of the impedance approach. Since the exact boundary condition (1.22) is decomposed in a series in odd powers of the parameter $\overline{Z}_S$, the Shchukin-Leontovich condition linear in $\overline{Z}_S$ differs from the exact only by terms proportional to $\sim |\overline{Z}_S|^3$. That is, the fields obtained using condition (1.17) are correct up to $\sim |\overline{Z}_S|^2$. Thus, the accuracy of condition (1.17) application turns out to be higher than it could be supposed based on the results obtained in [4].

1.4 Uniqueness Theorem and Duality Principle for Material Bodies with Impedance Boundaries

Since later in the monograph electrodynamic problems will be investigated using impedance boundary conditions, it is useful to discuss some aspects of using such boundary conditions.

1.4.1 Features of the Application of the Uniqueness Theorem

The electromagnetic problems are usually classified into two classes: internal and external problems. The internal problems concern with fields in a limited part of the space surrounded by a surface S. Extraneous currents $\vec{j}^{E(M)}$ are specified inside the surface S, while the boundary conditions for the tangential components of the electric and magnetic fields should be fulfilled on the surface itself. The external problems deal with fields in an infinite space outside the surface S, where the boundary conditions for the components E_τ and H_τ are specified, and external currents are defined in the space surrounding the surface.

From a mathematical point of view, the requirements to satisfy boundary conditions at the surface S for the Helmholtz Eqs. (1.6) or (1.10) determine that the solution should be unique. In the electrodynamics theory this statement is proved in the framework of the uniqueness theorem.

For the mono-chromatic electromagnetic fields uniqueness of solution of the internal and external boundary value problems for the Maxwell equations in conducting media with the complex permittivity $\varepsilon_1 = \mathrm{Re}\varepsilon_1 + i\mathrm{Im}\varepsilon_1$ and permeability $\mu_1 = \mathrm{Re}\mu_1 + i\mathrm{Im}\mu_1$ ($\mathrm{Im}\varepsilon_1 > 0$, $\mathrm{Im}\mu_1 > 0$) are intensively studied (here, for convenience of comparison with literary sources, we choose a time dependence in the form $e^{-i\omega t}$). The uniqueness problems were thoroughly studied when the boundary conditions for electric and magnetic fields are given apart from each other [16]. Since the uniqueness theorem proof for boundary value problems with impedance surfaces satisfying the boundary conditions (1.17) are covered in literature contradictorily, additional consideration of this question can be found in [17].

A key point in the proof of the uniqueness theorem is a lemma asserting that the homogeneous Maxwell equations do not have solutions other than zero. The lemma is applied to difference electromagnetic fields $\left(\vec{E}, \vec{H}\right)$, and it is proved based on the complex Pointing theorem [16, 18] or the Lorentz lemma [8, 17].

If the tangential component of the electric and magnetic field are explicitly specified on the surface S: the electric field is defined on the surface S_1, the part of the surface S, and the tangential component of the magnetic field is defined on the surface S_2 ($S = S_1 + S_2$), the difference field $\left(\vec{E}, \vec{H}\right)$ should satisfy homogeneous Maxwell equations with uniform boundary conditions on the surface S $\vec{E}_\tau = 0$ on the surface S_1 and $\vec{H}_\tau = 0$ on the surface S_2.

If the Shchukin-Leontovich impedance boundary condition (1.17) are specified on the surface S, it is easy to verify that the difference field $\left(\vec{E}, \vec{H}\right)$ should also satisfy the homogeneous linear Maxwell equations with impedance boundary conditions (1.17) on the surface S. The distinction in the boundary conditions for difference fields dictates the need for the proof of the uniqueness theorem for the boundary value problems with impedance surfaces.

There exist several approaches to determining the surface impedance in the literature, different are used. In some works, the surface impedance defined as a linear operator connecting the tangential components $\vec{E}_\tau$ and $\left[\vec{H}_\tau, \vec{n}\right]$, and in other it is determined from the relation between $\vec{E}_\tau$ and $\left[\vec{H}_\tau, \vec{n}\right]$. Therefore, it is always necessary to keep a check on definitions of the surface impedance.

For the general case, we introduce definition of the anisotropic surface impedance in the form of a matrix

$$\hat{Z}_S = \left\| \begin{matrix} Z_{11} & Z_{12} \\ Z_{21} & Z_{22} \end{matrix} \right\|, \tag{1.24}$$

where $Z_{jk} = R_{jk} + X_{jk}$, indices j,k are equal to 1,2. To ensure the absence of additional energy sources on the impedance surface, or more precisely, the energy fluxes through the surface the inequalities should hold:

$$R_{11} \geq 0,\ R_{22} \geq 0,\ 4R_{11}R_{22} \geq \left|Z_{12} + Z_{21}^*\right|^2, \tag{1.25}$$

where Z_{21}^{*} is the complex conjugate of Z_{21}. It should be emphasized that when the normal vector to the surface S is directed inside the impedance body, the physically correct using the impedance boundary condition (1.17) is possible only if the requirements $\mathrm{Re}Z_S \geq 0$ and (1.25) are met for the isotropic and anisotropic surface impedance Z_S.

The authors of this book has proved the uniqueness theorem applying the Lorentz lemma to the difference electromagnetic field $\left\{\vec{E}_d, \vec{H}_d\right\} = \left\{\vec{E}, \vec{H}\right\} - \left\{\vec{E}^{*}, \vec{H}^{*}\right\}$ assuming that electrical and magnetic losses are nonzero at each point of the spatial region D_e, and applying the impedance boundary condition (1.20) on the given surface. The final expression for the case of isotropic surface impedance can be presented as

$$\oint\limits_S \mathrm{Re}Z_S\left|\left[\vec{n}, \vec{H}_d\right]\right|^2 \mathrm{d}s + \omega \int\limits_{D_e} \left(\mathrm{Im}\varepsilon_1\left|\vec{E}_d\right|^2 + \mathrm{Im}\mu_1\left|\vec{H}_d\right|^2\right)\mathrm{d}v = 0, \tag{1.26}$$

where $\mathrm{d}v$ is the volume element. As can be seen, the expression (1.26) includes the integral terms taking into account the electric and magnetic losses in the medium and the surface integral, proportional to $\mathrm{Re}Z_S$. Obviously, to satisfy condition (1.21) with and, and also when, everywhere in the region there should be and. Obviously, on the assumption that the inequalities $\mathrm{Im}\varepsilon_1 > 0$,$\mathrm{Im}\mu_1 > 0$, and $\mathrm{Re}Z_S \geq 0$ hold, the condition (1.21) can be fulfilled if everywhere in the region D_e the equalities $\vec{E}_d = 0$ and $\vec{H}_d = 0$ are valid. The expression (1.26) turns out to be valid even if the time dependence of monochromatic wave processes is chosen in the form $e^{i\omega t}$. In this case, the signs before ω, $\mathrm{Im}\varepsilon_1$,and $\mathrm{Im}\mu_1$ in (1.26) should be opposite.

Thus, it can be stated that the solutions of internal and external boundary value problems of the Maxwell system of equations for areas bounded by impedance surfaces in conducting media where the imaginary parts of permittivity and permeability satisfy the relations $\mathrm{Im}\varepsilon_1 > 0$ and $\mathrm{Im}\mu_1 > 0$ are unique if the conditions $\mathrm{Re}Z_S \geq 0$ and (1.25) are met in the case of an isotropic and anisotropic impedance surfaces.

1.4.2 Features of the Application of the Principle of Duality

The duality principle commonly used in the formulation of Weinstein [16], establish a relation between two diffraction problems: (1) diffraction in free space on a flat infinitely thin perfectly conducting plate; (2) diffraction in free space on a flat, infinitely thin perfectly conducting screen with an aperture that coincides with the plate in the first problem. The duality theorem is analogous to the Babinet theorem which relates diffraction phenomena on mutually complementary screens within the framework of Huygens scalar principle. An analysis of the duality principle for areas with impedance surfaces was absent in references prior to publication [17].

The duality principle is based on the property of permutation symmetry or permutation invariance of Maxwell equations with respect to external electric $\vec{j}^E$ and magnetic $\vec{j}^M$ currents. This property allows the following two-sided permutation

$$\vec{j}^E \Leftrightarrow -\vec{j}^M;\ \vec{H}_1 \Leftrightarrow \vec{E}_2;\ \vec{E}_1 \Leftrightarrow \vec{H}_2;\ \varepsilon \Leftrightarrow -\mu. \tag{1.27}$$

The choice of one or other pair of coupled electrodynamic problems with various geometry, differing by excitation conditions $\left(\vec{j}^E \Leftrightarrow -\vec{j}^M\right)$, which in fact determine the content of the reciprocity principle for each of the possible options can be justified on the basis of these permutations. Problems with only one type of the excitation currents should be solved by using the Maxwell equations, asymmetrical with respect to extraneous currents. In these cases, it is difficult to directly use the property of permutation invariance, and separate proofs of the duality principle are required, for example, as it was done in [16].

The bilateral permutations (1.27) are valid for solution of electrodynamic problems in infinite space if the field boundedness condition at infinity is satisfied. The boundary value problems for impedance surfaces have one significant feature based on the permutation principle (1.27)—the boundary conditions should be also transformed.

Since the impedance boundary condition (1.17) are asymmetric with respect to replacing only the fields $\vec{E}_1 \Leftrightarrow \vec{H}_2$ and $\vec{H}_1 \Leftrightarrow \vec{E}_2$, it is also necessary to make the replacement of the surface impedance. It is easy to verify that if the impedance is isotropic, the permutations (1.27) must be supplemented by

$$Z_S \Leftrightarrow -1/Z_S. \tag{1.28}$$

To verify the legitimacy of requirement (1.28), let us consider the problem of the normal incidence of a plane electromagnetic wave on a flat interface between the free half-space and a half-space with a homogeneous medium characterized by complex permittivity ε_1 and permeability μ_1. The formula of the wave impedance in the form $Z_0 = \sqrt{\mu_1/\varepsilon_1}$ was obtained as solution of this problem. In this case, the permutation (1.95) is directly resulted from permutations for the fields from (1.27) and $\varepsilon_1 \Leftrightarrow -\mu_1$.

Analyzing the permutation (1.28), the following conclusion can be made: in any electrodynamic volumes containing impedance boundaries, the duality principle can be realized only using permutation (1.28) in related problems. Since the relation $\mathrm{Re}\, Z_S > 0$ is a condition of physical feasibility for passive impedance structures, the of this permutation is possible only for surfaces characterized by the purely imaginary impedance. This conclusion remains valid if the surface impedance is represented in the matrix form (1.19). Since the algebra of rank-2 matrices coincides with the algebra of linear operators, the condition (1.28) can be written as

$$\hat{Z}_S \Leftrightarrow -\hat{Z}_S/\det\left[\hat{Z}_S\right], \tag{1.29}$$

where $\det\left[\hat{Z}_S\right]$ is the determinant of the impedance matrix $\hat{Z}_S$. In this case, the permutation (1.29) is possible only when $Z_{12} = Z_{21}$, i.e. matrix $\hat{Z}_S$ is symmetric.

Expressions for specific realizations of surface impedances for various surfaces, as well as for electrically thin vibrators, are detailed in [17, 19, 20] and Appendix C.

1.5 Tensor Green's Functions of the Vector Helmholtz Equation for Hertz Potentials in Electrodynamic Volumes

1.5.1 Properties of the Tensor Green's Function

As mentioned earlier, the vector Helmholtz equations (Sect. 1.1) and boundary conditions (Sects. 1.2 and 1.3) are a single boundary value problem. This boundary value problem can be analytically solved by the eigenfunctions and Green's function method.

The of eigenfunctions approach (see, for example, [16]), is based on solving differential equations by the separation of variables. The eigenfunctions are solutions of homogeneous ordinary differential equations which a separation constants satisfy the corresponding boundary conditions at the interval ends of the independent variables. The values of the separation constant, valid for specified boundary conditions, are called eigenvalues. This method will not be used in this book, therefore it is not discussed here in detail.

The second method is more physically obvious, since Green's function represent a field at the observation point generated by a point source. To obtain the field generated by sources distributed in space, one should find the corresponding volume integral over the entire region where the sources are given.

Let us consider the basic properties of the tensor Green's function of the vector Helmholtz equation [21]:

(a) In the system of orthogonal curvilinear coordinates, the tensor Green's function $\hat{G}\left(q_1, q_2, q_3; q_1', q_2', q_3'\right) = \hat{G}\left(\vec{q}, \vec{q}'\right)$ satisfies the inhomogeneous Helmholtz equation

$$\Delta\hat{G}\left(\vec{q}, \vec{q}'\right) + k^2\hat{G}\left(\vec{q}, \vec{q}'\right) = -4\pi\hat{I}\frac{\delta\left(q_1 - q_1'\right)\delta\left(q_2 - q_2'\right)\delta\left(q_3 - q_3'\right)}{h_1 h_2 h_3}, \quad (1.30)$$

where $\hat{I}$ is the unit tensor, $\left(q_1', q_2', q_3'\right)$ are coordinates of the source, $\delta\left(q - q'\right)$ is the Dirac delta function, h_n are Lame coefficients, and the Laplacian is applied to all components of the tensor;

(b) the Green's function is the symmetric tensor (affinor) of the second rank defined by nine components, since each of the three components of a vector source can create three field components. The tensor symmetry ensures the equality

$$\vec{F}(\vec{q})\hat{G}(\vec{q}, \vec{q}') = \hat{G}(\vec{q}, \vec{q}')\vec{F}(\vec{q}). \tag{1.31}$$

holds;

(c) all components of the Green's tensor are invariant with respect to the mutual change of the of the source $(\vec{q}')$ and observation $(\vec{q})$ coordinates;

(d) when the source and observation points coincide, all components of the Green's tensor have an integrable singularity

$$1\Big/\sqrt{(q_1 - q_1')^2 + (q_2 - q_2')^2 + (q_3 - q_3')^2};$$

(e) the solution of the vector inhomogeneous Helmholtz equation can be obtained in integral form by using the Green's tensor.

The solution for vector Hertz potentials is presented in the following form [21]

$$\vec{\Pi}^{E(M)}(\vec{q}) = \frac{1}{i\omega\varepsilon_1\mu_1}\int\limits_V \vec{j}^{E(M)}(\vec{q}')\ \hat{G}^{E(M)}(\vec{q}, \vec{q}')\,\mathrm{d}v$$
$$+\oint\limits_S \Big\{\mathrm{div}\vec{\Pi}^{E(M)}(\vec{q}')\ \hat{G}^{E(M)}(\vec{q}, \vec{q}')\ \vec{n} - \mathrm{div}\hat{G}^{E(M)}(\vec{q}, \vec{q}')\vec{\Pi}^{E(M)}(\vec{q}')\ \vec{n}$$
$$+\big[\vec{n},\ \hat{G}^{E(M)}(\vec{q}, \vec{q}')\big]\,rot\,\vec{\Pi}^{E(M)}(\vec{q}') - \Big[\vec{n},\ \vec{\Pi}^{E(M)}(\vec{q}')\Big]\,rot\hat{G}^{E(M)}(\vec{q}, \vec{q}')\Big\}\,ds', \tag{1.32}$$

where $\vec{n}$ is the unit vector of the outward normal to the surface S, $\mathrm{d}v$ is the volume element ($\mathrm{d}s'$ is the area element in the dashed coordinates). The volume integral is taken over the entire volume bounded by the surface S, the surface integral is taken over the entire surface S, and differentiation is performed with respect to operations the dashed coordinates Note that the expression in braces is a vector.

Thus, the solution of the inhomogeneous Helmholtz equation can be represented as the sum of the volume and surface integrals. The integrands of surface integrals include boundary values of the unknown function and its derivatives, that reduces the range of solvable problems. If the boundary values of thc unknown function are not known in advance, the equalities (1.32) are converted into integral equations whose solution is no less difficult than the initial differential equations.

The surface integrals can be excluded from (1.32) by constructing the Green's function in a special way. If components of the Green's function $\hat{G}(\vec{q}, \vec{q}')$ satisfy the boundary conditions on the surface S coincide with components of the vector potentials $\vec{\Pi}^{E(M)}(\vec{q})$, the surface integrals disappear, since the components of the integrand vector vanish. Hence, the solution for $\vec{\Pi}^{E(M)}(\vec{q})$ will be represented only as the volume integral.

Substituting this solution into Eq. (1.8), we obtain the following expressions for the electromagnetic field

$$\begin{aligned}\vec{E}(\vec{q}) &= \frac{1}{4\pi i\omega\varepsilon_1}\left[\text{graddiv} + k_1^2\right]\int\limits_V \vec{j}^E(\vec{q}')\,\hat{G}^E(\vec{q},\vec{q}')\,\text{d}v \\ &-\frac{1}{4\pi}rot\int\limits_V \vec{j}^M(\vec{q}')\,\hat{G}^M(\vec{q},\vec{q}')\,\text{d}v, \\ \vec{H}(\vec{q}) &= \frac{1}{4\pi i\omega\mu_1}\left[\text{graddiv} + k_1^2\right]\int\limits_V \vec{j}^M(\vec{q}')\,\hat{G}^M(\vec{q},\vec{q}')\,\text{d}v \\ &+\frac{1}{4\pi}rot\int\limits_V \vec{j}^E(\vec{q}')\,\hat{G}^E(\vec{q},\vec{q}')\,\text{d}v.\end{aligned} \tag{1.33}$$

Let us now represent of the electromagnetic field through the dyadic Green's functions

$$\begin{bmatrix}\vec{E}(\vec{q})\\ \vec{H}(\vec{q})\end{bmatrix} = \int\limits_V \begin{bmatrix}\hat{g}^E(\vec{q},\vec{q}') & \hat{g}^{EM}(\vec{q},\vec{q}')\\ \hat{g}^{ME}(\vec{q},\vec{q}') & \hat{g}^M(\vec{q},\vec{q}')\end{bmatrix} \times \begin{bmatrix}\vec{j}^E(\vec{q}')\\ \vec{j}^M(\vec{q}')\end{bmatrix}\text{d}v \tag{1.34}$$

where $\hat{g}^E(\vec{q},\vec{q}')$,$\hat{g}^M(\vec{q},\vec{q}')$,$\hat{g}^{EM}(\vec{q},\vec{q}')$, and $\hat{g}^{ME}(\vec{q},\vec{q}')$ are electric, magnetic, electromagnetic, and magneto-electric functions. Then, comparing expressions (1.33) and (1.34), we can define the dyadic Green's functions through the Green's tensors of the Hertz vector potentials as

$$\begin{aligned}\hat{g}^E(\vec{q},\vec{q}') &= \frac{1}{4\pi i\omega\varepsilon_1}\left[\text{graddiv} + k_1^2\right]\hat{G}^E(\vec{q},\vec{q}')\ , \\ \hat{g}^{EM}(\vec{q},\vec{q}') &= -\frac{1}{4\pi}rot\hat{G}^M(\vec{q},\vec{q}')\ , \\ \hat{g}^{ME}(\vec{q},\vec{q}') &= \frac{1}{4\pi}\text{rot}\hat{G}^E(\vec{q},\vec{q}')\ , \\ \hat{g}^M(\vec{q},\vec{q}') &= \frac{1}{4\pi i\omega\mu_1}\left[\text{graddiv} + k_1^2\right]\hat{G}^M(\vec{q},\vec{q}')\ .\end{aligned} \tag{1.35}$$

As indicated above, the Green's functions for vector potentials $\hat{G}^{E(M)}(\vec{q},\vec{q}')$ in the source region have an integrable singularity. In the case of the field Green's functions (1.35) the singularity is non-integrable. Therefore, the apparatus of generalized functions should be used to regularize the singularity which significantly complicates the Green's tensors constructing.

In this connection, the excitation problems of various electrodynamic volumes by external currents will be solved by using the tensor Green's function of the Helmholtz equations for Hertz vector potentials $\vec{\Pi}^{E(M)}(\vec{q})$. When the potentials $\vec{\Pi}^{E(M)}(\vec{q})$ is determined, the electromagnetic fields $\vec{E}(\vec{q})$ and $\vec{H}(\vec{q})$, can be found from (1.33).

1.5.2 Construction of the Green's Tensor in Orthogonal Curvilinear Coordinate Systems

The constructing the tensor Green's function of the Helmholtz vector equation, developed in the monograph [21], is based on the possibility of representing its components as series expansions in the system of one longitudinal and two transverse Hansen vector functions, which are expressed through scalar eigenfunctions. This approach imposes several requirements to the coordinate systems that limit the number of systems in which the possibility of constructing Green's tensors exists. These requirements will be briefly analyzed bellow.

First of all, the coordinate system should ensure that the variable separations in the three-dimensional Helmholtz equation which allows determining the scalar eigenfunctions $\varphi(\vec{q})$,$\psi(\vec{q})$ and $\chi(\vec{q})$. There exist eleven types of coordinate systems which according to Robertson condition allow the variable separations. The scalar functions in each coordinate system are represented as products of functional multipliers, each of which depends only on one variable, for example, $\varphi(\vec{q}) = X_1(q_1)X_2(q_2)X_3(q_3)$.

However, in the Helmholtz vector equations, the variables cannot always be separated but even if they are separated, the components of the vector field may be admixed so that they are included in all three equations. To avoid this admixing, the method of vector field separation into one longitudinal component presented as a grad operator acting on the scalar potential and two transverse or solenoidal components given as rot operator acting on the vector potential [21]. Then, if the Lame coefficient $h_1 = 1$, the complete system of Hansen eigenvector functions can be obtained based on solutions of three scalar Helmholtz equation for the functions $\varphi(\vec{q})$, $\psi(\vec{q})$ and $\chi(\vec{q})$:

$$
\begin{aligned}
\vec{L} &= grad\varphi\left(\vec{q}\right), \\
\vec{M} &= rot\left(\vec{n}_1 w\psi\left(\vec{q}\right)\right), \\
\vec{N} &= \frac{1}{k} rotrot\left(\vec{n}_1 w\chi\left(\vec{q}\right)\right),
\end{aligned}
\tag{1.36}
$$

where $\vec{n}_1$ is the unit vector of the coordinate q_1, $1/k$ is the coefficient introduced to ensure that the dimension of the components $\vec{L}$, $\vec{M}$, $\vec{N}$ is the same.

The requirements that $h_1 = 1$ and the ratio of the other two Lame coefficients do not depend upon the coefficient corresponding to this coordinate allows only six coordinate systems: rectangular coordinates in which x, y or z can be selected as the coordinate q_1, three cylindrical coordinate systems in which the coordinate z corresponds to q_1, spherical and conical coordinate systems in which the radius r is q_1. In the expressions (1.36), $w = 1$ in the first four cases and $w = r$ in the last two cases.

The system of the eigenvector vector functions (1.36) constructed in this way can ensure the fulfillment of boundary conditions for vector fields only if the boundary surfaces coincide with coordinate surfaces of the coordinate system. That is, the

Green's functions can be constructed only for spatial domains with so-called coordinate boundaries. Thus, this requirement determines the choice of the coordinate system for the solutions of boundary-value problems of various geometries. The eigenfunctions of the vector Helmholtz equation can be defined based on (1.36) as

$$\vec{F}_n(\vec{q}) = \vec{L} + \vec{M} + \vec{N}, \tag{1.37}$$

where a four-digit subscript n is assign to each eigenvector $\vec{F}_n(\vec{q})$, so that the first digit of this index shows which of the three systems $\vec{L}$, $\vec{M}$ or $\vec{N}$ the vector belongs to, and the other three are numbers of the generating scalar eigenfunctions, $\varphi(\vec{q})$, $\psi(\vec{q})$ and $\chi(\vec{q})$. It can be shown that the eigenvector $\vec{F}_n(\vec{q})$ for all possible n form a complete orthogonal system of functions, that is

$$\iiint \vec{F}_n^*(\vec{q})\, \vec{F}_m(\vec{q})\, \mathrm{d}v = \begin{cases} 0\,, & m \neq n, \\ \lambda_n\,, & m = n, \end{cases} \tag{1.38}$$

where integration is performed over the entire spatial domain, $\vec{F}_n^*(\vec{q})$ is a vector function, complex-conjugate to the function $\vec{F}_n(\vec{q})$, λ_n is the norm of the corresponding eigenfunctions.

The Green's tensor can be represented by using the operation of tensor multiplication of two vectors as an expansion in eigenvectors $\vec{F}_n(\vec{q})$.

$$\hat{G}(\vec{q}, \vec{q}') = 4\pi \sum_n \frac{\vec{F}_n^*(\vec{q}) \otimes \vec{F}_n(\vec{q}')}{\lambda_n \left(k_n^2 - k^2\right)}, \tag{1.39}$$

where the eigenvector $\vec{F}_n(\vec{q})$ satisfy the same boundary conditions as the components $\hat{G}(\vec{q}, \vec{q}')$, k_n^2 are the corresponding eigenvalues, and $\otimes$ is the symbol of tensor multiplication. If $\vec{F}_n(\vec{q})$ is a complex vector, then the conjugate vector in the decomposition (1.39), therefore the tensor $\hat{G}(\vec{q}, \vec{q}')$ is the Hermitian tensor. It can be also seen that the tensor $\hat{G}(\vec{q}, \vec{q}')$ is symmetric relative to $\vec{q}$ and $\vec{q}'$.

The tensor $\hat{G}(\vec{q}, \vec{q}')$ as a function of k has poles for all eigenvalues k_n, where the residue at the n-th pole is equal to $2\pi \vec{F}_n^*(\vec{q})\vec{F}_n(\vec{q}\prime) \Big/ (\lambda_n k_n)$. This corresponds to an infinite amplitude of oscillations at the resonant frequency of the driving force in absence of attenuation.

Let us now construct using the described method, the Green's tensor for the most interesting, from a practical point of view, the generalized cylindrical coordinate systems.

1.5.3 Green's Tensor for Areas with Cylindrical Boundaries

Let us construct the tensor Green's function for the vector Helmholtz equation in the generalized cylindrical coordinate systems including rectangular coordinate systems by the proposed method which is also used in [21]. Let (q_1, q_2, z) and (q_1', q_2', z') be the coordinates of the observation and source points.

The tensor functions based on the formula (1.39) can be represented as series over the full system of vector functions (1.36). For example, the tensor $\hat{G}^E(\vec{q}, \vec{q}')$ can be decompose as

$$\begin{aligned}\hat{G}^E(\vec{q}, \vec{q}\prime) = \sum_{n=0}^{\infty}\sum_{m=0}^{\infty}\Bigg\{ & \frac{1}{\left(k_{nm}^E\right)^2 \lambda_{nm}^E}\left[\vec{z}^\circ,\ \nabla\psi_{nm}(q_1, q_2)\right] \otimes \left[\vec{z}^\circ, \nabla\psi_{nm}^*(q_1', q_2')\right] h_{nm}^E \\ & + \frac{1}{\left(k_{nm}^M\right)^2 \lambda_{nm}^M} \nabla\chi_{nm}(q_1, q_2) \otimes \chi_{nm}^*\left(q_1', q_2'\right) h_{nm}^M \\ & + \frac{1}{\lambda_{nm}^M}\vec{z}^\circ \chi_{nm}(q_1, q_2) \times \vec{z}^\circ \chi_{nm}^*\left(q_1', q_2'\right) h_{nm}^M \Bigg\},\end{aligned} \tag{1.40}$$

where $\nabla u = \vec{q}_1^\circ \frac{1}{h_1}\frac{\partial u}{\partial q_1} + \vec{q}_2^\circ \frac{1}{h_2}\frac{\partial u}{\partial q_2}$, q_1, q_2 are curvilinear coordinates in the cross section of the considered area with Lame coefficients h_1 and h_2,$\vec{q}_1^\circ$,$\vec{q}_2^\circ$ are the unit vectors, and $\lambda_{nm}^{E(M)}$ are norms of the eigenfunctions.

In expression (1.40), $\psi_{nm}(q_1, q_2)$ and $\chi_{nm}(q_1, q_2)$ are mutually orthogonal eigenfunctions of the following two-dimensional Helmholtz equations

$$\begin{aligned}\Delta\, \chi_{nm}(q_1, q_2) + \left(k_{nm}^M\right)^2 \chi_{nm}(q_1, q_2) = 0, \\ \Delta\, \psi_{nm}(q_1, q_2) + \left(k_{nm}^E\right)^2 \psi_{nm}(q_1, q_2) = 0\end{aligned} \tag{1.41}$$

with boundary conditions $\chi_{nm}(q_1, q_2) = \frac{\partial}{\partial n}\psi_{nm}(q_1, q_2) = 0$.

The functions h_{nm}^E and h_{nm}^M in (1.40) satisfy non-homogeneous ordinary differential equations

$$\frac{\partial^2}{\partial z^2} h_{nm}^{E(M)}(z, z') + \left(\gamma_{nm}^{E(M)}\right)^2 h_{nm}^{E(M)}(z, z') = -\,4\pi\delta\left(z - z'\right), \tag{1.42}$$

with predefined boundary conditions at the ends of the interval z. In the Eq. (1.42) $\gamma_{nm}^{E(M)} = \sqrt{k^2 - \left(k_{nm}^{E(M)}\right)^2}$.

In the cylindrical coordinate systems, the form of Eq. (1.42) for $h_{nm}^{E(M)}(z, z')$ and their solutions do not depend on the field distribution in the region cross section. The solutions should be found in accordance with the geometry of the boundary value problem specified in the longitudinal direction.

The tensor Green's function $\hat{G}^E(\vec{q}, \vec{q}')$ in the considered coordinate systems has the following form

$$\hat{G}^E(\vec{q}, \vec{q}') = \begin{pmatrix} G_{11}^E(\vec{q}, \vec{q}') & G_{12}^E(\vec{q}, \vec{q}') & 0 \\ G_{21}^E(\vec{q}, \vec{q}') & G_{22}^E(\vec{q}, \vec{q}') & 0 \\ 0 & 0 & G_{33}^E(\vec{q}, \vec{q}') \end{pmatrix}, \tag{1.43}$$

since only five nonzero components can be obtained from the tensor decomposition (1.40). For example, the explicit expression for the component of the tensor Green's function $G_{33}^E(\vec{q}, \vec{q}')$ can be written as

$$G_{33}^E(\vec{q}, \vec{q}') = G_{zz}^E(\vec{q}, \vec{q}') = \sum_{n=0}^{\infty}\sum_{m=0}^{\infty} \frac{1}{\lambda_{nm}^E} \chi_{nm}(q_1, q_2)\, \chi_{nm}^*(q_1, q_2)\, h_{nm}^E(z, z'). \tag{1.44}$$

The tensor Green's functions of the magnetic type are also determined by the expression (1.40) with the following mutual substitutions: $\chi_{nm}(q_1, q_2) \Leftrightarrow \psi_{nm}(q_1, q_2)$, $h_{nm}^E(z, z\prime) \Leftrightarrow h_{nm}^M(z, z\prime)$, $k_{nm}^E \Leftrightarrow k_{nm}^M$, $\lambda_{nm}^E \Leftrightarrow \lambda_{nm}^M$.

The tensor Green's functions of electric and magnetic types for the various electrodynamic volumes are presented in Appendix A.

References

1. Leontovich MA (1985) Theoretical physics. Selected works. Nauka, Moskow (in Russian)
2. Leontovich MA (1948) On approximate boundary conditions for an electromagnetic field on the surface of well-conducting bodies. Radio wave propagation studies, pp 5–12 (in Russian)
3. Miller MA, Talanov VI (1961) Using the concept of surface impedance in the theory of surface electromagnetic fields (review). Izv Vuz Radiophys 4(5): 795-830 (in Russian)
4. Rytov SM (1940) Calculation of the skin-effect by the perturbation method. J Exp Theor Phys 10(2):180–190 (in Russian)
5. Ilyinsky AS, Kravtsov VV, Sveshnikov AG (1991) Mathematical models of electrodynamics. Vysshaya shkola, Moskow (in Russian)
6. Physical Encyclopedia (1983) In: Prokhorov AM (ed) Soviet encyclopedia Moskow (in Russian)
7. Feinberg EL (1999) Propagation of radio waves along the Earth's Surface. Fizmatlit, Moskow (in Russian)
8. Il'inskiy AS, Slepyan GYa (1991) Impedance boundary conditions and their application for calculating the absorption of electromagnetic waves in conducting media. Soviet J Commun Technol Electron 36(6):1–18
9. Senior TBA, Volakis JL (1995) Approximate boundary conditions in electro-magnetics. IEE, London
10. Hoppe DJ, Rahmat-Samii Y (1995) Impedance boundary conditions in electro-magnetics. Taylor and Fransic, Boca Raton
11. Tretyakov S (2003) Analytical modeling in applied electromagnetics. Artech House, Norwood, MA

12. Yuferev SV, Ida N (2009) Surface impedance boundary conditions. A comprehensive approach. CRC Press, Boca Raton
13. Belkovich IV, Zheksenov MA, Kozlov DA (2001) Comparison of options for impedance boundary conditions when a plane electromagnetic wave is incident on a dielectric half-space. J Radio Electron (7) (in Russian)
14. Brekhovskikh LM (1980) Waves in layered media. Academic Press, New York
15. Alshits VI, Lyubimov VN (2009) Generalization of the Leontovich approximation for electromagnetic fields on a dielectric–metal interface. Uspekhi Fizicheskikh Nauk and PN Lebedev Physics Institute of the Russian Acad Sci 52(8):815–820
16. Weinstein LA (1988) Electromagnetic waves. Radio i svyaz', Moskow (in Russian)
17. Penkin YuM, Katrich VA, Nesterenko MV, Berdnik SL, Dakhov VM (2019) Electromagnetic fields excited in volumes with spherical boundaries. Springer, Cham, Swizerland
18. Feld YaN (1968) Basic equations, uniqueness theorem and boundary problems of electrodynamics. I-th All-Union Workshop on Diffraction and Wave Propagation, pp 93–109 (in Russian)
19. Nesterenko MV, Katrich VA, Penkin YuM, Berdnik SL, Dumin OM (2020) Combined vibrator-slot structures. Theory and Applications. Springer, Cham, Swizerland
20. Berdnik SL, Gomozov AV, Gretskih DV, Katrich VA, Nesterenko MV (2022) Approximate boundary conditions for electromagnetic fields in electromagnetics. Radioelectronik Comput Syst 103(3):141–160
21. Morse PM, Feshbach H (1953) Methods of theoretical physics. McGraw-Hill, New York

Chapter 2
General Aspects of the Theory of Multielement Systems of Thin Impedance Vibrators and Narrow Slots in a Quasi-One-Dimensional Approximation

2.1 Problem Formulation and Initial Integral Equations

Let us formulate in the most general form the problem of excitation, scattering and radiation of electromagnetic fields by material bodies of finite size in the presence of the coupling holes between the two electrodynamic volumes. Let there be an arbitrary volume V_1 bounded by a perfectly conducting, impedance, or partially impedance surface S_1, some parts of which can be removed to infinity. The volume V_1 is coupling with another arbitrary volume V_2 through holes Σ_n $(n = 1, 2 \ldots N)$ cut in the common surface between the two volumes. The boundary wall between the coupling volumes V_1 and V_2 has an infinitely small thickness in the region of the coupling holes. The permittivity's and permeability's of the media filling the volumes V_1 and V_2, are equal to ε_1, μ_1 and ε_2, μ_2. Let the volume V_1 contains material bodies enclosed in local volumes V_{m_1} $(m_1 = 1, 2 \ldots M_1)$, which are bounded by smooth closed surfaces S_{m_1}. The bodies are characterized by homogeneous material parameters: permittivity ε_{m_1}, permeability μ_{m_1}, and conductivity σ_{m_1}. Analogously, the volume V_2 contains in local volumes V_{m_2} $(m_2 = 1, 2 \ldots M_2)$, material bodies are limited by smooth closed surfaces S_{m_2}, which characterized by homogeneous material parameters: permittivity ε_{m_2}, permeability μ_{m_2} and conductivity σ_{m_2} (Fig. 2.1).

Fields of extraneous sources can be defined in several different ways: 1) as fields of electromagnetic waves incident on bodies and holes in the scattering problem, 2) as fields of electromotive forces applied to the bodies which are different from zero only in certain areas of the volumes V_{m_1} and V_{m_2} in the radiation problem, and 3) as combination of these two fields. Without loss of generality, we will assume that the electromagnetic field $\{\vec{E}_0(\vec{r}), \vec{H}_0(\vec{r})\}$ of specified extraneous sources exist only inside the volume V_1 ($\vec{r}$ is the radius-vector of the observation point). The fields depend on time t as $e^{i\omega t}$ ($\omega = 2\pi f$ is the circular frequency, and f is the frequency in Hz). The problem is formulated as follows: to determine the total electromagnetic fields $\{\vec{E}_{V_1}(\vec{r}), \vec{H}_{V_1}(\vec{r})\}$ and $\{\vec{E}_{V_2}(\vec{r}), \vec{H}_{V_2}(\vec{r})\}$ in the volumes V_1 and V_2, satisfying the

M. V. Nesterenko et al., *Vibrator and Slot Antenna Arrays for Modern Applications*, Lecture Notes in Electrical Engineering 1244,
https://doi.org/10.1007/978-3-031-68346-6_2

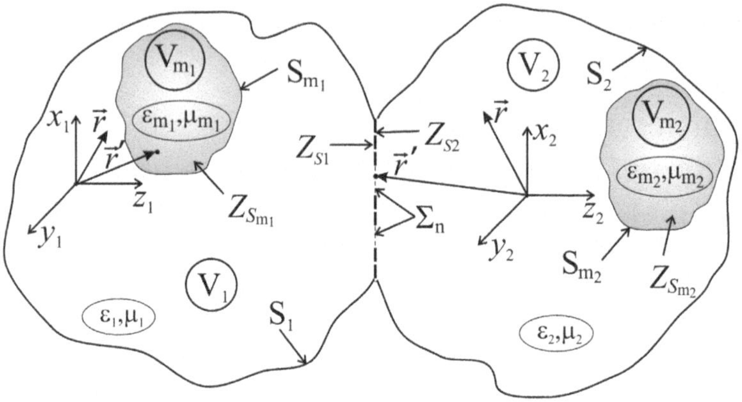

Fig. 2.1 The problem geometry and accepted notation

Maxwell equations and the boundary conditions on the surfaces S_{m_1}, S_{m_2}, Σ_n, S_1 and S_2.

The total electromagnetic fields in the volumes V_1 and V_2 can be expressed through the tangential field components on the surfaces S_{m_1}, S_{m_2} and Σ_n. In the Gaussian CGS system of units, the fields can be represented as Kirchhoff-Kotler integral equations [1, 2]

$$
\begin{aligned}
\vec{E}_{V_1}(\vec{r}) = \vec{E}_0(\vec{r}) &+ \frac{1}{4\pi ik\varepsilon_1}(\text{graddiv} + k_1^2) \sum_{m_1=1}^{M_1} \int\limits_{S_{m_1}} \hat{G}^e_{V_1}(\vec{r}, \vec{r}'_{m_1})[\vec{n}_{m_1}, \vec{H}_{V_1}(\vec{r}'_{m_1})]d\vec{r}'_{m_1} \\
&- \frac{1}{4\pi}\text{rot}\left\{ \sum_{m_1=1}^{M_1} \int\limits_{S_{m_1}} \hat{G}^m_{V_1}(\vec{r}, \vec{r}'_{m_1})[\vec{n}_{m_1}, \vec{E}_{V_1}(\vec{r}'_{m_1})]d\vec{r}'_{m_1} \right. \\
&\left. + \sum_{n=1}^{N} \int\limits_{\Sigma_n} \hat{G}^m_{V_1}(\vec{r}, \vec{r}'_n)[\vec{n}_n, \vec{E}_{V_1}(\vec{r}'_n)]d\vec{r}'_n \right\},
\end{aligned}
$$

$$
\begin{aligned}
\vec{H}_{V_1}(\vec{r}) = \vec{H}_0(\vec{r}) &+ \frac{1}{4\pi ik\mu_1}(\text{graddiv} + k_1^2) \left\{ \begin{aligned} &\sum_{m_1=1}^{M_1} \int\limits_{S_{m_1}} \hat{G}^m_{V_1}(\vec{r}, \vec{r}'_{m_1})[\vec{n}_{m_1}, \vec{E}_{V_1}(\vec{r}'_{m_1})]d\vec{r}'_{m_1} \\ &+ \sum_{n=1}^{N} \int\limits_{\Sigma_n} \hat{G}^m_{V_1}(\vec{r}, \vec{r}'_n)[\vec{n}_n, \vec{E}_{V_1}(\vec{r}'_n)]d\vec{r}'_n \end{aligned} \right\} \\
&+ \frac{1}{4\pi}\text{rot} \sum_{m_1=1}^{M_1} \int\limits_{S_{m_1}} \hat{G}^e_{V_1}(\vec{r}, \vec{r}'_{m_1})[\vec{n}_{m_1}, \vec{H}_{V_1}(\vec{r}'_{m_1})]d\vec{r}'_{m_1},
\end{aligned}
$$

$$\vec{E}_{V_2}(\vec{r}) = \frac{1}{4\pi i k \varepsilon_2}(\text{graddiv} + k_2^2)\sum_{m_2=1}^{M_2}\int\limits_{S_{m_2}} \hat{G}^e_{V_2}(\vec{r}, \vec{r}\,'_{m_2})[\vec{n}_{m_2}, \vec{H}_{V_2}(\vec{r}\,'_{m_2})]d\vec{r}\,'_{m_2}$$
$$-\frac{1}{4\pi}\text{rot}\left\{\sum_{m_2=1}^{M_2}\int\limits_{S_{m_2}} \hat{G}^m_{V_2}(\vec{r}, \vec{r}\,'_{m_2})[\vec{n}_{m_2}, \vec{E}_{V_2}(\vec{r}\,'_{m_2})]d\vec{r}\,'_{m_2}\right.$$
$$\left.+\sum_{n=1}^{N}\int\limits_{\Sigma_n} \hat{G}^m_{V_2}(\vec{r}, \vec{r}\,'_n)[\vec{n}_n, \vec{E}_{V_2}(\vec{r}\,'_n)]d\vec{r}\,'_n,\right.$$
$$\vec{H}_{V_2}(\vec{r}) = \frac{1}{4\pi i k \mu_2}(\text{graddiv} + k_2^2)\left\{\begin{array}{l}\sum\limits_{m_2=1}^{M_2}\int\limits_{S_{m_2}} \hat{G}^m_{V_2}(\vec{r}, \vec{r}\,'_{m_2})[\vec{n}_{m_2}, \vec{E}_{V_2}(\vec{r}\,'_{m_2})]d\vec{r}\,'_{m_2} \\ +\sum\limits_{n=1}^{N}\int\limits_{\Sigma_n} \hat{G}^m_{V_2}(\vec{r}, \vec{r}\,'_n)[\vec{n}_n, \vec{E}_{V_2}(\vec{r}\,'_n)]d\vec{r}\,'_n\end{array}\right\}$$
$$+\frac{1}{4\pi}\text{rot}\sum_{m_2=1}^{M_2}\int\limits_{S_{m_2}} \hat{G}^e_{V_2}(\vec{r}, \vec{r}\,'_{m_2})[\vec{n}_{m_2}, \vec{H}_{V_2}(\vec{r}\,'_{m_2})]d\vec{r}\,'_{m_2}, \tag{2.1}$$

where $k = 2\pi/\lambda$ is the wave number, λ is the free space wavelength, $k_1 = k\sqrt{\varepsilon_1\mu_1}$, $k_2 = k\sqrt{\varepsilon_2\mu_2}$ are the wave numbers in the media filling the volumes, $\vec{r}\,'_{m_1,m_2,n}$ are radius-vectors of source points located on surfaces S_{m_1}, S_{m_2} and Σ_n, $\vec{n}_{m1,m2,n}$ are the unit vectors of external normals to the surfaces, $\hat{G}^e_{V_1,V_2}(\vec{r}, \vec{r}\,')$ and $\hat{G}^m_{V_1,V_2}(\vec{r}, \vec{r}\,')$ are the electric and magnetic tensor Green's functions for the vector Hertz potentials of the coupling volumes, satisfying the Helmholtz vector equation and corresponding boundary conditions on the surfaces S_1 and S_2. The boundary conditions for $\hat{G}^{e,m}_{V_1,V_2}(\vec{r}, \vec{r}\,')$ at parts of the surfaces S_1 or S_2 displaced to infinity are transformed into the Sommerfeld radiation conditions.

The fields on the left-hand side of the Eqs. (2.1) are interpreted depending on the position of the observation point $\vec{r}$. If the observation point is located on the surfaces S_{m_1}, S_{m_2} of the bodies V_{m_1} and V_{m_2} or on the hole apertures Σ_n, the fields $\vec{E}(\vec{r})$ and $\vec{H}(\vec{r})$ coincide with fields in the integrands of the right-hand side of the Eq. (2.1) which is the inhomogeneous linear integral Fredholm equation of the second kind that has a unique solution. If the observation point lies outside the volumes V_{m_1}, V_{m_2} and surfaces Σ_n, the Eqs. (2.1) become equalities determining the total electromagnetic field by the field of the extraneous sources. These equalities allow us to solve, in the most general form, the excitation problem of electromagnetic fields by obstacles of finite size if the fields on the obstacle surfaces are known. Of course, these fields can be found if integral equations are solved.

The representation (2.1) can be also used to solve electrodynamic problems when some additional physical considerations for determining fields on the body surfaces are involved. For example, the current induced on well-conducting bodies ($\sigma \rightarrow$

∞), is concentrated near the body surface. Then, if the skin layer depth is small, the Shchukin-Leontovich approximate impedance boundary condition (1.17) can be used. If the impedance varies at the material body surface, the boundary condition can be presented as

$$[\vec{n},\vec{E}(\vec{r})]=\overline{Z}_S(\vec{r})[\vec{n},[\vec{n},\ \vec{H}(\vec{r})]], \tag{2.2}$$

where $\overline{Z}_S(\vec{r})=\overline{R}_S(\vec{r})+i\overline{X}_S(\vec{r})=Z_S(\vec{r})/Z_0$ is distributed variable surface impedance.

Based on the impedance boundary condition (2.2), the integral Eq. (2.1) can be converted to new variables, the density of the surface currents. Without loss of generality, let us convert the system (2.1) to the equation system relative to the surface currents for the configuration consisting of two material bodies in the volume V_1, coupling with the volume V_2 through a two holes. If the observation point is placed on the impedance surfaces S_{1_1} and S_{2_1} (indices 1, 2) and the continuity conditions for the tangential components of the magnetic fields on the holes Σ_1 and Σ_2 (indices 3, 4) are applied, the following system of integral equations can be obtained

$$\begin{aligned}
Z_{S1}(\vec{r}_1)\vec{J}_1^e(\vec{r}_1)&+\frac{k}{\omega}\text{rot}\left\{\int\limits_{\Sigma_1}\hat{G}_{V_1}^m(\vec{r},\vec{r}_3')\vec{J}_3^m(\vec{r}_3')d\vec{r}_3'+\int\limits_{\Sigma_2}\hat{G}_{V_1}^m(\vec{r},\vec{r}_4')\vec{J}_4^m(\vec{r}_4')d\vec{r}_4'\right\}\\
&=\vec{E}_0(\vec{r})\\
&+\frac{1}{i\omega\varepsilon_1}(\text{graddiv}+k_1^2)\left\{\int\limits_{S_{1_1}}\hat{G}_{V_1}^e(\vec{r},\vec{r}_1')\vec{J}_1^e(\vec{r}_1')d\vec{r}_1'+\int\limits_{S_{2_1}}\hat{G}_{V_1}^e(\vec{r},\vec{r}_2')\vec{J}_2^e(\vec{r}_2')d\vec{r}_2'\right\}\\
&+\frac{1}{4\pi}\text{rot}\left\{\int\limits_{S_{1_1}}\hat{G}_{V_1}^m(\vec{r},\vec{r}_1')Z_{S1}(\vec{r}_1')[\vec{n}_1,\ \vec{J}_1^e(\vec{r}_1')]d\vec{r}_1'\right.\\
&\left.+\int\limits_{S_{2_1}}\hat{G}_{V_1}^m(\vec{r},\vec{r}_2')Z_{S2}(\vec{r}_2')[\vec{n}_2,\ \vec{J}_2^e(\vec{r}_2')]d\vec{r}_2'\right\},
\end{aligned} \tag{2.3a}$$

$$\begin{aligned}
Z_{S2}(\vec{r}_2)\vec{J}_2^e(\vec{r}_2)&+\frac{k}{\omega}\text{rot}\int\limits_{\Sigma}\hat{G}_{V_1}^m(\vec{r},\vec{r}_3')\vec{J}_3^m(\vec{r}_3')d\vec{r}_3'=\vec{E}_0(\vec{r})\\
&+\frac{1}{i\omega\varepsilon_1}(\text{graddiv}+k_1^2)\left\{\int\limits_{S_{2_1}}\hat{G}_{V_1}^e(\vec{r},\vec{r}_2')\vec{J}_2^e(\vec{r}_2')d\vec{r}_2'+\int\limits_{S_{1_1}}\hat{G}_{V_1}^e(\vec{r},\vec{r}_1')\vec{J}_1^e(\vec{r}_1')d\vec{r}_1'\right\}\\
&+\frac{1}{4\pi}\text{rot}\left\{\int\limits_{S_{2_1}}\hat{G}_{V_1}^m(\vec{r},\vec{r}_2')Z_{S2}(\vec{r}_2')[\vec{n}_2,\ \vec{J}_2^e(\vec{r}_2')]d\vec{r}_2'\right.
\end{aligned}$$

$$
\left. + \int\limits_{S_{1_1}} \hat{G}^m_{V_1}(\vec{r}, \vec{r}_1{}')Z_{S1}(\vec{r}_1{}')[\vec{n}_1, \vec{J}^e_1(\vec{r}_1{}')]d\vec{r}_1{}' \right\}, \tag{2.3b}
$$

$$
\begin{aligned}
&\vec{H}_0(\vec{r}) + \frac{1}{i\omega\mu_1}(\text{graddiv} + k_1^2)\left\{ \int\limits_{\Sigma_1} \hat{G}^m_{V_1}(\vec{r}, \vec{r}_3{}')\vec{J}^m_3(\vec{r}_3{}')d\vec{r}_3{}' + \int\limits_{\Sigma_2} \hat{G}^m_{V_1}(\vec{r}, \vec{r}_4{}')\vec{J}^m_4(\vec{r}_4{}')d\vec{r}_4{}' \right\} \\
&+ \frac{1}{i\omega\mu_2}(\text{graddiv} + k_2^2)\left\{ \int\limits_{\Sigma_1} \hat{G}^m_{V_1}(\vec{r}, \vec{r}_3{}')\vec{J}^m_3(\vec{r}_3{}')d\vec{r}_3{}' + \int\limits_{\Sigma_2} \hat{G}^m_{V_1}(\vec{r}, \vec{r}_4{}')\vec{J}^m_4(\vec{r}_4{}')d\vec{r}_4{}' \right\} \\
&= \frac{1}{i\omega\varepsilon_1}(\text{graddiv} + k_1^2)\left\{ \begin{aligned} &\int\limits_{S_{1_1}} \hat{G}^m_{V_1}(\vec{r}, \vec{r}_1{}')Z_{S1}(\vec{r}_1{}')[\vec{n}_1, \vec{J}^e_1(\vec{r}_1{}')]d\vec{r}_1{}' \\ &+ \int\limits_{S_{2_1}} \hat{G}^m_{V_1}(\vec{r}, \vec{r}_2{}')Z_{S2}(\vec{r}_2{}')[\vec{n}_2, \vec{J}^e_2(\vec{r}_2{}')]d\vec{r}_2{}' \end{aligned} \right\} \\
&- \frac{k}{\omega}\text{rot}\left\{ \int\limits_{S_{1_1}} \hat{G}^e_{V_1}(\vec{r}, \vec{r}_1{}')\vec{J}^e_1(\vec{r}_1{}')d\vec{r}_1{}' + \int\limits_{S_{2_1}} \hat{G}^e_{V_1}(\vec{r}, \vec{r}_2{}')\vec{J}^e_2(\vec{r}_2{}')d\vec{r}_2{}' \right\},
\end{aligned} \tag{2.3c}
$$

with respect to electric surface current densities $\vec{J}^e_{1,2}(\vec{r}_{1,2})$ on $S_{1_1,2_1}$ and equivalent magnetic currents $\vec{J}^m_{3,4}(\vec{r}_{3,4})$ on $\Sigma_{1,2}$. These currents are equal to

$$
\vec{J}^e_{1,2}(\vec{r}_{1,2}) = \frac{c}{4\pi}[\vec{n}_{1,2}, \vec{H}(\vec{r}_{1,2})], \quad \vec{J}^m_{3,4}(\vec{r}_{3,4}) = -\frac{c}{4\pi}[\vec{n}_{3,4}, \vec{E}(\vec{r}_{3,4})]. \tag{2.4}
$$

Thus, the problem of electromagnetic fields excitation by the system of impedance bodies of finite sizes and by the system of coupling holes cut in the wall separating the electrodynamic volumes can be formulated as a boundary value problem of macroscopic electrodynamics which is reduced to integral equations for currents. The solution of this problem is an independent problem, often associated with significant mathematical difficulties. If the characteristic dimensions of the obstacles are much greater than the wavelength, that is, in the high-frequency approximation, the solution of the problem is usually sought in an inverse power of the wave number. In the low-frequency region, when the obstacles dimensions are much smaller than the wavelength, the representation of unknown functions in the form of power series expansion of the wavenumber reduces to solving a sequence of electrostatic problems. In the resonant region, where at least one of the dimensions of the obstacles is commensurable with the wavelength, the problem is the most difficult to analyze, since it requires a rigorous solution of the field equations. From a practical point of view, such obstacles as thin impedance vibrators and narrow slots in the resonant frequency domain are of particular interest.

2.1.1 Integral Equations for Electric and Magnetic Currents in Thin Vibrators and Narrow Slots

Solution of the equations system (2.3) for material bodies with complex surface forms S and coupling holes Σ with arbitrary geometry certain mathematical difficulties may encounter. However, the equation system solution for cylinders and holes whose transvers dimensions are small compared to free space wavelength, that is, for thin vibrators and narrow slots is considerably simplified [3, 4]. In addition, the boundary condition (2.2) can be extended to the vibrator surfaces with an arbitrary distribution of complex impedance on their surfaces, regardless of the exciting field structure and electrical characteristics of the vibrator material [5].

Let us transform the equations system (2.3) so that it can be applicable for thin vibrators made of circular cylindrical wires and rectilinear narrow slots. If the wire radii $r_{1,2}$ and lengths $2L_{1,2}$ and slot widths $d_{3,4}$ and lengths $2L_{3,4}$ satisfy the inequalities

$$\frac{r_{1,2}}{2L_{1,2}} \ll 1, \quad \frac{r_{1,2}}{\lambda_{1,2}} \ll 1, \quad \frac{d_{3,4}}{2L_{3,4}} \ll 1, \quad \frac{d_{3,4}}{\lambda_{1,2}} \ll 1, \tag{2.5}$$

($\lambda_{1,2}$ are the wavelengths in corresponding media), the induced currents on the vibrators and the equivalent magnetic currents in the slots can be represented as follows

$$\vec{J}^{e}_{1(2)}(\vec{r}_{1(2)}) = \vec{e}_{s_{1(2)}} J_{1(2)}(s_{1(2)}) \psi_{1(2)}(\rho_{1(2)}, \varphi_{1(2)}), \quad \vec{J}^{m}_{3(4)}(\vec{r}_{3(4)}) = \vec{e}_{s_{3(4)}} J_{3(4)}(s_{3(4)}) \chi_{3(4)}(\xi_{3(4)}), \tag{2.6}$$

where $\vec{e}_{s_{1(2)}}$ and $\vec{e}_{s_{3(4)}}$ are the unit vectors directed along the vibrator and slot axes, $s_{1(2)}$ and $s_{3(4)}$ are local coordinates associated with the vibrator and slot axes, $\psi_{1(2)}(\rho_{1(2)}, \varphi_{1(2)})$ are the functions of transverse polar coordinates $(\rho_{1(2)}, \varphi_{1(2)})$ for vibrators, $\chi_{3(4)}(\xi_{3(4)})$ are the functions of a transverse coordinates $\xi_{3(4)}$ for slots. The functions $\psi_{1(2)}(\rho_{1(2)}, \varphi_{1(2)})$ and $\chi_{3(4)}(\xi_{3(4)})$ satisfy the normalization conditions

$$\int\limits_{\perp_{1(2)}} \psi_{1(2)}(\rho_{1(2)}, \varphi_{1(2)}) \rho_{1(2)} d\rho_{1(2)} d\varphi_{1(2)} = 1, \quad \int\limits_{\xi_{3(4)}} \chi_{3(4)}(\xi_{3(4)}) d\xi_{3(4)} = 1. \tag{2.7}$$

The unknown currents $J_{1(2)}(s_{1(2)})$ and $J_{3(4)}(s_{3(4)})$ must satisfy the boundary conditions

$$J_{1(2)}(\pm L_{1(2)}) = 0, \quad J_{3(4)}(\pm L_{3(4)}) = 0, \tag{2.8}$$

where superscripts $e(m)$ are omitted. Let us project the Eqs. (2.3a), (2.3b) and (2.3c) on the axes of the vibrators and slots. Then, since $[\vec{n}_{1(2)}, \vec{J}_{1(2)}(\vec{r}_{1(2)})] \ll 1$ according to Eq. (2.5), the system of integral equations for currents on the vibrators and in the slot, accounting for mutual interactions between them can be presented as

$$
\begin{aligned}
&\left(\frac{d^2}{ds_1^2}+k_1^2\right)\left\{\int\limits_{-L_1}^{L_1} J_1(s_1')G_{s_1}^{V_1}(s_1,s_1')ds_1' + \int\limits_{-L_2}^{L_2} J_2(s_2')G_{s_2}^{V_1}(s_1,s_2')ds_2'\right\}\\
&\quad - ik\vec{e}_{s_1}\mathrm{rot}\left\{\int\limits_{-L_3}^{L_3} J_3(s_3')G_{s_3}^{V_1}(s_1,s_3')ds_3' + \int\limits_{-L_4}^{L_4} J_4(s_4')G_{s_4}^{V_1}(s_1,s_4')ds_4'\right\}\\
&\quad = -i\omega\varepsilon_1\left[E_{0s_1}(s_1) - z_{i1}(s_1)J_1(s_1)\right],\\
&\left(\frac{d^2}{ds_2^2}+k_1^2\right)\left\{\int\limits_{-L_2}^{L_2} J_2(s_2')G_{s_2}^{V_1}(s_2,s_2')ds_2' + \int\limits_{-L_1}^{L_1} J_1(s_1')G_{s_1}^{V_1}(s_2,s_1')ds_1'\right\}\\
&\quad - ik\vec{e}_{s_2}\mathrm{rot}\left\{\int\limits_{-L_3}^{L_3} J_3(s_3')G_{s_3}^{V_1}(s_2,s_3')ds_3' + \int\limits_{-L_4}^{L_4} J_4(s_4')G_{s_4}^{V_1}(s_2,s_4')ds_4'\right\}\\
&\quad = -i\omega\varepsilon_1\left[E_{0s_2}(s_2) - z_{i2}(s_2)J_2(s_2)\right],\\
&\frac{1}{\mu_1}\left(\frac{d^2}{ds_3^2}+k_1^2\right)\left\{\int\limits_{-L_3}^{L_3} J_3(s_3')G_{\mathrm{rot}s3}{}^{V_1}(s_3,s_3')ds_3' + \int\limits_{-L_4}^{L_4} J_4(s_4')G_{s_3}^{V_1}(s_3,s_4')ds_4'\right\}\\
&\quad + \frac{1}{\mu_2}\left(\frac{d^2}{ds_3^2}+k_2^2\right)\left\{\int\limits_{-L_3}^{L_3} J_3(s_3')G_{s_3}^{V_2}(s_3,s_3')ds_3' + \int\limits_{-L_4}^{L_4} J_4(s_4')G_{s_3}^{V_2}(s_3,s_4')ds_4'\right\}\\
&\quad + ik_1\vec{e}_{s_3}\mathrm{rot}\left\{\int\limits_{-L_1}^{L_1} J_1(s_1')G_{s_1}^{V_1}(s_3,s_1')ds_1' + \int\limits_{-L_2}^{L_2} J_2(s_2')G_{s_2}^{V_1}(s_3,s_2')ds_2'\right\} = -i\omega H_{0s_3}(s_3),\\
&\frac{1}{\mu_1}\left(\frac{d^2}{ds_4^2}+k_1^2\right)\left\{\int\limits_{-L_4}^{L_4} J_4(s_4')G_{s_4}^{V_1}(s_4,s_4')ds_4' + \int\limits_{-L_3}^{L_3} J_3(s_3')G_{s_4}^{V_1}(s_4,s_3')ds_3'\right\}\\
&\quad + \frac{1}{\mu_2}\left(\frac{d^2}{ds_4^2}+k_2^2\right)\left\{\int\limits_{-L_4}^{L_4} J_4(s_4')G_{s_4}^{V_2}(s_4,s_4')ds_4' + \int\limits_{-L_3}^{L_3} J_3(s_3')G_{s_4}^{V_2}(s_4,s_3')ds_3'\right\}\\
&\quad + ik_1\vec{e}_{s_4}\mathrm{rot}\left\{\int\limits_{-L_1}^{L_1} J_1(s_1')G_{s_1}^{V_1}(s_4,s_1')ds_1' + \int\limits_{-L_2}^{L_2} J_2(s_2')G_{s_2}^{V_1}(s_4,s_2')ds_2'\right\} = -i\omega H_{0s_4}(s_4),
\end{aligned}
\tag{2.9}
$$

where $z_{i1(2)}(s_{1(2)})$ ([Ohm/m]) are the internal linear vibrator impedances $Z_{S1(2)}(\vec{r}_{1(2)}) = 2\pi r_{1(2)} z_{i1(2)}(\vec{r}_{1(2)})$, $E_{0s_{1(2)}}(s_{1(2)})$, and $H_{0s_{3(4)}}(s_{3(4)})$ are the projections of the extraneous source fields on the vibrators and slots axes, and $G_{s_{1,2}}^{V_1}(s_{1,2,3,4}, s'_{1(2)})$, $G_{s_{3,4}}^{V_{1(2)}}(s_{1,2,3,4}, s'_{3(4)})$ are the corresponding components of the tensor Green's functions of the volumes V_1 and V_2.

If the interaction between the vibrators and the slots is absent, the system of Eqs. (2.9) is greatly simplified and has the following form

$$
\begin{aligned}
&\left(\frac{d^2}{ds_1^2}+k_1^2\right)\left\{\int_{-L_1}^{L_1} J_1(s'_1)G_{s_1}^{V_1}(s_1,s'_1)ds'_1+\int_{-L_2}^{L_2} J_2(s'_2)G_{s_2}^{V_1}(s_1,s'_2)ds'_2\right\}\\
&=-i\omega\varepsilon_1\left[E_{0s_1}(s_1)-z_{i1}(s_1)J_1(s_1)\right],\\
&\left(\frac{d^2}{ds_2^2}+k_1^2\right)\left\{\int_{-L_2}^{L_2} J_2(s'_2)G_{s_2}^{V_1}(s_2,s'_2)ds'_2+\int_{-L_1}^{L_1} J_1(s'_1)G_{s_1}^{V_1}(s_2,s'_1)ds'_1\right\}\\
&=-i\omega\varepsilon_1\left[E_{0s_2}(s_2)-z_{i2}(s_2)J_2(s_2)\right],\\
&\frac{1}{\mu_1}\left(\frac{d^2}{ds_3^2}+k_1^2\right)\left\{\int_{-L_3}^{L_3} J_3(s'_3)G_{s_3}^{V_1}(s_3,s'_3)ds'_3+\int_{-L_4}^{L_4} J_4(s'_4)G_{s_3}^{V_1}(s_3,s'_4)ds'_4\right\}\\
&\quad+\frac{1}{\mu_2}\left(\frac{d^2}{ds_3^2}+k_2^2\right)\left\{\int_{-L_3}^{L_3} J_3(s'_3)G_{s_3}^{V_2}(s_3,s'_3)ds'_3+\int_{-L_4}^{L_4} J_4(s'_4)G_{s_3}^{V_2}(s_3,s'_4)ds'_4\right\}\\
&=-i\omega H_{0s_3}(s_3),\\
&\frac{1}{\mu_1}\left(\frac{d^2}{ds_4^2}+k_1^2\right)\left\{\int_{-L_4}^{L_4} J_4(s'_4)G_{s_4}^{V_1}(s_4,s'_4)ds'_4+\int_{-L_3}^{L_3} J_3(s'_3)G_{s_4}^{V_1}(s_4,s'_3)ds'_3\right\}\\
&\quad+\frac{1}{\mu_2}\left(\frac{d^2}{ds_4^2}+k_2^2\right)\left\{\int_{-L_4}^{L_4} J_4(s'_4)G_{s_4}^{V_2}(s_4,s'_4)ds'_4+\int_{-L_3}^{L_3} J_3(s'_3)G_{s_4}^{V_2}(s_4,s'_3)ds'_3\right\}\\
&\quad=-i\omega H_{0s_4}(s_4).
\end{aligned}
\tag{2.10}
$$

If configuration consists only of one vibrator and one slot, the equations system (2.10) splits into two independent equations

$$
\left(\frac{d^2}{ds_1^2}+k_1^2\right)\int_{-L_v}^{L_v} J_v(s'_1)G_v^{V_1}(s_1,s'_1)ds'_1=-i\omega\varepsilon_1\left[E_{0s_1}(s_1)-z_i(s_1)J_v(s_1)\right],\tag{2.11}
$$

$$
\frac{1}{\mu_1}\left(\frac{d^2}{ds_2^2}+k_1^2\right)\int_{-L_{sl}}^{L_{sl}} J_{sl}(s'_2)G_{sl}^{V_1}(s_2,s'_2)ds'_2
$$

$$+\frac{1}{\mu_2}\left(\frac{d^2}{ds_2^2}+k_2^2\right)\int\limits_{-L_{sl}}^{L_{sl}} J_{sl}(s_2')G_{sl}^{V_2}(s_2,s_2')ds_2' = -i\omega H_{0s_2}(s_2), \tag{2.12}$$

where indices $\{v, 1\}$ and $\{sl, 2\}$ are related to the vibrator and slot, respectively, and in the free space V_1 (half-space $V_{1,2}$)

$$G_v^{V_1}(s_1,s_1') = \int\limits_{-\pi}^{\pi} \frac{e^{-ik\sqrt{(s_1-s_1')^2+[2r\sin(\varphi/2)]^2}}}{\sqrt{(s_1-s_1')^2+[2r\sin(\varphi/2)]^2}}\psi(r,\varphi)r\,d\varphi, \tag{2.13a}$$

$$G_{sl}^{V_{1,2}}(s_2,s_2') = \int\limits_{-d/2}^{d/2} \frac{e^{-ik\sqrt{(s_2-s_2')^2+(\xi)^2}}}{\sqrt{(s_2-s_2')^2+(\xi)^2}}\chi(\xi)d\xi. \tag{2.13b}$$

The problem solution based on integral equation kernels (2.13) may encounter serious difficulties, therefore we will use approximate expressions, known as "quasi-one-dimensional" kernels [3, 4, 6, 7]

$$G_v^{V_1}(s_1,s_1') = \frac{e^{-ik_1\sqrt{(s_1-s_1')^2+r^2}}}{\sqrt{(s_1-s_1')^2+r^2}}, \tag{2.14a}$$

$$G_{sl}^{V_{1,2}}(s_2,s_2') = \frac{e^{-ik_{1,2}\sqrt{(s_2-s_2')^2+(d/4)^2}}}{\sqrt{(s_2-s_2')^2+(d/4)^2}}, \tag{2.14b}$$

where the source points are assumed to be located on the geometric axes of the vibrator and slot. The observation points are placed on the physical vibrator surface and on the slot axis with the coordinates $\{s_2, \xi/2\}$. In this case, the functions $G_v^{V_1}(s_1,s_1')$ and $G_{sl}^{V_{1,2}}(s_2,s_2')$ are continuous everywhere, and the equations for the currents are significantly simplified without noticeable accuracy deterioration [8, 9].

Since the form of the Green's functions in the above equations was not specified, the Eqs. (2.9)–(2.12) are valid for any electrodynamic volumes if the corresponding Green's functions are known or can be constructed for these volumes.

Thus, the problem of the electromagnetic fields excitation by thin impedance vibrators and narrow slots connecting the two electrodynamic volumes has been reduced to the integral equations with respect to electric currents in the vibrators and equivalent magnetic currents in the slots. The solution of these equations is the final stage of the problem solution, since the total electromagnetic fields in the volumes can be easily found by using the expressions (2.1) and (2.4).

2.2 Methods for Solving Integral Equations for Currents

Direct numerical, numerical-analytical and analytical methods are used to solve boundary value problems on the excitation of electromagnetic fields by vibrator radiators (scatterers), slot holes of the coupling of electrodynamic volumes, as well as combined vibrator-slot structures. A detailed bibliography on these issues is presented in the monographs of the authors [2–4, 10]. Despite the fact that analytical methods are more vivid and effective as compared with numerical methods, various modifications the method of moment are used in the overwhelming majority of publications. In this regard, the method of moments, as the most universal in the theory of thin vibrators, slots, and systems based on these elements will be detailly consider below.

2.2.1 Basics of the Method of Moments

The essence of the method of moments (MoM) consists in reducing the integral equation to a system of linear algebraic equations (SLAE) by substituting the unknown functions as a linear combination of predefined functions with an unknown complex amplitudes. Thus, the solution for the electric or magnetic currents can be sought in the form:

$$\vec{J}(s) = \sum_{m=-M}^{M} J_m \vec{\varphi}_m(s), \tag{2.15}$$

where s is a coordinate along the vibrator or slot axis, J_m is unknown complex amplitudes, $\vec{\varphi}_m(s)$ are the vector basis functions. If the system of functions $\{\vec{\varphi}_m(s)\}$ is complete in the corresponding functional space, then an exact solution of the problem can be obtained under condition $|M| \to \infty$. To find an approximate solution, M is chosen finite based on a preliminary justification.

The right-hand side of the original equation is also represented as a sum $\sum_{\mu=-L}^{L} E_\mu \vec{\psi}_\mu(s)$, where vector of the linearly independent functions $\vec{\psi}_\mu(s)$ are called by weighted or trial functions. The problem solution can be subdivided into several stages: 1) both sides of the equation are scalar multiplied by weight functions $\vec{\psi}_\mu(s)$ and integrated over the surface of the vibrator or slot; 2) the inner products, known as moments, are calculated, and the equations are reduced to the matrix form of the SLAE; 3) the solution of the SLAE for unknown complex amplitudes J_m included in the representation (2.27) are found. At the first stage of the solution, the SLAE consisting of (2L + 1) equations with (2M + 1) unknown amplitudes is obtained. When M = L, the SLAE matrix turns out to be quadratic and the amplitudes J_m can be easily found. If the weight functions coincide with the basis functions $\left(\vec{\psi}_\mu = \vec{\varphi}_m\right)$, the method of moments is called the Galerkin method. The solution thus obtained has stationary properties, and the approach itself is equivalent to the variational Rayleigh–Ritz method.

In the theory of vibrator radiators, the Galerkin method is known as the generalized method of induced electromotive forces (EMF). This term is associated with the radio engineering, where the unknown function in the original integral equation corresponds to the electric current, and the equation kernel has the meaning of resistance. Therefore, the inner products, that is, the SLAE matrix coefficients, represent the EMF induced at some point on the antenna by the currents of other antenna sections or by other radiators. The right-hand side of the original equation usually corresponds to the extrinsic EMF. The term generalized method of the induced EMF reveals the physical meaning of the formulas used, therefore, the Galerkin method in the theory of slot radiators and coupling holes, is known as the generalized method of induced magnetomotive forces.

The SLAE can be solved numerically by using the Cramer's rule or matrix inversion, which is widely used for diffraction problem solutions. However, the matrix inversion may be inefficient and lengthy, especially when very large matrices are inverted or when a significant number of simulation options are required, for example, when studying the solution dependence upon many parameters. The inversion of ill-conditioned matrices, that is, matrices whose determinant is close to zero, can be difficult and can easily lead to large errors. Such situations can arise in problems related to perfectly conducting vibrators and slots in walls of electrodynamic volumes. These problems are based on Fredholm integral equations of the first kind. These problems are referred to as ill-posed electrodynamics problems whose numerical solution require additional procedures for its regularization.

In the method of moments, the choice of basic functions is of paramount importance. There exist an infinitely large number of basic and weighted function sets, however, for each specific electrodynamic problem, it is advisable to choose these functions so that they correspond to the physical entity and lead to efficient computational algorithms. The basic and weight functions, which for antenna radiators are called by spatial current harmonics, can be classified into two classes: 1) harmonics defined on the entire radiator, known as the basic functions of total domain; 2) harmonics that are non-zero only within some region of the radiator, known as the basic functions of subdomain or segment basic functions.

Among the most commonly used harmonics of the first group are:

- *harmonics of the Fourier series* $J(s) = \sum_m J_m e^{2\pi i m s/T}$. The Fourier series are often used for the approximation of functions, which properties are well studied. The formulas for SLAE coefficients are relatively simple, there are accuracy estimates based on a uniform and root-mean-square approximation. The harmonics of the Fourier series are often used to represent currents in rectilinear radiators, when the current on the radiator vanishes at its edges. In this case, the decomposition into sine or cosine functions can be quite natural if the coordinate origin coincides with the radiator center;

- *power polynomials* $J(s) = \sum_m J_m s^m$. This is also a class of approximating functions with known interpolation formulas and accuracy estimates. The power polynomials are often used with the Dirac delta functions as weighting functions, allowing to perform a pointwise function matching. This allows us to eliminate one of the two integrals in formulas for the mutual resistance. Polynomials $\left(1 - \frac{|s|}{L}\right)^m$ where L is the half length of the vibrator or slot can also be used;
- *Chebyshev polynomials* $J(s) = \sum_m J_m T_m(s)$. These polynomials are used as basis functions for representing currents in radiators with strong edge effect, for example, in strip vibrators or slots in infinitely thin walls of electrodynamic volumes. Since Chebyshev polynomials are characterized by the sharp increase at the edges of the interval, they allow us correctly describe the edge effect, that is, the diffraction at the radiator edge. This can accelerate convergence to the exact solution;
- *Bessel functions* $J(s) = \sum_m J_m I_m(s)$. They are usually used as basis functions for the radiators with axial symmetry, such as dielectric rods, spirals, etc.;

Consider some harmonics of the second group, which can be represented as $\varphi_m(s) = \begin{cases} f_m(s), & s \in \Delta s_m; \\ 0, & s \notin \Delta s_m. \end{cases}$. The most commonly used types of such harmonics are:

- *Dirac delta functions* $J(s) = \sum_m J_m \delta(s - s_m)$. These basis function can be used to satisfy the boundary conditions at separate points of the radiator. This procedure is also called pointwise stitching or the collocation method. Due to the properties of the delta function, calculation of the mutual resistances and the SLAE right-hand parts can be much-simplified. There are many examples of the successful application of this basis, however, it is often necessary to take into account quite a lot of stitching points and to solve a high order SLAE. A compromise between simplicity of the computational algorithms of the system matrix and the complexity of solving a SLAE with a large number of equations must be evaluated for specific problem solutions;
- *piecewise constant functions* $f_m(s) = J_m$. The properties of this basis functions is similar to that of the Dirac delta-functions. The integration universality and simplicity is compensated by slow convergence to the exact solution and the difficulty of solving high-order SLAE;
- *piecewise linear functions* $f_m(s)\Delta s_m = J_m(s_{m+1} - s) + J_{m+1}(s - s_m)$. This basis functions are characterized by comparative simplicity of integral calculating. It also accelerate convergence to the exact solution as compared with that for the piecewise constant functions;
- *piecewise sinusoidal functions* $f_m(s) \sin k\Delta s_m = J_m \sin k(s_{m+1} - s) + J_{m+1} \sin k(s - s_m)$. The advantage of this basis functions lies in the relatively small amount of computations, since the corresponding integrals can be presented in close forms allowing very fast convergence to the exact solution;

- *parabolic functions* $f_m(s) = A_m + B_m(s - s_m) + C_m(s - s_m)^2$. In this case, the subdomains can intersect. The properties of the approximation is close to the approximation by splines.

Recently, wavelets as basis and weight functions for numerical solution of electrodynamics problems, in particular, those concerning the theory of thin vibrators and narrow slots by the method of moments are increasingly used. In a number of cases, this approach can significantly improve the efficiency of numerical simulation, and, in addition, it optimizes the transition between the spatial-frequency and temporal representations of the solution. Wavelets are a generalized name for functions representing wave packets, localized along the axis of the independent variable, for example along the s coordinate in our case. The wave packets can be translated along the axis and scaled, that is, they can be compressed or stretched. These functions can be used in the method of moments both as basis functions of total domains and as segment basis functions. The Daubech orthogonal and Malvar wavelets are successfully used in the first and second cases.

2.2.2 Application of the Generalized Method of Induced EMMF for Multi-element Vibrator and Slot Structures

The integral equations for the multi-element vibrator-slot structures can be quite complicated, which significantly limits the application of various modifications of the numerical method of moments considered in Sect. 2.2.1. A promising method for analyzing vibrator-slot systems consists in the numerical-analytical implementation of the generalized method of induced electro-magneto-motive forces (EMMF) based on the previously built asymptotic solutions for currents are used as basic functions for the system elements.

Initially, the method of induced electromotive forces (EMF), as a special case of the EMMF was proposed for analyzing a system of linear perfectly conducting vibrators. In this approximations, the current on each m-th vibrator of the system can be described by basic functions $\vec{\varphi}_m(s_m) = \vec{e}_{s_m} \sin k(L_m - |s_m|)$, while functions $\vec{\psi}_\mu(s) = \vec{\varphi}_m(s)$ are used as weight functions. The number of equations in the system is equal to the number of vibrators, the complex amplitudes in the antinodes of the basic functions are unknown, the right-hand sides of the equations are source voltages applied at the center of each vibrator which are re-calculated to the current in the antinodes. Such a choice of weight functions is associated with the following physical considerations. If the current is approximated by one continuous basis function, the correspondence between the current at an antinode determining the vibrator radiation, and the current at the input of the vibrator defining the power supplied by the power source can be inevitably disrupted. The complex amplitude of the current can be chosen so that either the radiation power or other power calculated from this value is equal to or at least close to its true value. However, in this case, the balance between the power radiated by the vibrator and the power coming from the source can be violated. The choice of the weight functions $\vec{\psi}_\mu(s) = \vec{\varphi}_m(s)$ corresponds to such a selection of complex amplitudes that the balance of these powers is achieved, although both powers are determined with some error. The weight functions thus build cannot be optimal in all cases, although it is always desirable to use functions $\vec{\varphi}_m(s)$ and $\vec{\psi}_\mu(s)$ that do not differ substantially from exact solutions, especially if the vibrators have distributed surface impedance.

The most vulnerable point of the approach is that in real vibrator systems, due to the interaction of vibrators through surrounding space or asymmetrical self-excitation, the current distribution in each of them can be asymmetric. Therefore, using only the symmetric basis and weight functions considered above becomes inappropriate. That is, functions asymmetric relative to the vibrator center should be included into the system of basic functions representing the current in each vibrator; this require further generalizing the electromotive forces (EMF) method to use complex functional current distributions as adequate basic and weight functions. Similar arguments can be made for the induced magnetomotive forces (MMF) method, and, in general, for the hybrid method of induced EMMF.

2.3 Solution of Current Equations for Isolated Vibrator and Slot Structures

2.3.1 *Vibrator with Variable Surface Impedance in Free Space*

As is known, vibrators characterized by variable surface impedance can be used as an additional tool for controlling the antenna electrodynamic characteristics [2, 3, 10]. The results presented in this section have independent importance, besides, they will also be used through the monograph.

2.3.1.1 Solving the Current Equation by the Averaging Method

Consider the integral Eq. (2.11) defining the electric current on the vibrator located in free space ($\varepsilon_1 = \mu_1 = 1$)

$$\left(\frac{\mathrm{d}^2}{\mathrm{d}s^2} + k^2\right)\int\limits_{-L}^{L} J(s')\frac{e^{-ik\sqrt{(s-s')^2+r^2}}}{\sqrt{(s-s')^2+r^2}}\mathrm{d}s' = -i\omega[E_{0s}(s) - z_i(s)J(s)], \tag{2.16}$$

where $R(s, s') = \sqrt{(s-s')^2 + r^2}$. First, we single out the logarithmic singularity of the above equation kernel as it was done in [3, 6]. Then the vibrator current can be written as

$$\int_{-L}^{L} J(s')\frac{e^{-ikR(s,s')}}{R(s,s')}ds' = \omega(s)J(s) + \int_{-L}^{L} \frac{J(s')e^{-ikR(s,s')} - J(s)}{R(s,s')}ds', \tag{2.17}$$

where

$$\omega(s) = \int_{-L}^{L} \frac{ds'}{\sqrt{(s-s')^2 + r^2}} = \omega + \gamma(s), \tag{2.18}$$

$\omega = 2\ln\frac{2L}{r}$ is the large parameter, $\gamma(s) = \ln\frac{[(L+s)+\sqrt{(L+s)^2+r^2}][(L-s)+\sqrt{(L-s)^2+r^2}]}{4L^2}$ is a function equal to zero in the vibrator center and reaches the greatest value at the vibrator ends, where the current is zero according to the boundary conditions (2.8). Then, based on (2.18), the (2.16) can be transformed to the integral–differential equation with a small parameter

$$\frac{d^2J(s)}{ds^2} + k^2J(s) = \alpha\{i\omega E_{0s}(s) + F[s, J(s)] - i\omega z_i(s)J(s)\}, \tag{2.19}$$

where $\alpha = \frac{1}{2\ln[r/(2L)]}$ is the natural small parameter ($|\alpha| \ll 1$), and

$$F[s, J(s)] = -\frac{dJ(s')}{ds'}\frac{e^{-ikR(s,s')}}{R(s,s')}\bigg|_{-L}^{L} + \left[\frac{d^2J(s)}{ds^2} + k^2J(s)\right]\gamma(s)$$
$$+ \int_{-L}^{L} \frac{\left[\frac{d^2J(s')}{ds'^2} + k^2J(s')\right]e^{-ikR(s,s')} - \left[\frac{d^2J(s)}{ds^2} + k^2J(s)\right]}{R(s,s')}ds' \tag{2.20}$$

is the vibrator eigenfield in free space.

The approximate analytical solution of the Eq. (2.19) will be obtained by the asymptotic averaging method [3], Appendix B. To reduce (2.19) the equation system to the standard form with a small parameter, we make the change of variables and obtain by the method of variation of arbitrary constants

$$\begin{aligned}
&J(s) = A(s)\cos ks + B(s)\sin ks,\\
&\frac{dJ(s)}{ds} = -A(s)k\sin ks + B(s)k\cos ks,\\
&\left(\frac{dA(s)}{ds}\cos ks + \frac{dB(s)}{ds}\sin ks = 0\right),\\
&\frac{d^2J(s)}{ds^2} + k^2J(s) = -\frac{dA(s)}{ds}\sin ks + \frac{dB(s)}{ds}\cos ks,
\end{aligned} \tag{2.21}$$

where $A(s)$ and $B(s)$ are new unknown functions. Then, the Eq. (2.19) is transformed to the following system of integral–differential equations:

$$\frac{dA(s)}{ds} = -\frac{\alpha}{k}\left\{\begin{array}{l} i\omega E_{0s}(s) + F\left[s, A(s), \frac{dA(s)}{ds}, B(s), \frac{dB(s)}{ds}\right] \\ -i\omega z_i(s)[A(s)\cos ks + B(s)\sin ks] \end{array}\right\}\sin ks,$$

$$\frac{dB(s)}{ds} = +\frac{\alpha}{k}\left\{\begin{array}{l} i\omega E_{0s}(s) + F\left[s, A(s), \frac{dA(s)}{ds}, B(s), \frac{dB(s)}{ds}\right] \\ -i\omega z_i(s)[A(s)\cos ks + B(s)\sin ks] \end{array}\right\}\cos ks. \quad (2.22)$$

These equations are equivalent to the system of integral–differential Eq. (2.22) not resolved with respect to the derivative which are presented in the standard form. Since the right-hand sides of these equations are proportional to the small parameter α, the functions $A(s)$ and $B(s)$, on the right-hand sides of Eq. (2.22) are slowly varying functions, and the asymptotic method of averaging can be used to solve the equations system (2.22). The equations system (2.22) can be put into one-to-one correspondence with the simplified system, in which the derivatives $\frac{dA(s)}{ds}$ and $\frac{dB(s)}{ds}$ on right-hand sides of the equations are zeroes. After partial averaging the system (2.22) over the explicit variable s, we obtain the equations of the first approximation. The term partial means that the averaging operator acts on all terms, except those that contain the incident field $E_{0s}(s)$. Such averaging is valid for equation systems in the form:

$$\frac{d\overline{A}(s)}{ds} = -\alpha\left\{\frac{i\omega}{k}E_{0s}(s) + \overline{F}[s, \overline{A}(s), \overline{B}(s)]\right\}\sin ks + \chi_a\overline{A}(s) + \chi_s\overline{B}(s),$$

$$\frac{d\overline{B}(s)}{ds} = +\alpha\left\{\frac{i\omega}{k}E_{0s}(s) + \overline{F}[s, \overline{A}(s), \overline{B}(s)]\right\}\cos ks - \chi_s\overline{A}(s) - \chi_a\overline{B}(s), \quad (2.23)$$

where

$$\overline{F}[s, \overline{A}(s), \overline{B}(s)] = [\overline{A}(s')\sin ks' - \overline{B}(s')\cos ks']\left.\frac{e^{-ikR(s,s')}}{R(s,s')}\right|_{-L}^{L} \quad (2.24)$$

is the vibrator eigenfield averaged over its length. In the Eq. (2.23)

$$\chi_s = \lim_{2L\to\infty}\frac{1}{2L}\frac{\alpha i\omega}{k}\int_{-L}^{L} z_i^s(s)\sin^2 ks\{\cos^2 ks\}ds,$$

$$\chi_a = \lim_{2L\to\infty}\frac{1}{2L}\frac{\alpha i\omega}{k}\int_{-L}^{L}\frac{1}{2}z_i^a(s)\sin 2ksds, \quad (2.25)$$

where $z_i^s(s)$ and $z_i^a(s)$ are the symmetric and antisymmetric components of the impedance $z_i(s) = z_i^s(s) + z_i^a(s)$ distributed over the vibrator length. The solution of the equations system of (2.23) can be seek in the following form:

$$\begin{aligned}\overline{A}(s) &= \chi_s[C_1(s)\sin\chi s + C_2(s)\cos\chi s],\\ \overline{B}(s) &= -\chi_a[C_1(s)\sin\chi s + C_2(s)\cos\chi s] + \chi[C_1(s)\cos\chi s - C_2(s)\sin\cos\chi s],\end{aligned} \tag{2.26}$$

where $\chi^2 = \chi_s^2 - \chi_a^2$, $C_1(s)$ and $C_2(s)$ are new unknown functions. The functions $C_1(s)$, $C_2(s)$, and, hence, $\overline{A}(s)$, $\overline{B}(s)$ can be found by substituting the expressions (2.26) into the equations system (2.23). Thus, we obtain:

$$\begin{aligned}\bar{A}(s) &= \frac{1}{\chi}\left\{\bar{A}(-L)[\chi\cos\chi(L+s) + \chi_a\sin\chi(L+s)] + \bar{B}(-L)[\chi_s\sin\chi(L+s)]\right.\\ &\quad\left. + \alpha\int\limits_{-L}^{s}\begin{array}{l}\left\{\dfrac{i\omega}{k}E_{0s}(s') + \bar{F}[s',\bar{A},\bar{B}]\right\}\\ \times[\chi_s\cos ks'\sin\chi(s-s') - \chi\sin ks'\cos\chi(s-s')\\ +\chi_a\sin ks'\sin\chi(s-s')]ds'\end{array}\right\},\\ \bar{B}(s) &= \frac{1}{\chi}\left\{-\bar{A}(-L)[\chi_s\sin\chi(L+s)] + \bar{B}(-L)[\chi\cos\chi(L+s) - \chi_a\sin\chi(L+s)]\right.\\ &\quad\left. + \alpha\int\limits_{-L}^{s}\begin{array}{l}\left\{\dfrac{i\omega}{k}E_{0s}(s') + \bar{F}[s',\bar{A},\bar{B}]\right\}\\ \times[\chi_s\sin ks'\sin\chi(s-s') + \chi\cos ks'\cos\chi(s-s')\\ -\chi_a\cos ks'\sin\chi(s-s')]ds'\end{array}\right\}.\end{aligned} \tag{2.27}$$

Substituting the functions $\overline{A}(s)$ and $\overline{B}(s)$ as approximating functions for $A(s)$ and $B(s)$ in the formula $J(s) = A(s)\cos ks + B(s)\sin ks$, we obtain the most general asymptotic expression relative the parameter α for the current in the thin vibrator with impedance variable along its axis. The formula valid for arbitrary vibrator excitation can be written as

$$\begin{aligned}J(s) &= \overline{A}(-L)[\cos ks\cos\chi(L+s) - (\overline{\chi}_s\sin ks - \overline{\chi}_a\cos ks)\sin\chi(L+s)]\\ &\quad + \overline{B}(-L)[\sin ks\cos\chi(L+s) + (\overline{\chi}_s\cos ks - \overline{\chi}_a\sin ks)\sin\chi(L+s)]\\ &\quad + \alpha\int\limits_{-L}^{s}\left\{\frac{i\omega}{k}E_{0s}^{s}(s') + \overline{F}[s',\overline{A}(\pm L),\overline{B}(\pm L)]\right\}\{\sin k(s-s')\cos\chi(s-s')\\ &\quad + [\overline{\chi}_s\cos k(s-s') - \overline{\chi}_a\sin k(s+s')]\sin\chi(s-s')\}ds'\end{aligned} \tag{2.28}$$

where $\overline{\chi}_s = \chi_s/\chi$, $\overline{\chi}_a = \chi_a/\chi$.

When the vibrator impedance is constant, $\chi = \alpha\frac{i\omega}{2k}z_i$, $\overline{\chi}_s = 1$, and $\overline{\chi}_a = 0$, the expression (2.28) is reduced to

$$J(s) = \overline{A}(-L)\cos(\tilde{k}s + \chi L) + \overline{B}(-L)\sin(\tilde{k}s + \chi L)$$
$$+ \alpha \int_{-L}^{s} \left\{ \frac{i\omega}{k} E_{0s}(s') + \overline{F}[s', \overline{A}, \overline{B}] \right\} \sin \tilde{k}(s - s')ds', \tag{2.29}$$

where $\tilde{k} = k + \chi = k + i(\alpha/r)\overline{Z}_S$.

The constants $\overline{A}(\pm L)$ and $\overline{B}(\pm L)$ can be find by using the boundary conditions (2.8) for the current and symmetry conditions [6], that are uniquely related to the vibrator excitation method. If $E_{0s}(s) = E_{0s}^{s}(s)$, then $J(s) = J(-s) = J^{s}(s)$ and $\overline{A}(-L) = \overline{A}(+L), \overline{B}(-L) = -\overline{B}(+L)$; if $E_{0s}(s) = E_{0s}^{a}(s)$, then $J(s) = J(-s) = J^{a}(s)$ and $\overline{A}(-L) = -\overline{A}(+L)$, $\overline{B}(-L) = \overline{B}(+L)$. Then, for arbitrary vibrator excitation we obtain the following expression

$$J(s) = J^{s}(s) + J^{a}(s) = \alpha \frac{i\omega}{k} \left\{ \int_{-L}^{s} E_{0s}(s') \sin \tilde{k}(s - s')ds' \right.$$
$$- \frac{\sin \tilde{k}(L + s) + \alpha P^{s}[kr, \tilde{k}(L + s)]}{\sin 2\tilde{k}L + \alpha P^{s}(kr, 2\tilde{k}L)} \int_{-L}^{L} E_{0s}^{s}(s') \sin \tilde{k}(L - s')ds'$$
$$\left. - \frac{\sin \tilde{k}(L + s) + \alpha P^{a}[kr, \tilde{k}(L + s)]}{\sin 2\tilde{k}L + \alpha P^{a}(kr, 2\tilde{k}L)} \int_{-L}^{L} E_{0s}^{a}(s') \sin \tilde{k}(L - s')ds' \right\}, \tag{2.30}$$

where P^{s} and P^{a} are the vibrator eigenfield functions

$$P^{s}[kr, \tilde{k}(L + s)] = \int_{-L}^{s} \left[\frac{e^{-ikR(s', -L)}}{R(s', -L)} + \frac{e^{-ikR(s', L)}}{R(s', L)} \right] \sin \tilde{k}(s - s')ds' \Bigg|_{s=L} = P^{s}(kr, 2\tilde{k}L) \tag{2.31a}$$

$$P^{a}[kr, \tilde{k}(L + s)] = \int_{-L}^{s} \left[\frac{e^{-ikR(s', -L)}}{R(s', -L)} - \frac{e^{-ikR(s', L)}}{R(s', L)} \right] \sin \tilde{k}(s - s')ds' \Bigg|_{s=L} = P^{a}(kr, 2\tilde{k}L) \tag{2.31b}$$

Let the vibrator be excited by a normally incident plane electromagnetic wave with amplitude E_0 so that $E_{0s}^{s}(s) = E_0$, $E_{0s}^{a}(s) = 0$. Then, we obtain based on (2.30), (2.31a) the first approximation of the averaging method

$$J(s) = -\alpha \frac{i\omega}{k\tilde{k}} E_0 \frac{(\cos \tilde{k}s - \cos \tilde{k}L) + \alpha \left\{ \begin{array}{l} \sin \tilde{k}LP^{s}[kr, \tilde{k}(L + s)] \\ -[1 - \cos \tilde{k}(L + s)]P^{s}[kr, \tilde{k}L] \end{array} \right\}}{\cos \tilde{k}L + \alpha P^{s}[kr, \tilde{k}L]}, \tag{2.32}$$

where $P^s[kr, \tilde{k}L] = \int\limits_{-L}^{L} \frac{e^{-ik\sqrt{(L-s)^2+r^2}}}{\sqrt{(L-s)^2+r^2}} \cos\tilde{k}s\mathrm{d}s$.

Since the final expression for the current obtained from (2.28) is rather cumbersome and unsuitable for practical use, we will construct the solution of Eq. (2.16) by the generalized method of EMF by using formula (2.32).

2.3.1.2 Solving the Current Equation by the Generalized Method of Induced EMF

Let us apply the generalized method of induced EMF for the approximate analytical solution of Eq. (2.16) under symmetric vibrator excitation, $E_{0s}(s) = E_{0s}(-s)$, provided that $z_i(s) = z_i(-s)$. First, we approximate the distribution of electric current on the vibrator by

$$J(s) = J_0 f(s), \quad f(\pm L) = 0, \tag{2.33}$$

where J_0 is unknown current amplitude, $f(s)$ is the predefined function. Then, we multiply the left- and right-hand sides of Eq. (2.16) by $f(s)$ and integrate the resulting expressions over the vibrator length, and obtain

$$J_0 = \frac{-\frac{i\omega}{2k}\int_{-L}^{L} f(s)E_{0s}(s)\mathrm{d}s}{\frac{1}{2k}\int_{-L}^{L} f(s)\left[\left(\frac{\mathrm{d}^2}{\mathrm{d}s^2}+k^2\right)\int_{-L}^{L} f(s')\frac{e^{-ikR(s,s')}}{R(s,s')}\mathrm{d}s'\right]\mathrm{d}s - \frac{i\omega}{2k}\int_{-L}^{L} f^2(s)z_i(s)\mathrm{d}s}. \tag{2.34}$$

As is known, the accuracy of the integral equation solution for the vibrator current by the induced EMF method substantially depends on the choice of approximating functions for the current. If the impedance depends on the vibrator length, the most adequate to the real physical process, from our point of view, is a function containing information about the average value of the impedance along the vibrator. Such approach is quite obvious if the small fluctuations of the impedance around the average value are observed. The type of the functions defining impedance variation over the vibrator length considerable influences the current amplitude and, hence, the electrodynamic characteristics of the vibrator.

Consider normal incidence of the plane electromagnetic wave $E_{0s}(s) = E_0$ on the vibrator. Then, the first term in the numerator of (2.32) can be used as the function $f(s)$, i.e.,

$$f(s) = \cos\tilde{k}s - \cos\tilde{k}L, \tag{2.35}$$

where $\tilde{k} = k - \frac{i2\pi z_i^{av}}{Z_0\omega}$, $z_i^{av} = \frac{1}{2L}\int\limits_{-L}^{L} z_i(s)ds$ is the averaged internal impedance along the vibrator length, and $\omega = 2\ln(2L/r)$. The expression (2.35) contains information concerning the function $f(s)$ that is significantly different from the functions used in the literature. Argumentation for such function choice is presented in Subsection 2.3.1.3, were the function $f(s)$ is also proposed for the case $E_{0s}(s) = E_{0s}(-s)$, $z_i(s) \neq z_i(-s)$. After substituting (2.35) into (2.34) and calculating J_0, we obtain, based on (2.33), the expression

$$J(s) = -\frac{i\omega}{k\tilde{k}} E_0 \frac{(\sin\tilde{k}L - \tilde{k}L\cos\tilde{k}L)(\cos\tilde{k}s - \cos\tilde{k}L)}{Z(kr,\tilde{k}L) + F_z(\tilde{k}r,\tilde{k}L)}, \tag{2.36}$$

where

$$Z(kr,\tilde{k}L) = \frac{1}{2k}\int\limits_{-L}^{L}(\cos\tilde{k}s - \cos\tilde{k}L)\left(\frac{d^2}{ds^2} + k^2\right)F_f(s)ds, \tag{2.37}$$

$$F_f(s) = \int\limits_{-L}^{L}(\cos\tilde{k}s' - \cos\tilde{k}L)\frac{e^{-ikR(s,s')}}{R(s,s')}ds', \tag{2.38}$$

$$F_z(\tilde{k}r,\tilde{k}L) = -\frac{i}{r}\int\limits_{-L}^{L}(\cos\tilde{k}s - \cos\tilde{k}L)^2\overline{Z}_S(s)ds, \tag{2.39}$$

$\overline{Z}_S(s) = \frac{2\pi r z_i(s)}{Z_0}$ is the normalized surface complex impedance of the vibrator $\overline{Z}_S(s) = \overline{R}_S + i\overline{X}_S\phi(s)$, and $\phi(s)$ is the predefined function.

2.3.1.3 Justification of the Choice of Approximating Functions for Current

Let the vibrator with the distributed surface symmetric impedance, $z_i(s) = z_i(-s) = z_i^s(s)$, be excited by the symmetric field $E_{0s}(s) = E_{0s}(-s) = E_{0s}^s(s)$ created by external sources. Then we substitute the equalities $\chi_a = 0$, $\chi = \chi_s$, $J(s) = J(-s)$, $\overline{A}(-L) = \overline{A}(L) = C_1$, $\overline{B}(-L) = -\overline{B}(L) = C_2$ in the formula (2.28) and obtain that the current is

$$J(s) = C_1\cos[(k+\chi_s)s + \chi_s L] + C_2\sin[(k+\chi_s)s + \chi_s L]$$
$$+\alpha\int\limits_{-L}^{s}\left\{\frac{i\omega}{k}E_{0s}^s(s') + \overline{F}[s', C_1, C_2]\right\}\sin[(k+\chi_s)(s-s')]ds'. \tag{2.40}$$

The constants C_1 and C_2 are determined from the boundary conditions $J(\pm L) = 0$. Assuming $E_{0s}^s(s) = E_0$ in (2.40), we obtain

$$J(s) = J_c\left\{\alpha[\cos(k+\chi_s)s - \cos(k+\chi_s)L] + \alpha^2 P^s[kr, (k+\chi_s)s]\right\}, \tag{2.41}$$

where J_c is the current amplitude, $P^s[kr, (k+\chi_s)s]$ is the function defined in (2.31a). Assume that $z_i(s) = z_i\phi(s)$ where $\phi(s) = e^{-\beta\frac{|s|}{L}}$ Since the impedance $\overline{Z}_S = 2\pi r z_i/Z_0$, we obtain

$$k + \chi_s = k + i\frac{\alpha}{r}\overline{Z}_S\frac{1-e^{-\beta}}{\beta}. \tag{2.42}$$

Using the expression for $\tilde{k}$ defined in (2.35), we have

$$\tilde{k} = k - \frac{i2\pi z_i}{Z_0\omega}\frac{1}{2L}\int\limits_{-L}^{L}\phi(s)ds = k + i\frac{\alpha}{r}\overline{Z}_S\frac{1-e^{-\beta}}{\beta}. \tag{2.43}$$

As can be seen, expressions (2.42) and (2.43) are identically equal. When terms proportional to α^2 are discarded, the functions $f(s)$ in formulas (2.35) and (2.41) are also identical. The above conclusions confirm the validity of approximation (2.35) if the vibrator is symmetrically and the function $f(s)$ is also symmetric with respect to the vibrator feeding point.

The generalized method of induced EMF with adequate approximating functions can also be used to solve a more specific problem as compared to that defined in (2.32), namely: $E_{0s}(s) = E_{0s}^s(s)$, $z_i(s) = z_i^s(s) + z_i^a(s)$. In this case, the vibrator current will consist of symmetric and asymmetric parts

$$J(s) = J^s(s) + J^a(s) = J_0^s f^s(s) + J_0^a f^a(s). \tag{2.44}$$

The expression for the function $f^a(s)$ can be obtained from (2.28), assuming that $z_i(s) = -z_i(-s) = z_i^a(s)$. Finally, it can be written as

$$\begin{cases} f^a(s) = \sin ks sh\chi_a L - \sin kL sh\chi_a s, & \chi_a \neq 0, \\ f^a(s) = \sin 2ks - 2\sin ks\cos kL, & \chi_a = 0. \end{cases} \tag{2.45}$$

2.3.2 Impedance Vibrator Excited in an Arbitrary Point

In general case, the excitation field of extraneous sources E_{0s} and internal linear vibrator impedance $z_i(s)$, can be divided into symmetric and antisymmetric components relative to the vibrator geometrical center. These components marked by the indices "s" and "a" are presented as $E_{0s}(s) = E_{0s}^s(s)+E_{0s}^a(s)$ and $z_i(s) = z_i^s(s)+z_i^a(s)$. Quite naturally, the vibrator currents can also have also two components $J(s) = J^s(s) + J^a(s)$. Then the initial equation relative the vibrator current (2.11) has the following form

$$\left(\frac{d^2}{ds^2}+k_1^2\right)\int\limits_{-L}^{L}\left[J^s(s')+J^a(s')\right]G_s(s,s')ds'$$
$$=-i\omega\varepsilon_1\left\{\left[E_{0s}^s(s)+E_{0s}^a(s)\right]-\left[z_i^s(s)+z_i^a(s)\right]\left[J^s(s)+J^a(s)\right]\right\}, \tag{2.46}$$

where $E_{0s}^{s,a}(s)$ are the projections of extraneous source fields on the vibrator axis, $G_s(s,s')=\frac{e^{-ik_1\sqrt{(s-s')^2+r^2}}}{\sqrt{(s-s')^2+r^2}}$, s and s' are the local coordinates related to the vibrator axis and surface, $k_1=k_1'-ik_1''=k\sqrt{\varepsilon_1\mu_1}$, $k=2\pi/\lambda$, λ is the wavelength in free space, ε_1 and μ_1 are the permittivity and the permeability of the medium.

The Eq. (2.46) can be represented by the system of two coupled integral equations, relative to the unknown currents $J^s(s)$ and $J^a(s)$

$$\begin{cases}\left(\frac{d^2}{ds^2}+k_1^2\right)\int\limits_{-L}^{L}J^s(s')G_s(s,s')ds'=-i\omega\varepsilon_1\{E_{0s}^s(s)-[z_i^s(s)J^s(s)+z_i^a(s)J^a(s)]\},\\ \left(\frac{d^2}{ds^2}+k_1^2\right)\int\limits_{-L}^{L}J^a(s')G_s(s,s')ds'=-i\omega\varepsilon_1\{E_{0s}^a(s)-[z_i^s(s)J^a(s)+z_i^a(s)J^s(s)]\}.\end{cases} \tag{2.47}$$

The vibrator currents (unlike (2.35)) can be presented as product of the unknown complex amplitudes $J_n^{s,a}$ and weight functions $f_n^{s,a}(s')$ $(n=0,1)$ as

$$J^{s,a}(s')=J_0^{s,a}f_0^{s,a}(s')+J_1^{s,a}f_1^{s,a}(s'), f_n^{s,a}(\pm L)=0. \tag{2.48}$$

The solution of the equations system (2.47) can be obtained by the generalized method of induced EMF. To do so, let us multiply the left- and right- hand parts of the Eq. (2.47) by the functions $f_n^s(s)$ and $f_n^a(s)$, and integrate the resulting expressions over the vibrator length. Thus, the following SLAE is obtained

$$\begin{cases}J_0^sZ_{00}^{s\Sigma}+J_1^sZ_{01}^{s\Sigma}+J_0^a\tilde{\tilde{Z}}_{00}^{sa}+J_1^a\tilde{\tilde{Z}}_{01}^{sa}=-(i\omega/2k)E_0^s,\\ J_0^sZ_{10}^{s\Sigma}+J_1^sZ_{11}^{s\Sigma}+J_0^a\tilde{\tilde{Z}}_{10}^{sa}+J_1^a\tilde{\tilde{Z}}_{11}^{sa}=-(i\omega/2k)E_1^s,\\ J_0^aZ_{00}^{a\Sigma}+J_1^aZ_{01}^{a\Sigma}+J_0^s\tilde{\tilde{Z}}_{00}^{as}+J_1^s\tilde{\tilde{Z}}_{01}^{as}=-(i\omega/2k)E_0^a,\\ J_0^aZ_{10}^{a\Sigma}+J_1^aZ_{11}^{a\Sigma}+J_0^s\tilde{\tilde{Z}}_{10}^{as}+J_1^s\tilde{\tilde{Z}}_{11}^{as}=-(i\omega/2k)E_1^a,\end{cases} \tag{2.49}$$

where $(m=0,1;\ n=0,1)$,

$$Z_{mn}^{s,a}=\frac{1}{2k}\left\{-\frac{df_m^{s,a}(s)}{ds}A_n^{s,a}(s)\Bigg|_{-L}^{L}+\int\limits_{-L}^{L}\left[\frac{d^2f_m^{s,a}(s)}{ds^2}+k_1^2f_m^{s,a}(s)\right]A_n^{s,a}(s)ds\right\},$$

$$A_n^{s,a}(s) = \int_{-L}^{L} f_n^{s,a}(s')G_s(s,s')ds',$$

$$\tilde{Z}_{mn}^{s,a} = -\frac{i\omega}{2k}\int_{-L}^{L} f_m^{s,a}(s)f_n^{s,a}(s)z_i^s(s)ds,\; Z_{mn}^{(s,a)\Sigma} = Z_{mn}^{s,a} + \tilde{Z}_{mn}^{s,a},$$

$$\tilde{\tilde{Z}}_{mn}^{\left\{ sa \atop as \right\}} = -\frac{i\omega}{2k}\int_{-L}^{L} f_m^{\left\{ s \atop a \right\}}(s)f_n^{\left\{ a \atop s \right\}}(s)z_i^a(s)ds,\; E_m^{s,a} = \int_{-L}^{L} f_m^{s,a}(s)E_{0s}^{s,a}(s)ds. \tag{2.50}$$

If the symmetric and asymmetric parts of the vibrator impedance are $z_i^s(s) = const$ and $z_i^a(s) = 0$, the SLAE (2.49) can be divided into two independent equation systems relative to the unknown currents J_n^s and J_n^a

$$\begin{cases} J_0^s Z_{00}^{s\Sigma} + J_1^s Z_{01}^{s\Sigma} = -(i\omega/2k)E_0^s, \\ J_0^s Z_{10}^{s\Sigma} + J_1^s Z_{11}^{s\Sigma} = -(i\omega/2k)E_1^s, \end{cases}$$
$$\begin{cases} J_0^a Z_{00}^{a\Sigma} + J_1^a Z_{01}^{a\Sigma} = -(i\omega/2k)E_0^a, \\ J_0^a Z_{10}^{a\Sigma} + J_1^a Z_{11}^{a\Sigma} = -(i\omega/2k)E_1^a, \end{cases} \tag{2.51}$$

and the vibrator current can be written as $J(s) = J^s(s) + J^a(s)$, where

$$J^{s,a}(s) = -\frac{i\omega\varepsilon_1}{2k}\left[\begin{array}{l} \dfrac{E_0^{s,a}Z_{11}^{(s,a)\Sigma} - E_1^{s,a}Z_{01}^{(s,a)\Sigma}}{Z_{00}^{(s,a)\Sigma}Z_{11}^{(s,a)\Sigma} - Z_{10}^{(s,a)\Sigma}Z_{01}^{(s,a)\Sigma}} f_0^{s,a}(s) \\ + \dfrac{E_1^{s,a}Z_{00}^{(s,a)\Sigma} - E_0^{s,a}Z_{10}^{(s,a)\Sigma}}{Z_{00}^{(s,a)\Sigma}Z_{11}^{(s,a)\Sigma} - Z_{10}^{(s,a)\Sigma}Z_{01}^{(s,a)\Sigma}} f_1^{s,a}(s) \end{array}\right]. \tag{2.52}$$

Let the vibrator with the impedance $z_i^s(s) = const$, $z_i^a(s) = 0$, be excited in the point $s = -s_\delta$ by the voltage generator with amplitude V_0, as shown in Fig. 2.2. Then

$$\begin{aligned} E_{0s}(s) &= V_0\delta(s + s_\delta) = E_{0s}^s(s) + E_{0s}^a(s), \\ E_{0s}^s(s) &= \frac{V_0}{2}[\delta(s + s_\delta) + \delta(s - s_\delta)], \\ E_{0s}^a(s) &= \frac{V_0}{2}[\delta(s + s_\delta) - \delta(s - s_\delta)], \end{aligned} \tag{2.53}$$

where δ is the Dirac delta function. In this case, the vibrator current defined by (2.52) can be written as $J(s) = J^s(s) + J^a(s)$, where

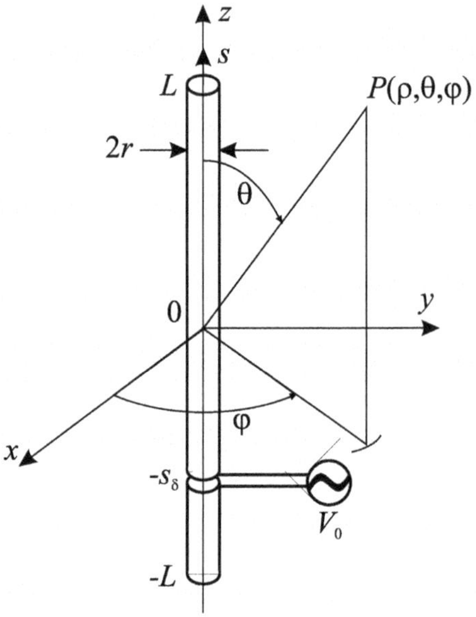

Fig. 2.2 The vibrator geometry and accepted notations

$$J^{s,a}(s) = -\frac{i\omega\varepsilon_1}{2k}V_0\left[\begin{array}{c}\dfrac{\tilde{E}_0^{s,a}Z_{11}^{(s,a)\Sigma} - \tilde{E}_1^{s,a}Z_{01}^{(s,a)\Sigma}}{Z_{00}^{(s,a)\Sigma}Z_{11}^{(s,a)\Sigma} - Z_{10}^{(s,a)\Sigma}Z_{01}^{(s,a)\Sigma}}f_0^{s,a}(s)\\ +\dfrac{\tilde{E}_1^{s,a}Z_{00}^{(s,a)\Sigma} - \tilde{E}_0^{s,a}Z_{10}^{(s,a)\Sigma}}{Z_{00}^{(s,a)\Sigma}Z_{11}^{(s,a)\Sigma} - Z_{10}^{(s,a)\Sigma}Z_{01}^{(s,a)\Sigma}}f_1^{s,a}(s)\end{array}\right]. \tag{2.54}$$

Let us choose the functions $f_0^{s,a}(s)$ obtained after substituting the expressions (2.53) into the general solution of the equation for the current by the averaging method in the form [3]:

$$\begin{aligned} f_0^s(s) &= \cos\tilde{k}_1 s_\delta \sin\tilde{k}_1 L\cos\tilde{k}_1 s - (1/2)\cos\tilde{k}_1 L(\sin\tilde{k}_1|s - s_\delta| + \sin\tilde{k}_1|s + s_\delta|),\\ f_0^a(s) &= \sin\tilde{k}_1 s_\delta \cos\tilde{k}_1 L\sin\tilde{k}_1 s + (1/2)\sin\tilde{k}_1 L(\sin\tilde{k}_1|s - s_\delta| - \sin\tilde{k}_1|s + s_\delta|),\end{aligned} \tag{2.55}$$

where $\tilde{k}_1 = k_1 - \frac{i\overline{Z}_S\sqrt{\varepsilon_1/\mu_1}}{2r\ln(2L/r)}$, $\overline{Z}_S = \overline{R}_S + i\overline{X}_S = 2\pi r z_i/Z_0$ is the distributed surface impedance, normalized on the free space wave impedance, $Z_0 = 120\pi\ Ohm$. The current distribution function $f_1^s(s)$ is defined in [3], and the function $f_1^a(s)$ can be found as solution of the integral Eq. (2.46), obtained for the case $z_i = 0$ [11]. These functions can be written as:

$$f_1^s(s) = \cos\tilde{k}_1 s - \cos\tilde{k}_1 L, \tag{2.56a}$$

$$f_1^a(s) = \sin k_1 s - (s/L)\sin k_1 L. \tag{2.56b}$$

The coefficients $Z_{mn}^{s,a}(m = 0, 1; n = 0, 1)$ in the formula (2.54) can be obtained by substituting the expressions (2.55) and (2.56) into the formulas (2.50)

$$Z_{0n}^{s} = \frac{\tilde{k}_1}{k}\left[\cos\tilde{k}_1 s_\delta A_n^s(L) - \cos\tilde{k}_1 L A_n^s(s_\delta)\right] + \frac{(k_1^2 - \tilde{k}_1^2)}{2k}\int\limits_{-L}^{L} f_0^s(s) A_n^s(s)ds,$$

$$Z_{1n}^{s} = \frac{\tilde{k}_1}{k}\sin\tilde{k}_1 L A_n^s(L) - \frac{1}{2k}\left[k_1^2\cos\tilde{k}_1 L\int\limits_{-L}^{L} A_n^s(s)ds - (k_1^2 - \tilde{k}_1^2)\int\limits_{-L}^{L}\cos\tilde{k}_1 s A_n^s(s)ds\right],$$

$$Z_{0n}^{a} = -\frac{\tilde{k}_1}{k}\left[\sin\tilde{k}_1 s_\delta A_n^a(L) - \sin\tilde{k}_1 L A_n^a(s_\delta)\right] + \frac{(k_1^2 - \tilde{k}_1^2)}{2k}\int\limits_{-L}^{L} f_0^a(s) A_n^a(s)ds,$$

$$Z_{1n}^{a} = \frac{k_1}{k}\left(\frac{\sin k_1 L}{k_1 L} - \cos k_1 L\right)A_n^a(L) - \frac{k_1^2}{2kL}\sin k_1 L\int\limits_{-L}^{L} A_n^a(s)sds,$$

$$\tilde{Z}_{mn}^{s,a} = \frac{\overline{Z}_S}{ir}\int\limits_{-L}^{L} f_m^{s,a}(s) f_n^{s,a}(s)ds,$$

$$\tilde{E}_0^s = \cos\tilde{k}_1 s_\delta\sin\tilde{k}_1(L - |s_\delta|),\ \tilde{E}_1^s = \cos\tilde{k}_1 s_\delta - \cos\tilde{k}_1 L,$$
$$\tilde{E}_0^a = -\sin\tilde{k}_1|s_\delta|\sin\tilde{k}_1(L - |s_\delta|),\ \tilde{E}_1^a = \sin k_1 s_\delta - \left(s_\delta/L\right)\sin k_1 L.$$

The input impedance $Z_{in} = R_{in} + iX_{in}$ and admittance $Y_{in} = G_{in} + iB_{in}$ can be presented as:

$$Z_{in}[\text{Ohm}] = \frac{60i/\varepsilon_1}{J_0^s f_0^s(s_\delta) + J_1^s f_1^s(s_\delta) + J_0^a f_0^a(s_\delta) + J_1^a f_1^a(s_\delta)},\ Y_{in}[\text{millimhos}] = \frac{10^3}{Z_{in}}. \tag{2.57}$$

Then, the voltage standing wave ratio (VSWR) in the antenna feeder with the wave impedance W is equal to:

$$\text{VSWR} = \frac{1 + |S_{11}|}{1 - |S_{11}|}, \tag{2.58}$$

where $S_{11} = \frac{Z_{in} - W}{Z_{in} + W}$ is the reflection coefficient in the feeder.

The comparison of the calculated and the experimental results in [3] allows us to state that the approximating functions (2.55) and (2.56) adequately represent the real physical process if the electrical lengths of the vibrator $(2L/\lambda) \leq 1.4$. Two other approximating functions $f_2^s(s)$ and $f_2^a(s)$ valid in the range $1.4 < (2L/\lambda) \leq 2.5$ can also be substituted in the expressions (2.48).These functions were obtained in [12] in the form

$$f_2^s(s) = \cos\frac{k_1 s}{2} - \cos\frac{k_1 L}{2}, \quad f_2^a(s) = \sin\frac{k_1 s}{2} - \sin\frac{k_1 L}{2} \tag{2.59}$$

If the perfectly conducting vibrator coated by the magnetodielectric with permittivity $\varepsilon = \varepsilon' - i\varepsilon''$ and permeability $\mu = \mu' - i\mu''$, the formula for the inductive surface impedance is $\overline{Z}_S = i\sqrt{\mu_1/\varepsilon_1}kr\mu\ln(r/r_i)$, where r and r_i are the vibrator and coating radiuses. This formula is valid if condition $|(k\sqrt{\varepsilon\mu}r)^2\ln(k\sqrt{\varepsilon\mu}r_i)| \ll 1$ is satisfied [3]. If $\varepsilon_1 = \mu_1 = 1$, the formula for surface impedance reduces to $\overline{Z}_S = ikrC_L$, where constant C_L is determined by the vibrator geometrical dimensions and electrical-physical parameters of the vibrator material. Formulas for specific realizations of the surface impedance of vibrators are presented in Appendix C.

The mathematical model was verified by comparing calculated and experimental data. The comparison results for the medium with material parameters $\varepsilon_1 = \mu_1 = 1$ and $s_\delta = 0.0$ are presented in Fig. 2.3.

To illustrate the efficiency of using impedance vibrators in practical applications, we present some calculated results [14]. On the basis of rigorous electrodynamic modeling, we will show the possibility of tuning an impedance vibrator with asymmetric excitation at several predetermined resonant frequencies. The multiparameter calculations, have allowed us to define the vibrator antenna design with the required input resistance based on the perfectly conducting cylinder coated by a uniform magnetic layer ($\mu = 4.7$). The vibrator parameters are as follow: total length $2L = 120.0$ mm, radius is 1.37 mm, external coating radius is 2.0 mm, and the local coordinate of the excitation point $s_\delta = -0.395L$. That is, the length of the vibrator arms should be equal to 36.3 mm and 83.7 mm. The frequency plots (the functions (2.55), (2.56) and (2.59)) of the modulus of the reflection coefficient $|S_{11}|$ for the vibrator antenna are presented in Fig. 2.4, where the boundaries of standard frequency ranges of the GSM 900 (880 ÷ 960 MHz), GSM 1800 (1710 ÷ 1880 MHz), WiMAX (3400 ÷ 3600 MHz) bands are marked by color strips. The dimensions of the proposed antenna are compared with dimensions of a typical smartphone case are also shown in Fig. 2.4.

As can be seen from Fig. 2.4, the impedance antenna is indeed characterized by three resonant frequencies $f_1 = 0.9\,GHz$, $f_2 = 1.79\,GHz$, and $f_3 = 3.48\,GHz$ with operating bands {0.866, 0.935}, {1.750, 1.835} and {3.446, 3.512} GHz at the VSWR level 2.0. It should be noted that if $VSWR = 2.5$ the operating bands of the impedance antenna are almost completely cover the frequency ranges of GSM 900 and GSM 1800, and 46% of the WiMAX range.

The simulation results have also confirmed that the electrodynamic characteristics of the triple-band antenna satisfy the requirements of practice. The frequency plots of the main antenna characteristic: directivity D, efficiency $\eta = 1 - |S_{11}|^2$, and gain $G = \eta D$ are presented in Fig. 2.5. The directivity was calculated by the formula

$$D = 2/\int_0^\pi |F(\theta)|^2 \sin(\theta)d\theta, \tag{2.60}$$

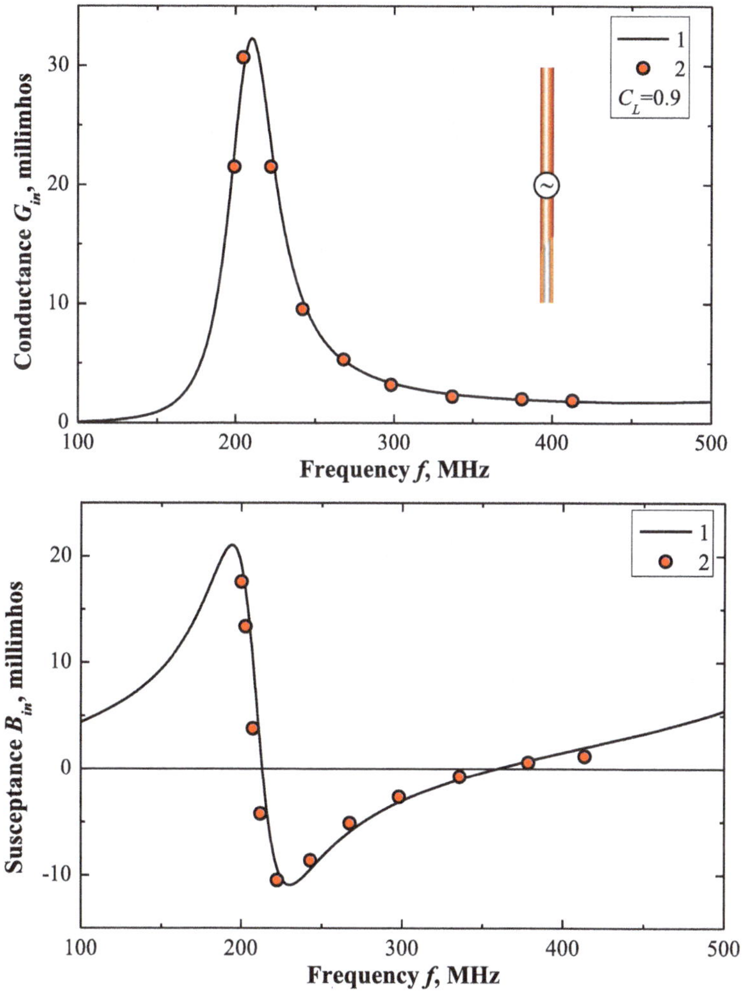

Fig. 2.3 The input admittance of the metallic conductor of the radius $r_i = 0.5175$ cm, covered by the ferrite ($\mu = 4.7$) shell of the radius $r = 0.6$ cm versus frequency at $2L = 30.0$ cm, $s_\delta = 0.0$: 1–calculation (the functions (2.55) and (2.56)), 2–experimental data [13]

where the antenna radiation pattern $F(\theta)$ in the denominator (2.60) was calculated by the formula $F(\theta) = \frac{\sin(\theta)}{F_{\max}} \int\limits_{-L}^{L} J(s)e^{iks\cos(\theta)}ds$, where $F_{\max}$ is the normalizing factor.

Note that the proposed method can be extended to vibrators of variable cross section without any particular fundamental difficulties [15].

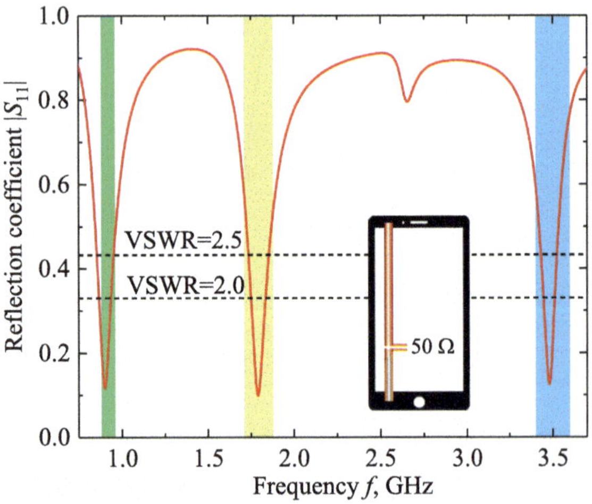

Fig. 2.4 The modulus of the reflection coefficient $|S_{11}|$ of the metallic conductor of the radius $r_i = 1.37$ mm, covered by the ferrite ($\mu = 4.7$) shell of the radius $r = 2.0$ mm ($C_L = 1.776$) versus frequency at $2L = 120.0$ mm, $s_8 = -0.395L$

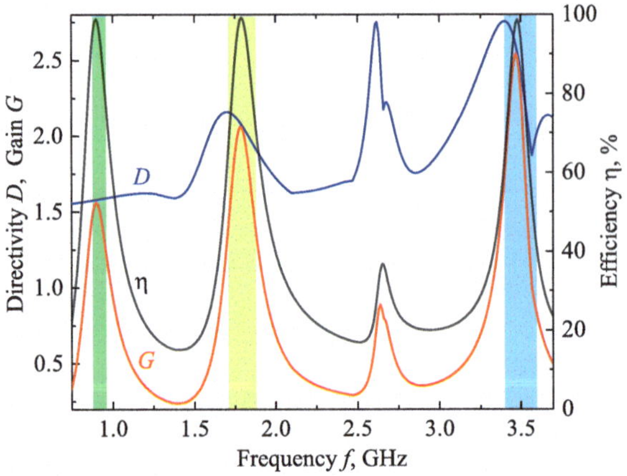

Fig. 2.5 The directivity D, gain G, and efficiency η of the triple-band antenna

2.3.3 *Narrow Slots in Rectangular Waveguide Walls*

2.3.3.1 Solving the Current Equation by the Averaging Method

Consider a narrow rectilinear slot cut in a broad wall of a hollow ($\varepsilon_2 = \mu = 1$) rectangular waveguide with a cross-section $\{a \times b\}$ (the region index is Wg). The slot radiates a free half-space placed over an infinite perfectly conducting plane (the region index is Hs) (Fig. 2.6). The slot length and width, $2L$ and d satisfy the inequalities $[d/(2L)] \ll 1$, $[d/\lambda] \ll 1$.

The structure analysis, according to (2.12), can be carried out based on the integral equation for the equivalent magnetic current in the slot

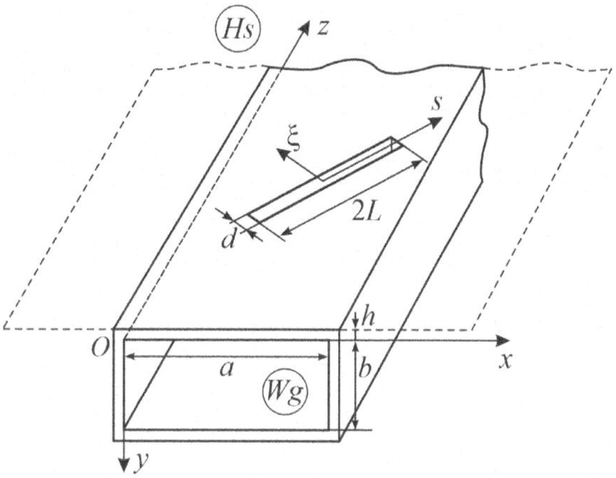

Fig. 2.6 The problem geometry and accepted notations

$$\left(\frac{\mathrm{d}^2}{\mathrm{d}s^2}+k^2\right)\int_{-L}^{L} J(s')\left[G_s^{Hs}(s,s')+G_s^{Wg}(s,s')\right]\mathrm{d}s' = -i\omega H_{0s}(s), \tag{2.61}$$

where $H_{0s}(s)$ is the projection of the field of extraneous sources to the slot axis, s is the local coordinate related to the slot axis, and $G_s^{Hs}(s,s') = 2\frac{e^{-ikR(s,s')}}{R(s,s')}$, $G_s^{Wg}(s,s') = 2\frac{e^{-ikR(s,s')}}{R(s,s')} + g_s^{Wg}(s,s')$, $R(s,s') = \sqrt{(s-s')^2+(d/4)^2}$, $g_s^{Wg}(s,s')$ is the s-component of the regular part of the magnetic Green's function of the waveguide, obtained as a result of the procedure for separating the Green's function of the free half-space from the Green's function of a rectangular waveguide [4].

The Eq. (2.61) will be solved by the averaging method. First, we made a change of variables according to the formulas (2.21). Then, if the slot is excited by the symmetric and antisymmetric field components $H_{0s}(s) = H_{0s}^s(s) + H_{0s}^a(s)$ mark by indices s and a, the current components in the slot can be obtained with an accuracy up to terms α^2 as

$$J(s) = J_0^s f^s(s) + J_0^a f^a(s) = \alpha\frac{i\omega}{k}\left\{\int_{-L}^{s} H_{0s}(s')\sin k(s-s')\mathrm{d}s' - \frac{\sin k(L+s)\int_{-L}^{L} H_{0s}^s(s')\sin k(L-s')\mathrm{d}s'}{\sin 2kL + \alpha N^s(kd_e, 2kL)} - \frac{\sin k(L+s)\int_{-L}^{L} H_{0s}^a(s')\sin k(L-s')\mathrm{d}s'}{\sin 2kL + \alpha N^a(kd_e, 2kL)}\right\}, \tag{2.62}$$

where $\alpha = 1/[8\ln(d/2L)]$ and $N^s(kd_e, 2kL)$, $N^a(kd_e, 2kL)$ are functions of the slot eigenfield determined by the magnetic Green's functions $G_s^{\Sigma} = G_s^{Hs} + G_s^{Wg}$ for the coupling volumes equal to

$$N^s(kd_e, 2kL) = \int_{-L}^{L} [G_s^{\Sigma}(d_e; s, -L) + G_s^{\Sigma}(d_e; s, L)] \sin k(L - s) ds,$$

$$N^a(kd_e, 2kL) = \int_{-L}^{L} [G_s^{\Sigma}(d_e; s, -L) - G_s^{\Sigma}(d_e; s, L)] \sin k(L - s) ds, \quad (2.63)$$

where d_e is equivalent slot width, which takes into account the waveguide wall thickness h. The functions d_e and d are related as [4]:

$$d_e = \mathrm{d}\frac{\mathrm{K}}{\mathrm{E(K)}}, \quad (2.64)$$

where E(K) is a complete elliptic integral of the second kind and K is the integral modulus. Convenient relations can be obtained from (2.64) in the two limiting cases [4]:

$$d_e = \mathrm{d}\left(1 - \frac{h}{\pi d}\ln\frac{d}{h}\right) if \frac{h}{d} \ll 1 \quad (2.65a)$$

$$d_e = \mathrm{d}\left(\frac{8}{\pi}e^{-\left(\frac{\pi h}{2d}+1\right)}\right) if \frac{h}{d} \ll 1 \quad (2.65b)$$

The formulas (2.65a) and (2.65b) can be combined with a sufficient accuracy into one expression [16], which is valid in the region $0 \leq (h/d) < 1 < 1$

$$d_e = de^{-\frac{\pi h}{2d}}. \quad (2.66)$$

First, we will find from (2.62) the current distribution functions for various slot positions relative to the waveguide walls. The functions $f^s(s)$ and $f^a(s)$ can be presented as (see Fig. 2.6):

(1) If the slot axis oriented at the angle φ relative to the longitudinal waveguide axis at the distance x_0 between the narrow waveguide wall and slot center ($s = 0$)

$$f^s(s) = \frac{\cos ks \cos k_2 L - \cos kL \cos k_2 s}{(\sin\varphi + (k_c/k_g)\cos\varphi)^2} e^{ik_c x_0} - \frac{\cos ks \cos k_1 L - \cos kL \cos k_1 s}{(\sin\varphi - (k_c/k_g)\cos\varphi)^2} e^{-ik_c x_0},$$

$$f^a(s) = \frac{\sin ks \sin k_2 L - \sin kL \sin k_2 s}{(\sin\varphi + (k_c/k_g)\cos\varphi)^2} e^{ik_c x_0} + \frac{\sin ks \sin k_1 L - \sin kL \sin k_1 s}{(\sin\varphi - (k_c/k_g)\cos\varphi)^2} e^{-ik_c x_0}, \quad (2.67)$$

where $k_1 = k_c \sin\varphi + k_g \cos\varphi$, $k_2 = k_c \sin\varphi - k_g \cos\varphi$, $k_g = \sqrt{k^2 - k_c^2}$, $k_c = \pi/a$;

(2) If the slot axis coincides with the longitudinal waveguide axis (longitudinal slot, $\varphi = 0^O$)

$$\begin{aligned} f^s(s) &= \cos ks \cos k_g L - \cos kL \cos k_g s, \\ f^a(s) &= \sin ks \sin k_g L - \sin kL \sin k_g s; \end{aligned} \quad (2.68)$$

(3) If the slot axis is perpendicular to the longitudinal waveguide axis (transverse slot, $\varphi = 90^O$)

$$\begin{aligned} f^s(s) &= \cos ks \cos k_c L - \cos kL \cos k_c s, \\ f^a(s) &= \sin ks \sin k_c L - \sin kL \sin k_c s. \end{aligned} \quad (2.69)$$

If $2L \approx \lambda/2$, $x_0 = a/2$, and $\varphi = 90^O$ (symmetric transverse slot), expressions (2.69) can be satisfactorily approximated by one symmetric function

$$f(s) = f^s(s) = \cos ks - \cos kL. \quad (2.70)$$

2.3.3.2 Symmetric Transverse Slot in the Broad Waveguide Wall

Consider the transverse slot cut in the broad wall of rectangular waveguide symmetrically to its longitudinal axis which radiates into free half-space above infinite perfectly conducting plane $\{x0y\}$. The Eq. (2.61) relative to the magnetic current in the slot will be solved by using the generalized method of induced MMF. Then we introduce the Cartesian coordinate system related to the waveguide, as in Fig. 2.6. Assume that the longitudinal slot axis and the local coordinate s coincide with the axis $\{0x\}$.

If the slot is excited by the wave $H_{0s}(s) = H_{0s}^s(s) = H_0 \cos k_c s$, where H_0 is the wave amplitude, and the slot current is approximated by the formula $J(s) = J_0(\cos ks - \cos kL)$ (2.70), the current amplitude according to expression (2.34) can be obtain in the form

$$\begin{aligned} J_0 &= -\frac{i\omega}{k} H_0 \frac{(1/k) f(kL)}{Z^{Hs}(kd_e, kL) + Z^{Wg}(kd_e, kL)} = -\frac{i\omega}{k}\tilde{J}_0, \\ f(kL) &= \sin kL - kL \cos kL. \end{aligned} \quad (2.71)$$

where

$$Z^{Hs}(kd_e, kL) = (Si4kL - iCin4kL) - 2\cos kL \times \left[2(\sin kL - kL\cos kL)\left(\ln\frac{16L}{d_e} - Cin2kL - iSi2kL\right) + \sin 2kLe^{-ikL}\right], \tag{2.72}$$

$$Z^{Wg}(kd_e, kL) = \frac{8\pi}{ab}\sum_{m=1,3..}^{\infty}\sum_{n=0,1..}^{\infty}\frac{\varepsilon_n k}{k_z(k^2 - k_x^2)}e^{-k_z\frac{d_e}{4}} [\sin kL\cos k_xL - (k/k_x)\cos kL\sin k_xL]^2, \tag{2.73}$$

Si and Cin are the integral sine and cosine, $\varepsilon_n = \begin{cases} 1, n = 0 \\ 2, n \neq 0 \end{cases}$, $k_x = \frac{m\pi}{a}$, $k_y = \frac{n\pi}{b}$, $k_z = \sqrt{k_x^2 + k_y^2 - k^2}$, m and n are integer.

If the formulas (2.70) and (2.71)are taken into account, the current in the slot is

$$J(s) = -H_0\frac{i\omega}{k}\tilde{J}_0(\cos ks - \cos kL). \tag{2.74}$$

The energy characteristics of the vibrator-slot structure, the field reflection and transmission coefficients S_{11} and S_{12}, and the power radiation coefficient $|S_\Sigma|^2$ under condition that the amplitude of the exciting H_0-wave is equal to unity, can be determined by the following expressions

$$S_{11} = \frac{8\pi ik_g}{abk^2}\tilde{J}_0 f(kL)e^{2ik_g z}, \tag{2.75}$$

$$S_{12} = 1 + \frac{8\pi ik_g}{abk^2}\tilde{J}_0 f(kL), \tag{2.76}$$

$$|S_\Sigma|^2 = 1 - |S_{11}|^2 - |S_{12}|^2. \tag{2.77}$$

The numerical and experimental results for the symmetric transverse slot radiator are presented in Fig. 2.7.

As can be seen from the plot, the theoretical and experimental curves are in satisfactory agreement with one another, confirming the legitimacy of applying the generalized method of induced MMF with the current distribution function found by the averaging method.

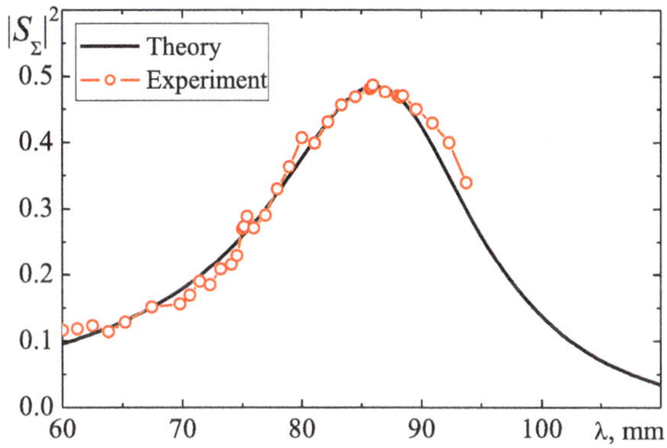

Fig. 2.7 The slot radiation coefficient $|S_\Sigma|^2$ as the function of wavelength: $a = 58.0$ mm, $b = 25.0$ mm, $h = 0.5$ mm, $d = 4.0$ mm, $2L = 40.0$ mm

2.3.3.3 Longitudinal Slot in the Broad Waveguide Wall

Consider a longitudinal slot cut in a broad wall of a rectangular waveguide at a distance x_0 from a narrow wall (in the Cartesian coordinate system shown in Fig. 2.6). The solution of Eq. (2.61) by the generalized method of induced MMF with approximating current functions $f^5(s)$ and $f^a(s)$ define in (2.68) can be written as

$$J(s) = -\frac{i\omega}{2k^2} H_0 \cos\frac{\pi x_0}{a}\left[J_0^s f^s(s) + iJ_0^a f^a(s)\right], \tag{2.78}$$

where $J_0^s = \frac{F_k^s}{Z_s^{Hs}+Z_s^{Wg}}$, $J_0^a = \frac{F_k^a}{Z_a^{Hs}+Z_a^{Wg}}$,

$$F_k^s = 2\cos k_g L\frac{\sin kL\cos k_g L - (k_g/k)\cos kL\sin k_g L}{1-(k_g/k)^2} - \cos kL\frac{\sin 2k_g L + 2k_g L}{2(k_g/k)},$$

$$F_k^a = 2\sin k_g L\frac{\cos kL\sin k_g L - (k_g/k)\sin kL\cos k_g L}{1-(k_g/k)^2} - \sin kL\frac{\sin 2k_g L - 2k_g L}{2(k_g/k)},$$

$$Z_s^{Hs} = 2\left\{\begin{array}{l}\left[\left(\dfrac{k_g}{k}\right)\cos kL\sin k_g L - \sin kL\cos k_g L\right]F^s(L) \\ +k\cos k_g L\displaystyle\int_{-L}^{L}\cos ks F^s(s)ds - \dfrac{k^2+k_g^2}{2k}\cos kL\displaystyle\int_{-L}^{L}\cos k_g s F^s(s)ds\end{array}\right\},$$

$$Z_a^{Hs} = 2\left\{\begin{array}{l}\left[\left(\dfrac{k_g}{k}\right)\sin kL\cos k_g L - \cos kL\sin k_g L\right]F^a(L) \\ +k\sin k_g L\displaystyle\int_{-L}^{L}\sin ks F^a(s)ds - \dfrac{k^2-k_g^2}{2k}\sin kL\displaystyle\int_{-L}^{L}\sin k_g s F^a(s)ds\end{array}\right\},$$

$$F^{s(a)}(s) = \int_{-L}^{L} f^{s(a)}(s') \frac{e^{-ik\sqrt{(s-s')^2+(d_e/4)^2}}}{\sqrt{(s-s')^2+(d_e/4)^2}} ds',$$

$$Z_s^{Wg} = \frac{2\pi}{ab} \sum_{m=0}^{\infty} \sum_{n=0}^{\infty} \frac{\varepsilon_m \varepsilon_n}{k^2} \cos k_x x_0 \cos k_x \left(x_0 + \frac{d_e}{4} \right)$$
$$\times \left\{ \begin{aligned} & \left[\cos k_g L \left(\frac{k}{k_z} \sin kL - \cos kL \right) \right] F_e^s \\ & - \frac{\cos kL}{k_z^2 + k_g^2} \left[(k_z^2 + k^2) \left(\frac{k_g}{k_z} \sin k_g L - \cos k_g L \right) F_e^s + k_c^2 F_k^s \right] \end{aligned} \right\},$$

$$F_e^s = \frac{k \cos k_g L}{k_z^2 + k^2} [k_z \cos kL (1 - e^{-2k_z L}) + k \sin kL (1 + e^{-2k_z L})]$$
$$- \frac{k \cos kL}{k_z^2 + k_g^2} [k_z \cos k_g L (1 - e^{2k_z L}) + k_g \sin k_g L (1 + e^{-2k_z L})],$$

$$Z_a^{Wg} = \frac{2\pi}{ab} \sum_{m=0}^{\infty} \sum_{n=0}^{\infty} \frac{\varepsilon_m \varepsilon_n}{k^2} \cos k_x x_0 \cos k_x \left(x_0 + \frac{d_e}{4} \right)$$
$$\times \left\{ \begin{aligned} & \left[- \sin k_g L \left(\frac{k}{k_z} \cos kL + \sin kL \right) \right] F_e^a \\ & + \frac{\sin kL}{k_z^2 + k_g^2} \left[(k_z^2 + k^2) \left(\frac{k_g}{k_z} \cos k_g L + \sin k_g L \right) F_e^a + k_c^2 F_k^a \right] \end{aligned} \right\},$$

$$F_e^a = \frac{k \sin k_g L}{k_z^2 + k^2} [k_z \sin kL (1 + e^{-2k_z L}) - k \cos kL (1 - e^{-2k_z L})]$$
$$- \frac{k \sin kL}{k_z^2 + k_g^2} [k_z \sin k_g L (1 + e^{2k_z L}) - k_g \cos k_g L (1 - e^{-2k_z L})],$$

$\varepsilon_m = \begin{cases} 1, m = 0 \\ 2, m \neq 0 \end{cases}$ and the axis of the current coordinate s coincides with the axis $\{0z\}$. These relations can be used to simulate the electrodynamic characteristics of longitudinal slot cut in the broad wall of rectangular waveguide which radiates into free half-space above perfectly conducting plane. Under condition $H_0 = 1$, the field reflection S_{11} and transmission S_{12} coefficients in the waveguide and the power radiation coefficient $|S_\Sigma|^2$ can be presented as

$$S_{11} = -\frac{2\pi k_c^2 \cos^2(\pi x_0/a)}{iabk_g k^3} (J_0^s F_k^s + J_0^a F_k^a) e^{2ik_g z}, \tag{2.79}$$

$$S_{12} = 1 - \frac{2\pi k_c^2 \cos^2(\pi x_0/a)}{iabk_g k^3} (J_0^s F_k^s - J_0^a F_k^a), \tag{2.80}$$

$$|S_\Sigma|^2 = 1 - |S_{11}|^2 - |S_{12}|^2. \tag{2.81}$$

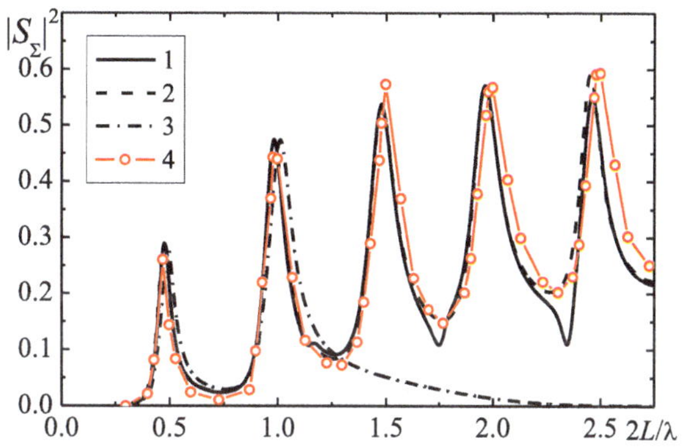

Fig. 2.8 The plots of the radiation coefficient $|S_\Sigma|^2$ as function of the slot electrical length: $a = 23.0$ mm, $b = 10.0$ mm, $d = 1.5$ mm, $\lambda = 30.0$ mm, $h = 1.0$ mm, $x_0 = a/4$; **1**–simulation by the formulas (2.69), **2**–simulation by the Galerkin method, **3**–simulation by the formula (2.82), **4**–experimental data

The plots of the experimental and simulated radiation coefficient $|S_\Sigma|^2$ as the functions of the slot length are shown in Fig. 2.8. The simulation results were obtained by using formulas (2.79) – (2.81), the moment (Galerkin) method with the current approximated by trigonometric functions $J(s) = \sum_{P=1}^{P} J_p \sin\frac{p\pi(L+s)}{2L}$ defined in Subsection 2.2.1 and by approximating function consisting of the first two basis functions [17]

$$J(s) = J_1 \cos\frac{\pi s}{2L} + J_2 \cos\frac{\pi s}{L} \tag{2.82}$$

As follows from the plots, the current approximating functions (2.69) and the Galerkin method can be successfully applied up to $(2L/\lambda) = 2.75$, while the approximating function (2.82) can be applied only $(2L/\lambda) = 1.25$.

References

1. Khizhnyak NA (1986) Integral equations of macroscopical electrodynamics. Naukova dumka, Kiev (in Russian)
2. Nesterenko MV, Katrich VA, Penkin YuM, Berdnik SL, Dumin OM (2020) Combined vibrator-slot structures. Theory and applications. Springer, Cham, Switzerland
3. Nesterenko MV, Katrich VA, Penkin YuM, Dakhov VM, Berdnik SL (2011) Thin impedance vibrators. Theory and applications. Springer, New York
4. Nesterenko MV, Katrich VA, Penkin YuM, Berdnik SL (2008) Analytical and hybrid methods in theory of slot-hole coupling of electrodynamic volumes. Springer, New York
5. Miller MA, Talanov VI (1961) Using the concept of surface impedance in the theory of surface electromagnetic fields (review). Izv Vuz Radiophysics 4(5):795–830 (in Russian)
6. King RWP (1956) The theory of linear antennas. Harv Univ Press, Cambr-Mass
7. Leontovich M (1946) On a theorem of the theory of diffraction and its application to diffraction by a narrow slot of arbitrary length. J Exp Theor Phys 16(6):474–479 (in Russian)
8. King RWP, Aronson EA, Harrison CW (1966) Determination of the admittance and effective length of cylindrical antennas. Radio Sci 1:835–850
9. Naiheng Y, Harrington R (1983) Electromagnetic coupling to an infinite wire through a slot in a conducting plane. IEEE Trans Antennas Propag 31:310–316

10. Penkin YuM, Katrich VA, Nesterenko MV, Berdnik SL, Dakhov VM (2019) Electromagnetic fields excited in volumes with spherical boundaries. Springer, Cham, Swizerland
11. King RWP, Wu TT (1965) The cylindrical antenna with arbitrary driving point. IEEE Trans Antennas Propag 13:710–718
12. King RWP, Smith GS (1981) Antennas in matter. MIT Press, Cambridge, USA
13. Bretones R, Martin RG, García IS (1995) Time-domain analysis of magnetic-coated wire antennas. IEEE Trans Antennas Propag 43:591–596
14. Nesterenko MV, Katrich VA, Berdnik SL, Dumin OM, Antonenko YO (2021) Asymmetric impedance vibrator for multi-band communication systems. Prog Electromagn Res M 102:81–89
15. Nesterenko MV, Gomozov AV, Katrich VA, Berdnik SL, Kijko VI (2022) Scattering of electromagnetic waves by impedance biconical vibrators in a free space and in a rectangular waveguide. Prog Electromagn Res C 119:275–285
16. Warne LK (1995) Eddy current power dissipation at sharp corners: closely spaced rectangular conductors. Journal of Electromagnetic Waves and Applications 9:1441–1458
17. Garb HL, Levinson IB, Fredberg PS (1968) Effect of wall thickness in slot problems of electrodynamics. Radio Eng Electron Phys 13:1888–1896

Chapter 3
Impedance Vibrators Antenna Arrays

3.1 System of Vibrators with Constant Surface Impedance in Free Space

3.1.1 Problem Formulation and Solution

Let us consider the system, consisting of N of the parallel impedance vibrators, located in free space. We introduce numeration of the vibrators and designate the length and the radius of the vibrator with the number n via $2L_n$ and r_n, correspondingly, and the corresponding coordinates of the vibrator centre in the Cartesian coordinate system—via z_n, x_n, y_n.

The projection of the $E_{0s_n}(s_n)$ impressed sources electrical field on the vibrator axis with the number n ($n = 1, 2...N$) can be represented in the form of the sum of two components: symmetrical $E^s_{0s_n}(s_n)$ and antisymmetrical $E^a_{0s_n}(s_n)$ relatively to the vibrator geometrical centre (s_n is the local coordinate along the vibrator axis with the number n): $E_{0s_n}(s_n) = E^s_{0s_n}(s_n) + E^a_{0s_n}(s_n)$ as in the case of the single vibrator (see Sect. 5.1). Satisfying the requirements of the impedance boundary condition for the electrical field on the surfaces of each of the vibrators, we obtain the integral–differential equations system relatively to the currents of the vibrators $J_n(s_n)$ in the form of ($m = 1, 2...N$):

$$\sum_{n=1}^{N}\left(\frac{d^2}{ds_m^2}+k^2\right)\int_{-L_n}^{L_n} J_n(s_n')G_{s_m}(s_m, s_n')\,ds_n'$$
$$= -i\omega[E^s_{0s_m}(s_m) + E^a_{0s_m}(s_m)] + i\omega z_{i_m}(s_m)J_m(s_m), \qquad (3.1)$$

where $z_{i_m}(s_m)$ is the internal impedance per unit length of the vibrator with the number m, changing along its length in a general case.

M. V. Nesterenko et al., *Vibrator and Slot Antenna Arrays for Modern Applications*, Lecture Notes in Electrical Engineering 1244,
https://doi.org/10.1007/978-3-031-68346-6_3

Because the impressed sources field is represented in the form of two components, the currents of each vibrator will also consist of two functions—symmetrical and antisymmetrical relatively to the centres of the vibrators—$J_n(s_n) = J_n^s(s_n) + J_n^a(s_n)$. Let us introduce the currents components in the vibrators in the form of the product of the $J_{nq}^{s,a}$ unknown complex amplitudes and the set scalar functions of distribution $f_{nq}^{s,a}(s_n)$ $(q = 0, 1...Q)$ farther:

$$J_n^{s,a}(s) = \sum_{q=0}^{Q} J_{nq}^{s,a} f_{nq}^{s,a}(s_n), \quad f_{nq}^{s,a}(\pm L_n) = 0. \tag{3.2}$$

Let us note, that as the system, consisting of the parallel vibrators, is considered, then it is not necessary to introduce the vector functions of the form (3.2) into the problem solution, the use of which is characteristic for the vibrators systems with arbitrary orientation in space. Then the equations system (3.1) can be written in the form:

$$\sum_{n=1}^{N}\sum_{q=0}^{Q}\left(\frac{d^2}{ds_m^2} + k^2\right)\int_{-L_n}^{L_n}\left[J_{nq}^s f_{nq}^s(s_n') + J_{nq}^a f_{nq}^a(s_n')\right]G_{s_m}(s_m, s_n')\,ds_n'$$
$$- i\omega z_{i_m}(s_m)\sum_{p=0}^{Q}\left[J_{mq}^s f_{mq}^s(s_m) + J_{mq}^a f_{mq}^a(s_m)\right] = -i\omega[E_{0s_m}^s(s_m) + E_{0s_m}^a(s_m)]. \tag{3.3}$$

We multiply the left and the right parts of the Eq. (3.3) on $f_{mp}^s(s_m)$ and $f_{mp}^a(s_m)$ $(p = 0, 1...Q)$ alternately and integrate them along the vibrators lengths due to the generalized method of the induced EMF. As a result, we obtain the system of the algebraic equations relatively to the J_{nq}^s and J_{nq}^a currents unknown amplitudes:

$$\begin{cases} \sum_{n=1}^{N}\sum_{q=0}^{Q}\left[J_{nq}^s\left(Z_{mn,pq}^{ss} + \delta_{mn}\tilde{Z}_{m,pq}^{ss}\right) + J_{nq}^a\left(Z_{mn,pq}^{sa} + \delta_{mn}\tilde{Z}_{m,pq}^{sa}\right)\right] = -\frac{i\omega}{2k}E_{0\,mp}^s, \\ \sum_{n=1}^{N}\sum_{q=0}^{Q}\left[J_{nq}^s\left(Z_{mn,pq}^{as} + \delta_{mn}\tilde{Z}_{m,pq}^{as}\right) + J_{nq}^a\left(Z_{mn,pq}^{aa} + \delta_{mn}\tilde{Z}_{m,pq}^{aa}\right)\right] = -\frac{i\omega}{2k}E_{0\,mp}^a. \end{cases} \tag{3.4}$$

The following symbols are accepted in (3.4):

$$Z_{mn,pq}^{\begin{Bmatrix} ss \\ aa \\ sa \\ as \end{Bmatrix}} = \frac{1}{2k} \int_{-L_m}^{L_m} f_{mp}^{\begin{Bmatrix} s \\ a \\ s \\ a \end{Bmatrix}}(s_m) \left[\left(\frac{d^2}{ds_m^2} + k^2 \right) \int_{-L_n}^{L_n} f_{nq}^{\begin{Bmatrix} s \\ a \\ a \\ s \end{Bmatrix}}(s_n') G_{s_m}(s_m, s_n') ds_n' \right] ds_m,$$

$$\tilde{Z}_{m,pq}^{\begin{Bmatrix} ss \\ aa \\ sa \\ as \end{Bmatrix}} = -\frac{i\omega}{2k} \int_{-L_m}^{L_m} f_{mp}^{\begin{Bmatrix} s \\ a \\ s \\ a \end{Bmatrix}}(s_m) f_{mq}^{\begin{Bmatrix} s \\ a \\ s \\ a \end{Bmatrix}}(s_m) z_{i\,m}(s_m) ds_m,$$

$$E_{0\,mp}^{s,a} = \int_{-L_m}^{L_m} f_{mp}^{s,a} E_{0s_m}^{s,a}(s_m) ds_m, \ \delta_{mn} = \begin{cases} 1 \text{ at } m = n, \\ 0 \text{ at } m \neq n. \end{cases}$$

Let us consider the Yagi-Uda array, named after its inventors [1, 2] and representing itself a linear system of the similar oriented vibrators, the axes of which are perpendicular to the line of their location, applied in practice widely, in details further. The main principles of operating of such an array are following. The supply by generator is obtained by only one of the vibrators, called active. The rest vibrators represent themselves passive radiators, the currents on which are induced at the expense of interconnection. Such a system is assigned for the operation only as the array with axial radiation. The required characteristics of directivity of the antenna are obtained by regulation of the passive vibrators lengths and the distances between them. The passive elements, located in the direction of maximal radiation (directors) must be a bit shorter than the active vibrator in order to obtain axial radiation. The vibrator, which is usually active, is resonant with the length less than $\lambda/2$ ($(0.45 \div 0.49)\lambda$ in more cases), and the directors lengths—about $(0.4 \div 0.45)\lambda$. Besides it is not obligatory to have a similar length and (or) a diameter for the directors. The reflector (the vibrator, located in the direction of minimal radiation) is a bit longer than the active vibrator. The distances between the directors are usually equal $(0.3 \div 0.4)\lambda$ and not similar to obtain optimal characteristics. The distance between the active vibrator and the reflector is usually a bit less than the distance between the active element and the nearest director, and it equals to 0.25λ approximately.

Let us locate the vibrators in space so, that their central points will be on the $\{0z\}$ axis of the Cartesian coordinate system, and the longitudinal axes of the vibrators will be oriented parallely to the $\{0x\}$ axis (Fig. 3.1).

The vibrator with the $n = 1$ number is active, with the $n = 2$ number is a reflector, and the rest vibrators ($n = 3, 4...N$) are directors. We consider, that the surface impedance of each of the vibrators is constant along its length, that is, $z_{i_n}(s_n) =$

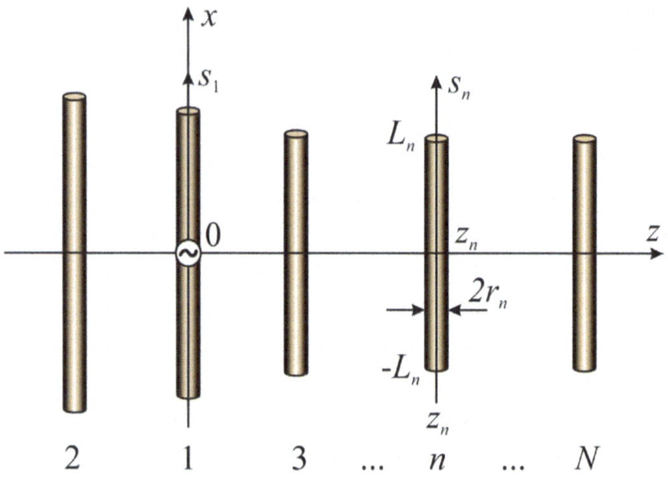

Fig. 3.1 The configuration of the Yagi-Uda antenna

$z_{i_n} = const$. The active vibrator ($n = 1$) is excited in the central point ($s_1 = 0$) from the δ-generator of harmonic oscillations with the amplitude of voltage V_0. Thus the projection of the impressed sources electrical field on the longitudinal axis of the first vibrator has only a symmetrical component relatively to its centre: $E_{0s_1}(s_1) = E^s_{0s_1}(s_1) = V_0\delta(s_1)$ and the field $E_{0s_n}(s_n) = 0$ at $n = 2, 3...N$.

As the system is symmetrical relatively to the $\{y0z\}$ plane, then the currents of each vibrator will have only symmetrical components:

$$J_n(s_n) = J^s_n(s_n) = \sum_{q=0}^{Q} J^s_{nq} f^s_{nq}(s_n)$$

and the equations system (5.23) is simplified sufficiently:

$$\sum_{n=1}^{N}\sum_{q=0}^{Q} J^s_{nq}\left(Z^s_{mn,pq} + \delta_{mn}\tilde{Z}^s_{m,pq}\right) = -\frac{i}{60}E^s_{0\,1p}\delta_{m1}, \quad m = 1, 2...N; \;\; p = 0, 1...Q. \tag{3.5}$$

Here

$$Z^s_{mn,pq} = \frac{1}{2k}\int_{-L_m}^{L_m}\int_{-L_n}^{L_n} f^s_{mp}(s_m) f^s_{nq}(s'_n)\tilde{G}_{s_m}(s_m, s'_n)\mathrm{d}s'_n\mathrm{d}s_m,$$

$$\tilde{Z}^s_{m,pq} = \frac{\overline{Z}_{S_m}}{ir_m}\int_{-L_m}^{L_m} f^s_{mp}(s_m) f^s_{mq}(s_m)\mathrm{d}s_m, \overline{Z}_{S_m} = \frac{r_m z_{i_m}}{60}, E^s_{0\,1p} = V_0 f^s_{1p}(0),$$

$$\tilde{G}_{s_m}(s_m, s'_n) = \left(\frac{d^2}{ds_m^2} + k^2\right) G(s_m, s'_n),\ G(s_m, s'_n) = \frac{e^{-ikR(s_m, s'_n)}}{R(s_m, s'_n)},$$

$R(s_m, s'_n) = \sqrt{(s_m - s'_n)^2 + z_{mn}^2}$, $z_{mn} = r_m$ at $m = n$ and $z_{mn} = (z_m - z_n)$ at $m \neq n$.

Let us make the following identical transformations for the convenience of calculations:

$$\tilde{G}(s_m, s'_n) = \frac{G(s_m, s'_n)}{R^4}\left\{(1 + ikR)\left(2R^2 - 3z_{mn}^2\right) + k^2 z_{mn}^2 R^2\right\}.$$

The distributed surface impedance is a complex value $\overline{Z}_{S_n} = \overline{R}_{S_n} + i\overline{X}_{S_n}$ in a general case. We obtain the impedance of an inductive kind, which, as it is shown in Chap. 2, can be represented in the form of $\overline{X}_{S_n} = kr_n C_{L_n}$ at $\overline{X}_{S_n} > 0$ and the impedance of a capacitive kind with its possible representation $\overline{X}_{S_n} = -\frac{C_{Cn}}{kr_n}$—at $\overline{X}_{S_n} < 0$.

We shall approximate the currents of the active vibrator (with the $n = 1$ number) by two functions due to the results of investigation of the characteristics of the single impedance vibrators, given in Chap. 2.

$$f_{10}^s(s_1) = \sin\tilde{k}_1(L_1 - |s_1|),\quad f_{11}^s(s_1) = \cos\tilde{k}_1 s_1 - \cos\tilde{k}_1 L_1, \tag{3.6}$$

and the currents of the passive vibrators ($n = 2, 3...N$)—by the functions:

$$f_{n1}^s(s_n) = \cos\tilde{k}_n s_n - \cos\tilde{k}_n L_n, \tag{3.7a}$$

$$f_{n2}^s(s_n) = \cos\frac{ks_n}{2} - \cos\frac{kL_n}{2}, \tag{3.7b}$$

where $\tilde{k}_n = k - \frac{i2\pi z_{in}}{Z_0\Omega_n}$, $\Omega_n = 2\ln(2L_n/r_n)$.

The array radiation pattern (RP) is defined by the following expression (θ and φ are the coordinates of the spherical coordinate system):

$$F(\theta, \varphi) = \frac{1}{F_N}\left|\sin(\arccos(\sin\theta\cos\varphi))\sum_{n=1}^{N} e^{ikz_n\cos\theta}\int_{-L_n}^{L_n} J_n(s_n)e^{iks_n\sin\theta\cos\varphi}ds_n\right|, \tag{3.8}$$

where, F_N is the normalizing multiplier, RP in the plane of the electrical field vector $\vec{E}$ equals to $F_E(\theta) = F(\theta, \varphi = 0)$ and RP in the plane of the magnetic field vector $\vec{H}$ equals to $F_H(\theta) = F(\theta, \varphi = \pi/2)$.

One can obtain the directive gain D by the formula, taking into consideration symmetry of RP of the array:

$$\mathrm{D} = \frac{\pi}{\int_0^{\pi/2}\int_0^{\pi} F^2(\theta,\varphi)\sin\theta\, d\theta\, d\varphi}. \tag{3.9}$$

It is possible to simplify calculation of D sufficiently, using the approximate formula $D = \sqrt{D_E D_H}$, where D_E, D_H are the directive gains, defined by radiation patterns in the planes of the vectors $\vec{E}$ and $\vec{H}$, correspondingly:

$$D_E = \frac{2}{\int_0^{\pi} F_E^2(\theta)\sin\theta\, d\theta}, \quad D_H = \frac{2}{\int_0^{\pi} F_H^2(\theta)\sin\theta\, d\theta}. \tag{3.10}$$

3.1.2 Numerical and Experimental Results

The reflector of the antenna must be created so, that the current, occurring in it, will lead the current of the active vibrator along the phase, and the current of the nearest director ($n = 3$) will lag along the phase in respect to the current of the active vibrator in order to amplify energy flux in one direction and to attenuate it in an opposite one. It is also necessary, that the current of the director 4 will lag along the phase in respect to the current of the director 3, the current in the director 5 will lag along the phase in respect to the current of the director 4 and so on.

The analysis, made on the ground of the generalized method of induced EMF shows, that it is necessary, that the complete impedance of the reflector, concerning the current antinode, has a positive (inductive) reactive component, and the complete impedance of the directors has a negative (capacitive) reactive component to obtain the phases indicated ratio. One can obtain a suitable reactive component of the vibrator impedance by the corresponding fitting of the vibrator length and (or) its surface impedance value (Fig. 3.2).

The reactive impedance with a positive sign is obtained at the vibrator lengthening in comparison with its resonant length or at the use of the inductive surface impedance and with a negative sign—at the vibrator shortening or the use of the capacitive surface impedance. The required reactive impedance, that one must obtain by the vibrator lengthening or shortening (the surface impedance value), also depends upon the dimensions of its cross section. The concrete values of the reactive impedances of the reflector and the directors and their lengths and values of the surface impedances, correspondingly, required for obtaining of one or another optimal characteristics, depend on the distance between the vibrators and also the number of directors.

The calculated distributions of the currents of the vibrators in comparison with the experimental data [3] for the array, consisting of two elements—the active and passive vibrators ($2L_1 = 2L_2 = \lambda$, $r_1 = r_2 = 0.007022\lambda$), located at the distance $\Delta z = 0.25\lambda$(Fig. 3.3) and $\Delta z = 0.125\lambda$ (Fig. 3.4) from each other, are given in Fig. 3.3 and Fig. 3.4 in order to prove adequacy of the proposed mathematical model to real physical process. It is necessary to approximate the currents for such a length of

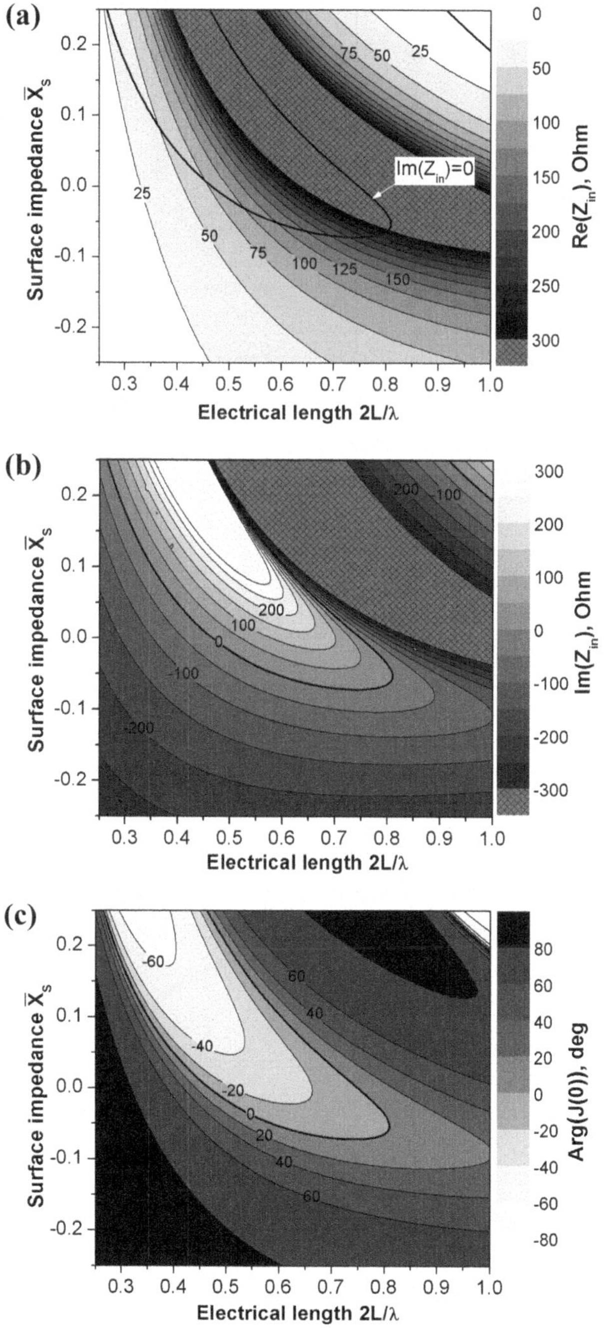

Fig. 3.2 The active (**a**) and reactive (**b**) parts of the vibrator input impedance, and also the vibrator current phase in the feeding point (**c**) in dependence from the vibrator electrical length and the value of the surface impedance reactive part at $r = 0.01\lambda$, $\lambda = 100$ mm

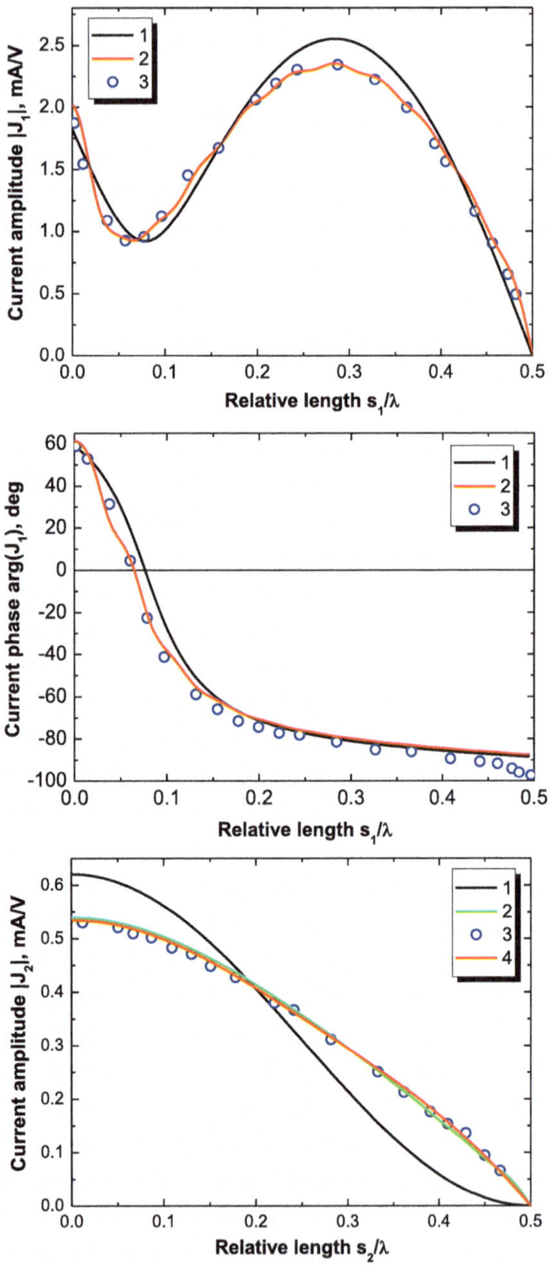

Fig. 3.3 The currents distribution in the array, consisting of two vibrators, at $\Delta z = 0.25\lambda$: 1–the calculation (the functions (3.6) and (3.7a)), 2–the calculation (the functions $J(s) = \sum_{n=1}^{20} J_n \sin \frac{n\pi(L+s)}{2L}$, moment method, Chap. 2), 3–the experimental data [3], 4–the calculation (the functions (3.7a) and (3.7b))

the passive vibrators by two functions (3.7), at this one observes a rather satisfactory agreement of the calculated and experimental data.

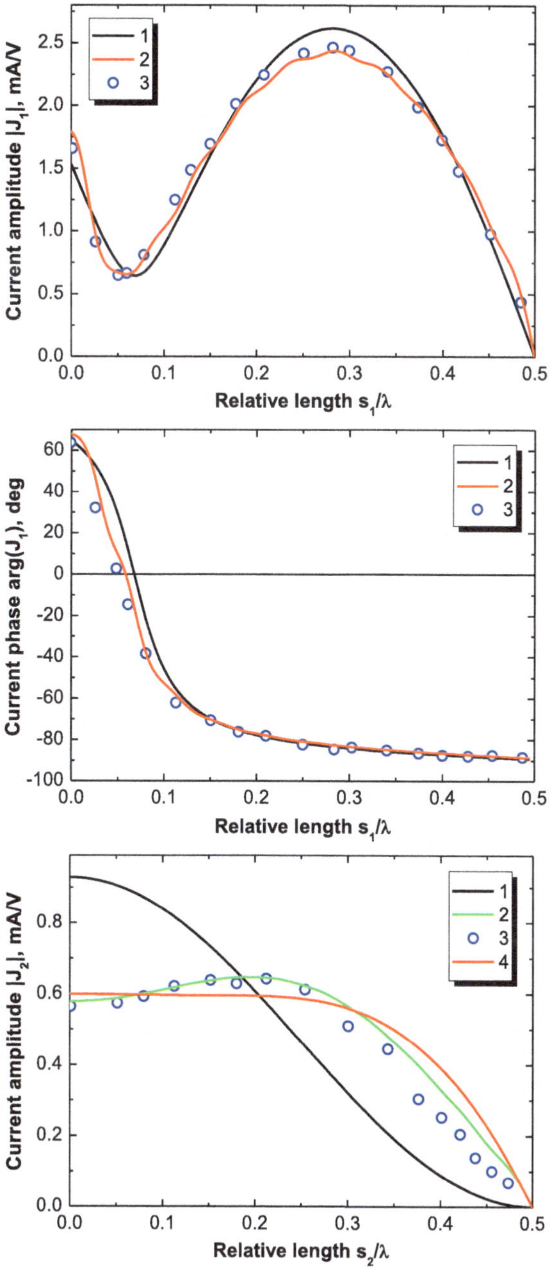

Fig. 3.4 The currents distribution in the array, consisting of two vibrators, at $\Delta z = 0.125\lambda$: 1–the calculation (the functions (3.6) and (3.7a)), 2–the calculation (the functions $J(s) = \sum_{n=1}^{20} J_n \sin\frac{n\pi(L+s)}{2L}$, moment method, Chap. 2), 3–the experimental data [3], 4–the calculation (the functions (3.7a) and (3.7b))

The use of the impedance vibrators in the Yagi-Uda antenna allows to increase the matching of the antenna with the feeding feeder. The change of the electrical length of the active vibrator with the corresponding change of $\overline{X}_S$ simultaneously gives the opportunity to vary the value of the active part of the input impedance R_{in} in wide ranges so, that at this the reactive part of the input resistance stays equal to null. At this the directed characteristics of the antenna radiation (D, the level of the side lobes) do not decrease. Figures 3.5 and 3.6 give the results of investigation of the array, consisting of 3 vibrators—an active vibrator, a reflector and a director. At this $2L_2 = 50$ mm, $2L_3 = 38$ mm, $r_{1,2,3} = 1$ mm, the distance between the vibrators $\Delta z = 25$ mm ($z_3 = -z_2 = 25$ mm). The surface impedances of the reflector and the director equal to null, and the value of the surface impedance of the active vibrator and its length are fitted so, that the antenna input impedance is active and equals to the wave resistance W of the feeder (50, 75 or 110 Ohm (Ω is omitted in the plots)) on the set wavelength ($\lambda = 100$ mm in this case). The parameters, corresponding to the system in question, are represented in Table 3.1. Figure 3.5 gives RP, Fig. 3.6 gives the dependences of the coefficients of reflection in the feeding feeder of the arrays in question from the wavelength for feeders with different wave resistances.

We note that, if the decrease of the value of the active vibrator input impedance real part is observed at the tuning of the Yagi-Uda antenna, then application of the capacitive impedance (and at this longer vibrators) permits to increase the R_{in} value. The tuning of the antenna on the set wavelength or the retuning of the antenna on another wavelength is also possible by means of the surface impedance change.

The required currents phases of the vibrators in the array to provide with the axial radiation can be obtained by fitting the suitable value of the surface impedance of each of the vibrators of the array at the unchangeable length of the vibrators. Figure 3.7 gives RP of the arrays, consisting of three vibrators ($z_2 = 0.25\lambda, z_3 = 0.2\lambda$, $r_{1,2,3} = 1$ mm, $\lambda = 100$ mm): 1—the surface impedances of all vibrators equal to null; 2—the lengths of all vibrators equal to $2L_n = 0.5\lambda$; 3—the lengths of all vibrators equal to $2L_n = 0.35\lambda$; 4—the lengths of all vibrators equal to $2L_n = 0.65\lambda$;

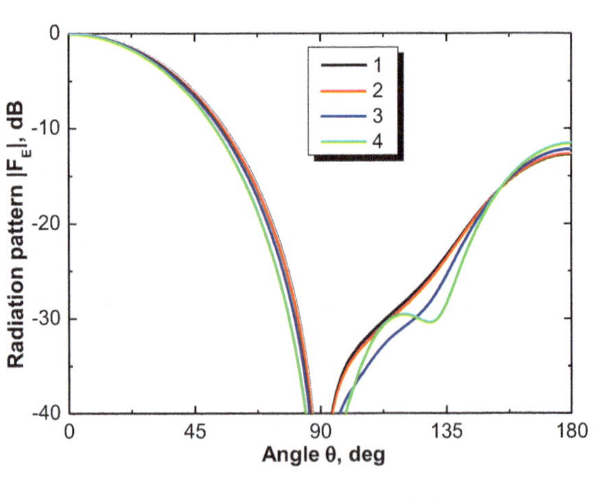

Fig. 3.5 RP of the Yagi-Uda arrays with the impedance active vibrator (from Table 3.1)

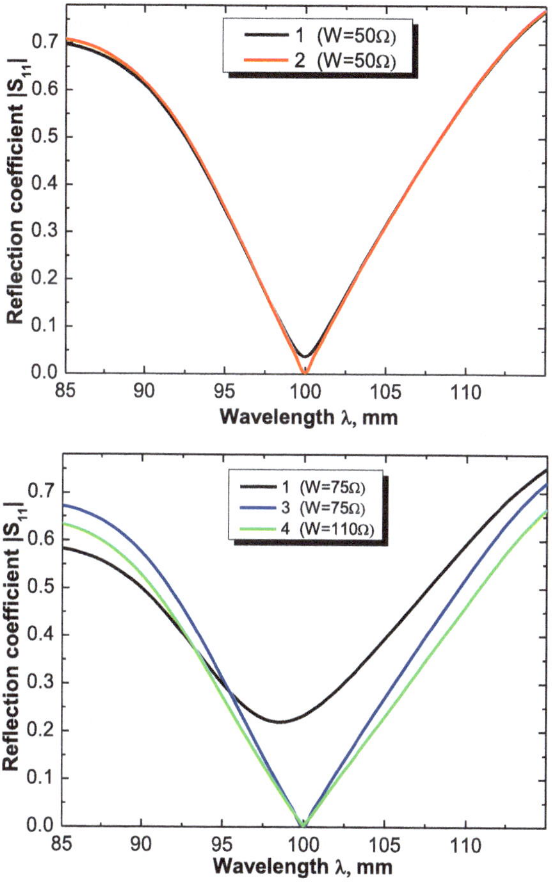

Fig. 3.6 The reflection coefficients dependencies in the feeding line of the Yagi-Uda arrays with the impedance active vibrator from the wavelength for the feeders with different wave resistance W (from Table 3.1)

Table 3.1 The parameters of the Yagi-Uda arrays with the impedance active vibrator

No.	$2L_1$, mm	$\overline{X}_{S_1}$	$C_{C_1} \times 10^3$	R_{in}, Ohm	D
1	43.8	0	0	46.6	6.4
2	45.75	−0.0145	0.911	50	6.4
3	56.9	−0.07075	4.445	75	6.5
4	70	−0.106	6.66	110	6.6

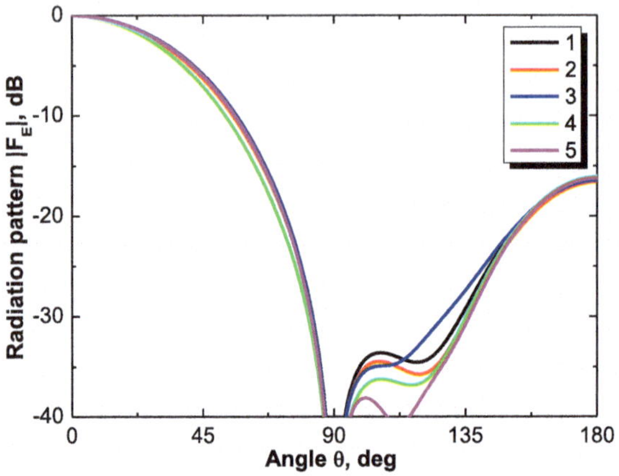

Fig. 3.7 The RP of the Yagi–Uda arrays from $N = 3$ elements (from Table 3.2)

Table 3.2 The parameters of the Yagi-Uda arrays from $N = 3$ elements

No.	$2L_n$, mm	$\overline{X}_{S_n}$	$C_{C_n} \times 10^3$	C_{L_n}	R_{in}, Ohm	D
1	44.85	0	–	–	43.37	6.36
	50		–	–		
	39		–	–		
2	50	−0.0303	1.9	–	52.62	6.5
		0	0	–		
		−0.09	5.655	–		
3	35	0.091	–	1.448	26.2	6.46
		0.122	–	1.942		
		0.054	–	0.859		
4	65	−0.084	5.28	–	84.6	6.78
		−0.053	3.3	–		
		−0.016	0.01	–		
5	44.7	0	–	0	42.03	6.44
	35	0.122	–	1.942		
	35	0.054	–	0.859		

5—the length of the reflector and the length of the director equal to $2L_{2,3} = 0.35\lambda$. The values of the vibrators surface impedances, providing a minimal level of the back lobe of RP (δ_{180}), the values of the input resistances and D of the arrays are represented in Table 3.2 for these cases. The energetic characteristics and the characteristics of directivity of these arrays in dependence from the wavelength are represented in Fig. 3.8.

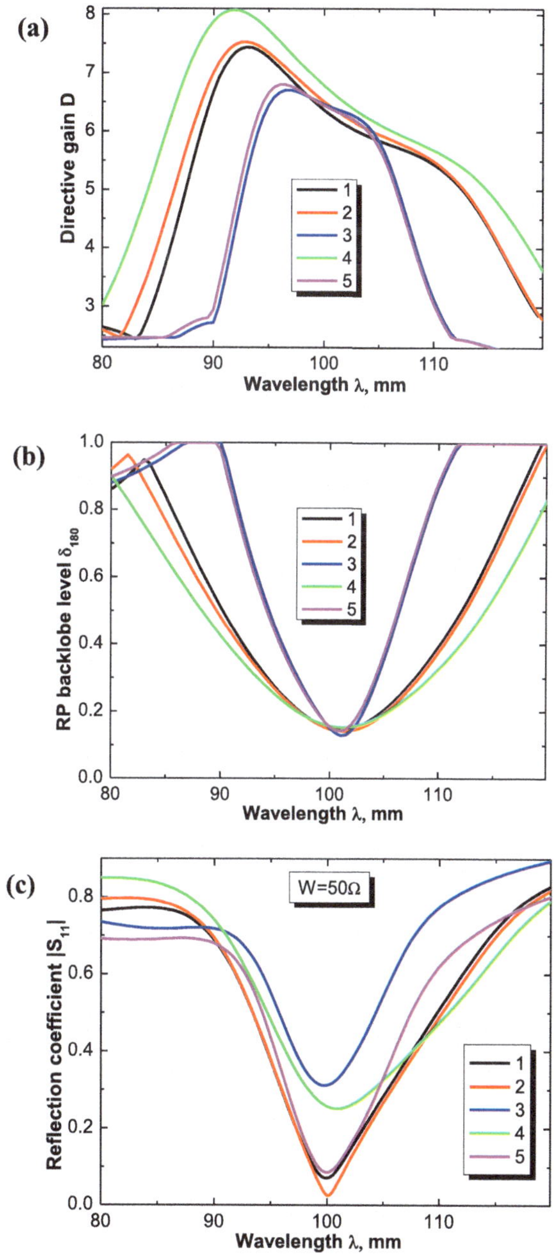

Fig. 3.8 D (**a**), the back lobe level of RP (**b**) and the coefficients of reflection in the feeding feeder (**c**, **d**) of the Yagi-Uda arrays from $N = 3$ elements in dependence from the wavelength (from Table 3.2)

Fig. 3.8 (continued)

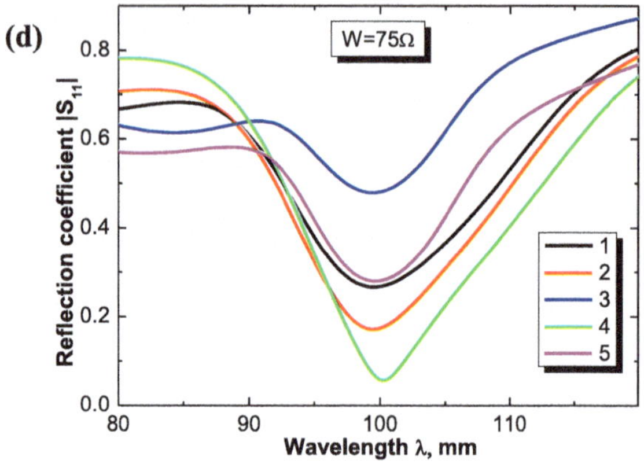

As it is seen from the plots, application of the vibrators with the inductive impedance (a variant 3) in the Yagi-Uda array allows to decrease the vibrators sizes, but at this the value of the input resistance R_{in} decreases, which can be increased by the length increase of either only the active vibrator (a variant 5) or at the length increase of all vibrators with simultaneous increase of the capacitive impedance values (the variants 2 and 4). The increase of the antenna band properties is observed at the use of longer vibrators with capacitive impedance (Fig. 3.8). One would note, that it is possible to change the direction of the antenna radiation (a function of the reflector is performed by the director, and of the director—by the reflector), having changed the surface impedances values of passive radiators suitably.

Figures 3.9 and 3.10 give RP and the band characteristics of the arrays, consisting of $N = 10$ elements (the distance between the vibrators $\Delta z = 0.25\lambda$ and their radiuses $r_{1-10} = 0.01\lambda$) at $\lambda = 100$ mm, correspondingly, as an example of the use of the distributed surface impedance for the arrays with a large quantity of vibrators. The lengths of the vibrators and the values of their surface impedances, chosen from the condition of attainment of a maximal D, are represented in Table 3.3.

3.2 System of Vibrators with Variable Surface Impedance in Free Space

3.2.1 Problem Formulation and Solution

In this case, we will assume that in formulas (3.1)–(3.7) $z_{i_{m(n)}}(s_{m(n)}) = r_{i_{m(n)}} + ix_{i_{m(n)}}(s_{m(n)})$ and $\tilde{k}_n = k - \frac{i2\pi z_{in}^{av}}{Z_0 \Omega_n}$, where $z_{i_n}^{av} = \frac{1}{2L_n}\int_{-L_n}^{L_n} z_{i_n}(s_n) \mathrm{d}s_n$ is the mean internal impedance along the vibrator length [4], $Z_0 = 120\pi$ Ohm and $\Omega = 2\ln(2L_n/r_n)$. The impedance distribution along the vibrators can be represented by $z_{in}(s_n) = z_{in}^{av}\,\phi_n(s_n)$,

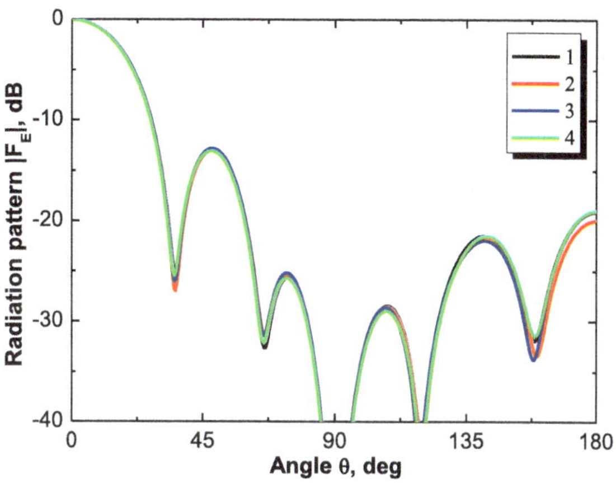

Fig. 3.9 The RP of the Yagi-Uda arrays from $N = 10$ elements (from Table 3.3)

where the distribution functions $\phi_n(s_n)$ are normalized so that mean values over the vibrator length were equal to unit.

3.2.2 Numerical Results

In Sect. 3.1 shows that the characteristics of the Yagi-Uda antenna can be varied by adjusting the lengths of vibrators and distances between them. To obtain axial radiation of the Yagi-Uda antenna composed of perfectly conducting vibrators, the directors should be slightly shorter and the reflector should be longer than the active vibrator (Fig. 3.11a). Recent studies have shown that the reactance of the vibrators in the Yagi-Uda array required to obtain the necessary input parameters and radiation characteristics can also be achieved by variable the magnitude and distribution function of the surface impedance of the fixed length vibrator. The electrical length of the impedance vibrators can be both shorter and longer than that of the perfectly conducting vibrator.

The inductive surface impedance reduces the vibrator resonant length, therefore the Yagi-Uda antenna with impedance vibrators can be used, when the weight and size parameters are significant. The capacitive impedance and, hence, longer vibrators can increase the antenna input resistance. Let us consider several functions defining distribution of the vibrator surface impedance, namely: constant distribution function (CDF), and decreasing (DDF) and increasing (IDF) distribution functions. All these function have equal mean and are symmetrical relative to the vibrator center. The inductive impedance with DDF and the capacitive impedance with IDF increase the vibrator resonant length relative to the impedance with CDF, while the inductive impedance with IDF and the capacitive impedance with DDF decrease the resonant length. That is, the DDF increases impedance while the IDF decreases it. This effect can be used during the design of antenna arrays. Let's illustrate this

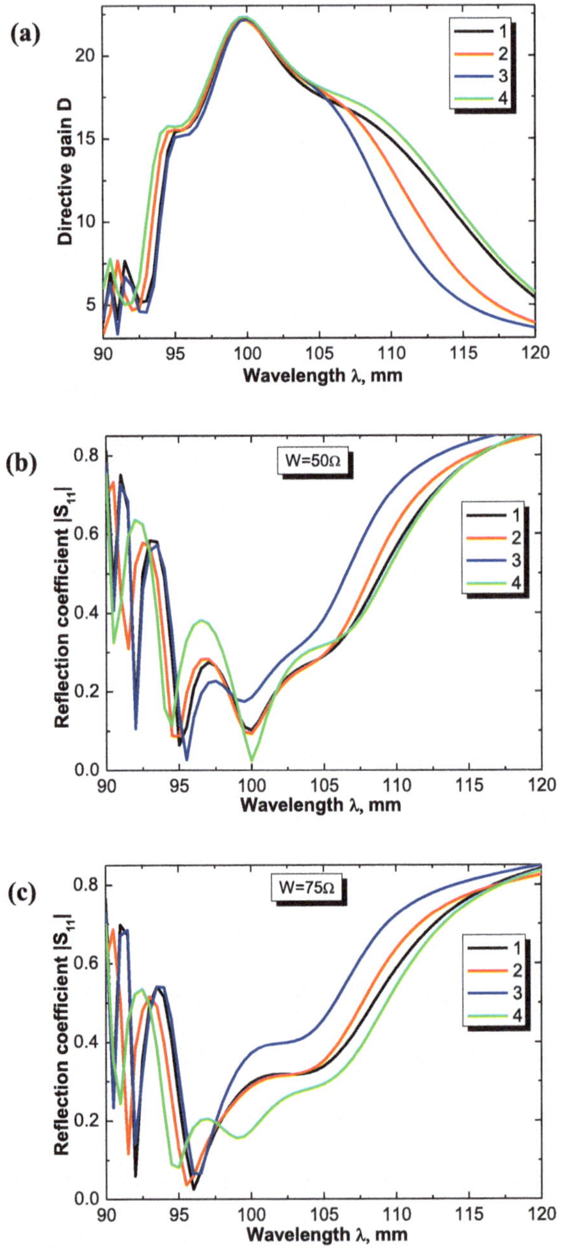

Fig. 3.10 D (**a**) and the coefficients of reflection in the feeding feeder (**b**, **c**) of the Yagi-Uda arrays from $N = 10$ elements in dependence from the wavelength (from Table 3.3)

Table 3.3 The parameters of the Yaga-Uda arrays from $N = 10$ elements

No.	$2L_1$, mm $2L_2$, mm $2L_{3-10}$, mm	$\overline{X}_{S_1}$ $\overline{X}_{S_2}$ $\overline{X}_{S_{3-10}}$	R_{in}, Ohm	D
1	43.6 50 39.6	0	40.76	22.2
2	43.6	0 0.038 −0.0394	51.5	22.1
3	39.6	0.0335 0.07 0	34.4	22.2
4	50	−0.0415 0 −0.086	52.6	22.4

Fig. 3.11 Three-element Yagi-Uda antenna array: **a** perfectly conducting vibrators ($2L_1 = 0.44\lambda_0$, $2L_2 = 0.5\lambda_0$, $2L_3 = 0.38\lambda_0$); **b** inductive impedance vibrators ($2L_n = 0.35\lambda_0$); **c** capacitive impedance vibrators ($2L_n = 0.65\lambda_0$)

possibility by numerical calculations of electrodynamic characteristics for a three-element Yagi-Uda antenna array with the equal-length vibrators and the variable inductive impedance $\overline{Z}_{Sn}(s_n) = i\overline{X}^{av}_{Sn}\phi_n(s_n)$ (Fig. 3.11b) and capacitive impedance $\overline{Z}_{Sn}(s_n) = -i\overline{X}^{av}_{Sn}\phi_n(s_n)$ (Fig. 3.11c). Since the normalized surface impedances $\overline{X}^{av}_{Sn}$ of all vibrators in the array are equal, phases of currents in the radiators to ensure the axial radiation can be obtained by selection of the surface impedance distribution functions. The geometric parameters of array elements and impedances $\overline{X}^{av}_{Sn}$ should be selected so as to match the antenna input impedance at operating wavelength λ_0 and the feeder line characteristic impedance and to provide a low voltage standing-wave ratio (VSWR) in the feeder line of the active vibrator.

Consider the exponentially decreasing and exponentially increasing distribution functions $\phi_n(s_n) = \alpha \exp[-\beta|s_n|/L_n]$ and $\phi_n(s_n) = \alpha \exp[-\beta(|s_n|/L_n - 1)]$, where $\alpha = \beta/[1-\exp(-\beta)]$ is the normalization factor, and β is the arbitrary dimensionless constant. The functions $\phi_n(s_n)$ for the three antennas shown in Fig. 3.11 are plotted in Fig. 3.12.

The plots of VSWR in feeder lines with wave resistance W = 50 Ohm (curves 1, 4), W = 25 Ohm (curve 2), and W = 75 Ohm (curve 3) and directivity D versus the wavelength are shown in Fig. 3.13 for the three-element Yagi-Uda arrays (Fig. 3.11) with the impedance distribution presented in the Fig. 3.12 ($C_{Ln}^{av} = 1.448$, $C_{Cn}^{av} = 5.466 \times 10^{-3}$, $r_n = 0.01\lambda_0$, $z_2 = -0.25\lambda_0$, $z_3 = 0.2\lambda_0$).

As can be seen from the plots in Fig. 3.13a, the antenna array with impedance vibrators and the feed line can be better matched as compared with the antenna with perfectly conducting vibrators. Figure 3.14 shows similar plots for seven-element arrays with perfectly conducting vibrators ($2L_1 = 0.45\lambda_0, 2L_2 = 0.5\lambda_0, 2L_{3-7} = 0.4\lambda_0$, W = 50 Ohm), with impedance vibrators (variable impedance of inductive type: $2L_n = 0.35\lambda_0$, $C_{Ln}^{av} = 1.464$, $\beta = 1.7$ for the reflector and directors, W = 25 Ohm), and with impedance vibrators (variable impedance of capacitive type: $2L_n = 0.65\lambda_0$, $C_{Cn}^{av} = 5.341 \times 10^{-3}$, $\beta = 1.7$ for the reflector and $\beta = 4.5$ for

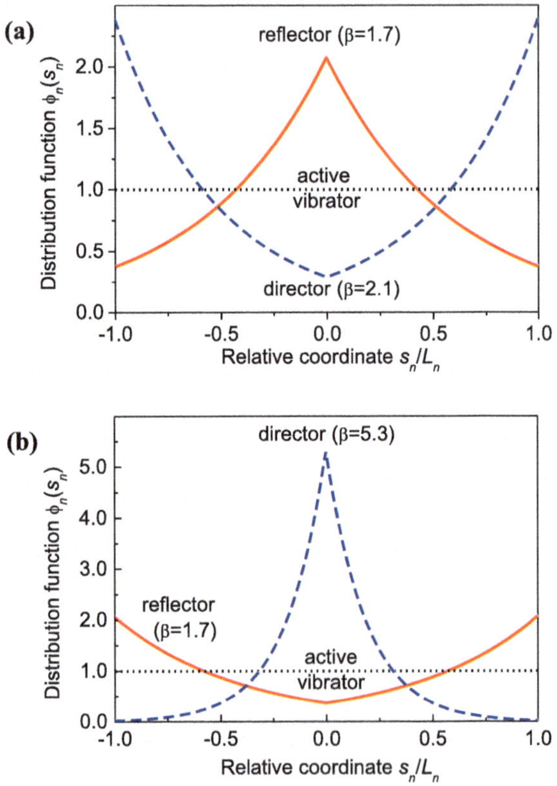

Fig. 3.12 The distribution functions of the surface impedance of vibrators in Yagi-Uda arrays: **a** the inductive impedance, **b** the capacitive impedance, (reflector, director, active vibrator)

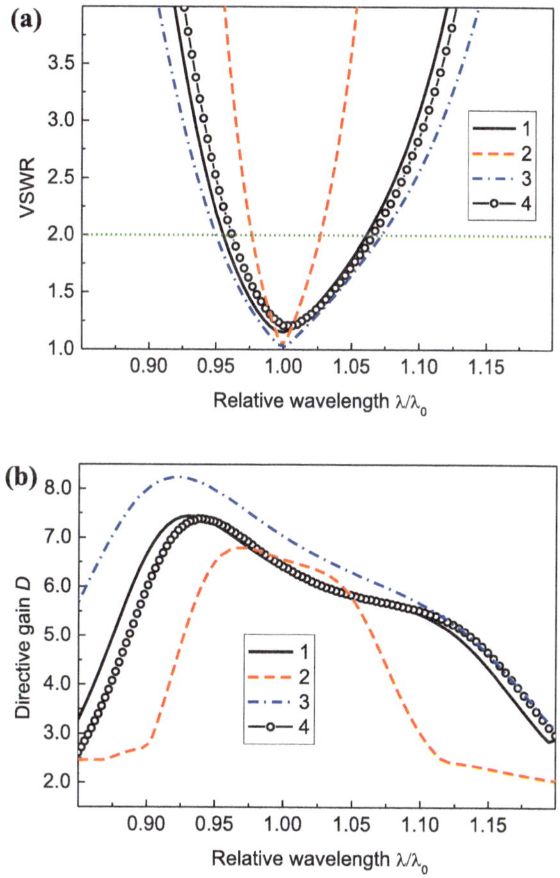

Fig. 3.13 VSWR and D versus the wavelength for the three-element Yagi-Uda antenna: curves 1, 4—perfectly conducting vibrators, curve 4 was obtained by the method of moments with piecewise constant basis vibrators; curve 2—vibrators with variable impedance of inductive type; curve 3—vibrators with variable capacitive impedance

the directors, W = 75 Ohm). The other parameters are as follows: $r_n = 0.01\lambda_0$, $z_2 = -0.25\lambda_0$, $z_3 = 0.2\lambda_0$, and the distances between the directors are $0.2\lambda_0$. As can be seen from the plots in Figs. 3.13a and 3.14a, the array with the impedance vibrators are better matched with feed line as compared with perfectly conducting vibrators. The inductive impedance decrease and capacitive impedance increases the array operating band defined using VSWR (for example, at the level VSWR = 2) and the directivity D as compared with the case $\overline{Z}_{Sn} = 0$.

The RP of the Yagi-Uda antennas at $\lambda = \lambda_0$ are shown in Fig. 3.15.

Thus, the problems solutions presented in this Chapter can be used as a basis for multi-parameter optimization of electrodynamic characteristics of radiating multi-element structures built on vibrators with constant and variable distributed surface impedance. The distinctive peculiarity of the method, proposed by the authors, is the use of the approximating functions, resulted from the integral equation solution for the current by the asymptotic averaging method, in the current distribution along the impedance vibrator. The ground of rightness and correctness of such an approach

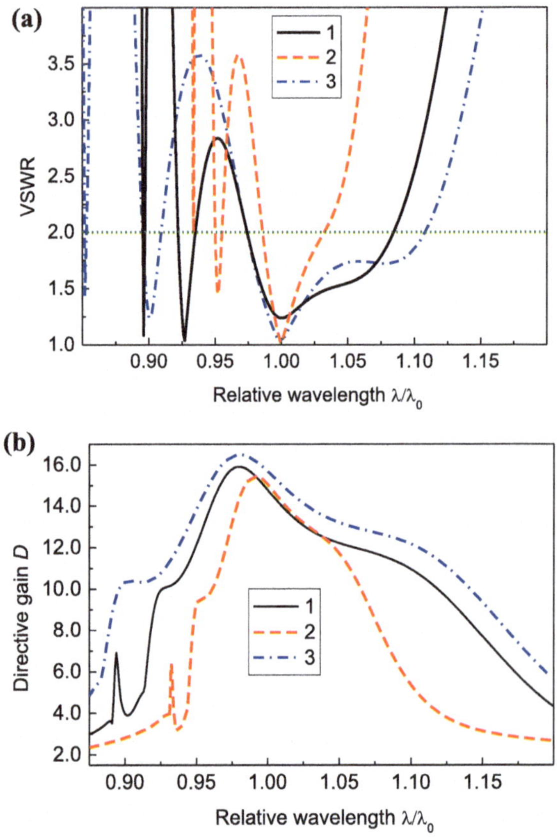

Fig. 3.14 VSWR and D versus the wavelength for the seven-element Yagi-Uda antenna: curve 1—perfectly conducting vibrators, curve 2—vibrators with variable impedance of inductive type; curve 3—vibrators with variable capacitive impedance

is represented in the format of comparative analysis with the calculated results by the method of moments. One would note, that the new conception of the generalized method of induced EMF, keeping all known advantages of numerical-analytical methods in comparison with direct numerical methods, extends to the cases of the vibrator with the impedance, variable along its length, and the impedance vibrators systems rather simply. Thus the proposed generalized method of induced EMF allows to widen the boundaries of numerical-analytical investigations of practically significant problems of the impedance vibrators application sufficiently.

In conclusion, we note that Yagi-Uda arrays have attracted the attention of researchers for almost 100 years. A huge number of publications have been devoted to them up to the present time (see, for example [5–8]). However, arrays of impedance vibrators (especially those with variable surface impedance) were first studied in the publications of the authors [4, 9–12].

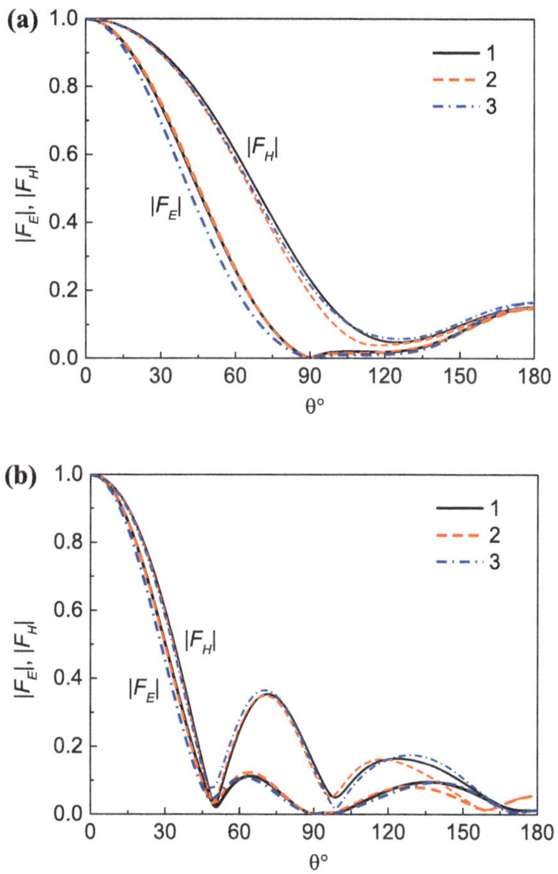

Fig. 3.15 The RP of the **a** three-element and **b** seven-element Yagi-Uda antennas: curves 1—perfectly conducting vibrators, curves 2—vibrators with variable impedance of inductive type; curves 3—vibrators with variable capacitive impedance

References

1. Uda S (1925) On the wireless beam of short electric waves. J IEE Japan 449:1128 (in Japanese)
2. Yagi H, Uda S (1926) Projector of the sharpest beam of electric waves. Proc Imperial Academy Japan 2:49–52
3. King RWP, Mack RB, Sandler SS (1968) Arrays of cylindrical dipoles. Cambridge University Press, New York
4. Nesterenko MV, Katrich VA, Penkin YuM, Dakhov VM, Berdnik SL (2011) Thin impedance vibrators. Theory and applications. Springer Science+Business Media, New York
5. Zhu W, Guo L (2021) A no-network matched high gain vertical array antenna for VHF band applications. Prog Electromagn Res L 100:1–8
6. Smith LP, Howell JC, Lim S (2021) A size-reduced, 15-element, planar Yagi antenna. IEEE Trans Antennas Propag 69:2410–2415
7. Donelli M, Manekiya M, Mulloni V, Marchi G, Mendicino R (2022) A compact ultra-wide band printed log-periodic antenna using a bar-tie structure. Prog Electromagn Res C 124:43–52
8. Karasawa Y (2022) Maximum gain of array antenna with closely placed dipole elements. IEEE Antennas Wireless Propag Lett 21:1674–1677

9. Nesterenko MV, Katrich VA, Berdnik SL, Penkin YuM, Dakhov VM (2010) Application of the generalized method of induced EMF for investigation of characteristics of thin impedance vibrators. Prog Electromagn Res B 26:149–178
10. Berdnik SL, Katrich VA, Nesterenko MV, Kiyko VI (2012) System of impedance vibrators in free space. In: Proceedings of the XVII-th international seminar/workshop on direct and inverse problems of electromagnetic and acoustic wave theory. Tbilisi, Georgia, pp 53–56
11. Berdnik SL, Katrich VA, Nesterenko MV, Penkin YuM, Pshenichnaya SV (2016) Yagi-Uda antennas with impedance vibrators. In: Proceedings of the XVI-th international conference on mathematical methods in electromagnetic theory. Lviv, Ukraine, pp 403–406
12. Berdnik SL, Katrich VA, Nesterenko MV, Penkin YuM (2016) Electromagnetic waves radiation by a vibrators system with variable surface impedance. Prog Electromagn Res M 51:157–163

Chapter 4
Slotted Waveguide Antenna Arrays

At present time, a widespread type of antenna systems is represented by waveguide-slotted antenna arrays (AA) [see, for example, 1–30] which have several advantages over other antenna systems. These AA are characterized by following features: absence of protruding parts, simple excitation of AA with large number of radiators, and possibility to realize a wide range of amplitude-phase distribution of fields and currents over the AA aperture. This Chapter presents results concerning electrodynamic characteristics of some waveguide-slotted AA obtained by the authors.

4.1 Leaky-Wave Slotted Waveguide Arrays

The problem solutions concerning electromagnetic wave scattering on a single slot with arbitrary lengths and various configurations presented in Chap. 2 can significantly expand applications of generalized method of induced magnetomotive forces (MMF) for calculating characteristics of multielement arrays (equidistant and nonequidistant) with slotted elements.

In this section, calculation results and experimental studies concerning characteristics of electromagnetic fields radiated by multielement slotted arrays are presented. Analysis of radiated field formation and search for required amplitude-phase distributions on the AA aperture to obtain their frequency-energy characteristics are also offered.

M. V. Nesterenko et al., *Vibrator and Slot Antenna Arrays for Modern Applications*,
Lecture Notes in Electrical Engineering 1244,
https://doi.org/10.1007/978-3-031-68346-6_4

4.1.1 Problem Formulation and Solution

If the generalized method of induced MMF is combined with analytical problem solutions for single slot radiators, it can be used to simulate multielement waveguide-slotted arrays and to study characteristics of AAs consisting of transverse slots displaced relative one another. The slots of arbitrary length and width are cut in wide walls of rectangular waveguides or resonators (Fig. 4.1). Consider a system consisting of N narrow rectilinear slots cut in a wide wall of rectangular waveguide symmetrically relative to its longitudinal axis shown in Fig. 4.2. The system radiates into an external half-space over an infinite perfectly conducting plane. The permittivity and permeability of the half-space are $\varepsilon_1 = \mu_1 = 1$, the waveguide cross section is $\{a \times b\}$, and wide wall thickness is h. The parameters of the $n-$ th slot satisfy inequalities: $\{d_n/(2L_n)\} << 1$, $\{d_n/\lambda\} << 1$ (d_n and $2L_n$ are the slot width and length, and λ is the wavelength in free space).

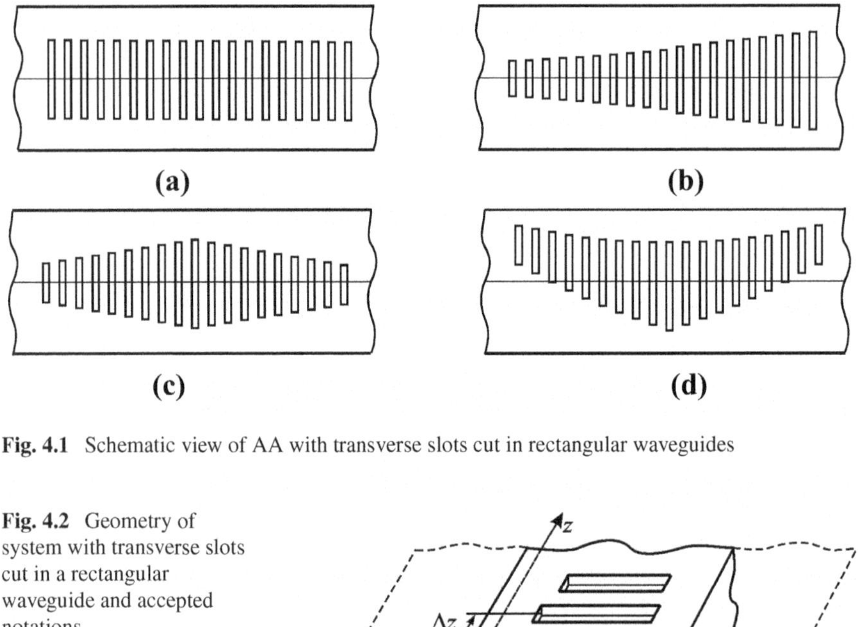

Fig. 4.1 Schematic view of AA with transverse slots cut in rectangular waveguides

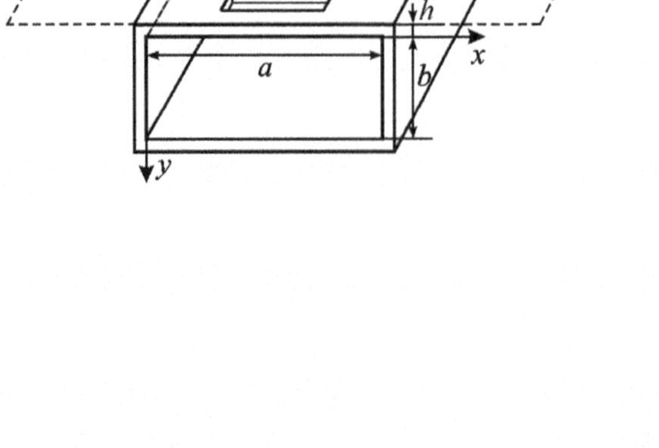

Fig. 4.2 Geometry of system with transverse slots cut in a rectangular waveguide and accepted notations

When the slot currents are approximated by the relations: $J_n(s_n) = J_{0n} f_n(s_n)$ and $f_n(\pm L_n) = 0$ ($f_n(s_n)$ are the predefined functions) considered in Chap. 2, the system of linear algebraic equation (SLAE) relative to the unknown current amplitudes J_{0n} can be obtained using the boundary conditions for the continuity of tangential components of magnetic fields on both surfaces of each slot. Thus, the following expressions can be written as

$$\sum_{n=1}^{N} J_{0n} Y_{mn}^{Wg,R,Hs}(kL_m, kL_n) = -\frac{i\omega}{2k} M_m(kL_m), \quad m = 1, 2 \cdots N, \tag{4.1}$$

where

$$y_{mn}^{wg,R,Hs}(kL_m, kL_n) = \frac{1}{2k} \int_{-L_m}^{L_m} f_m(s_m) \left[\left(\frac{d^2}{ds_m^2} + k^2 \right) \int_{-L_n}^{L_n} f_n(s_n') G_{s_m}^{Wg,R,Hs}(s_m, s_n) ds_n' \right] ds_m \tag{4.2}$$

are normalized intrinsic ($m = n$) and mutual ($m \neq n$) slot conductivities,

$$M_m(kL_m) = \int_{-L_m}^{L_m} f_m(s_m) H_{0s_m}(s_m) ds_m \tag{4.3}$$

are magnetomotive forces, $G_{s_m}^{Wg,R,Hs}$ are quasi-one-dimensional magnetic Green's functions for an infinite or semi-infinite rectangular waveguide (Wg), rectangular resonator (R) formed by a slot cavity (Appendix A), and half-space above the infinite screen (Hs), and H_{0s_m} are projections of the magnetic field generated by extrinsic sources on the slot axes.

Let us assume that the system is excited by the H_{10}-wave with amplitude H_0 propagating in a waveguide from the region $z = -\infty$ and choose functions $f_n(s_n)$ found as solutions of integral equation for magnetic currents in a single transverse slot cut in a wide wall of infinite rectangular waveguide by the asymptotic averaging method (2.69). These functions can be written as

$$f_n(s_n) = \cos ks_n \cos k_c L_n - \cos kL_n \cos k_c s_n, \tag{4.4}$$

where, $k_c = 2\pi/\lambda_c$ is the critical wavelength of the H_{10}-wave.

When all SLAE coefficients (4.1) are found by substituting (4.4) into the expressions (4.2) and (4.3), the SLAE solution allows us to find energy and spatial characteristics of waveguide-slotted AA. Thus, the expressions for the slot conductivities and MMF can be obtained in the following form:

$$Y_{mm}^{Wg}(kL_m, kL_m) = \frac{2\pi}{ab} \sum_{m=1,3\ldots}^{\infty} \sum_{n=0}^{\infty} \frac{\varepsilon_n(k^2 - k_x^2)}{kk_z} e^{-k_z \frac{d_m}{4}} I_{Wg}^2(kL_m), \tag{4.5}$$

$$Y_{mn}^{Wg}(kL_m, kL_n) = \frac{2\pi}{ab} \sum_{m=1,3\ldots}^{\infty} \sum_{n=0}^{\infty} \frac{\varepsilon_n(k^2 - k_x^2)}{kk_z} e^{-k_z|z_m - z_n|} I_{Wg}(kL_m) I_{Wg}(kL_n), \tag{4.6}$$

$$\begin{aligned} Y_{\left\{ mm \atop mn \right\}}^{R}(kL_m) = {} & \frac{2\pi}{L_m d_m} \sum_{m=1,3\ldots}^{\infty} \sum_{n=0}^{\infty} \frac{\varepsilon_n(k^2 - k_{xR}^2)}{kk_{zR}} \left\{ \begin{matrix} \coth(k_{zR}h) \\ 1/sh(k_{zR}h) \end{matrix} \right\} \cos\frac{n\pi}{2} \\ & \cos\left(k_{yR}\frac{3d_m}{4}\right) I_R^2(kL_m), \end{aligned} \tag{4.7}$$

$$\begin{aligned} Y_{mn}^{Hs}(kL_m, kL_n) = {} & \frac{1}{2k}\{(k\cos k_c L_m \sin kL_m - k_c \cos kL_m \sin k_c L_m) \\ & - \int_{-L_n}^{L_n} f_n(s_n\prime)[G_{s_m}^{Hs}(L_m, s_n\prime) + G_{s_m}^{Hs}(-L_m, s_n\prime)]\mathrm{d}s_n\prime \\ & - k_g \cos kL_m \int_{-L_m}^{L_m} \cos k_c s_m \left[\int_{-L_n}^{L_n} f_n(s_n')G_{s_m}^{Hs}(s_m, s_n')\mathrm{d}s_n\prime \right] \mathrm{d}s_m \bigg\}, \end{aligned} \tag{4.8}$$

$$M_m(kL_m) = H_0 \frac{1}{k} e^{-ik_g z_m} F(kL_m) \tag{4.9}$$

In formulas (4.5)–(4.9), the following notations are accepted:

$$\begin{aligned} I_{Wg(R)}(kL_m) = 2 & \left\{ \frac{k \sin kL_m \cos k_{x(xR)}L_m - k_{x(xR)} \cos kL_m \sin k_{x(xR)}L_m}{k^2 - k_{x(xR)}^2} \cos k_c L_m \right. \\ & \left. - \frac{k_c \sin k_c L_m \cos k_{x(xR)}L_m - k_{x(xR)} \cos k_c L_m \sin k_{x(xR)}L_m}{k_c^2 - k_{x(xR)}^2} \cos kL_m \right\}, \end{aligned}$$

$$\begin{aligned} F(kL_m) = {} & 2\cos k_c L_m \frac{\sin kL_m \cos k_c L_m - (k_c/k)\cos kL_m \sin k_c L_m}{1 - (k_c/k)^2} \\ & - \cos kL_m \frac{\sin 2k_c L_m + 2k_c L_m}{2k_c/k} \end{aligned}$$

$k_x = \frac{m\pi}{a}$, $k_y = \frac{n\pi}{b}$, $k_z = \sqrt{k_x^2 + k_y^2 - k^2}$, $\varepsilon_n = (2 - \delta_{0n})$ is Neumann multiplier, $k_{xR} = \frac{m\pi}{2L_m}$, $k_{yR} = \frac{n\pi}{d_m}$, $k_{zR} = \sqrt{k_{xR}^2 + k_{yR}^2 - k^2}$, m, n are integers, $z_m(z_n)$ is coordinate of $m-$th ($n-$th) slot axis. It should be noted that the proposed approximation to the distribution functions of the slot currents has allowed us to obtain expressions for Y_{mn}^{Wg}, Y_{mn}^{R} in the form of rapidly converging double series. Thus, the double integral in the formula (4.8) can be reduced to a single integral. As a result, the computation time required for obtaining characteristics of AAs with large number of radiators (up the order of several hundred) can be significantly reduced.

When the H_{10}-wave propagates in the waveguide, the reflection, transmission, and the power radiation coefficients S_{11}, S_{12}, and $|S_\Sigma|^2$ for the AA can be expressed as:

$$S_{11} = \frac{2\pi i k_g}{abk^3}\left\{\sum_{n=1}^{N} J_{0n} F(kL_n) e^{-ik_g z_n}\right\} e^{2ik_g z},\ S_{12} = 1 + \frac{2\pi i k_g}{abk^3}\left\{\sum_{n=1}^{N} J_{0n} F(kL_n) e^{ik_g z_n}\right\} \tag{4.10}$$

$$|S_\Sigma|^2 = 1 - |S_{11}|^2 - |S_{12}|^2 \tag{4.11}$$

The field RP of the AA in the E-plane is determined by the following expression

$$F(\theta) = \sum_{n=1}^{N} J_{0n} \frac{\sin\left(\frac{kd_n}{2}\cos\theta\right)}{\frac{kd_n}{2}\cos\theta} e^{ikz_n \cos\theta} \tag{4.12}$$

where the angle θ is measured from the axis $\{0z\}$ (see Fig. 4.2).

4.1.2 Numerical and Experimental Results

The theoretical and experimental plots of the power radiation coefficient $|S_\Sigma|^2$ as function of wavelength for the system consisting of 20 identical transverse slots cut at equal distances from one another in the wide wall of the rectangular waveguide are shown in Fig. 4.3. The system parameters were selected as follows: the waveguide cross section and wall thickness are $\{23.0 \times 10.0\}$ mm^2 and $h = 1.0$ mm (Fig. 4.2.), the slot length and width are $2L_n/\lambda = 0.4$ and $d_n = 1$ mm. Comparison of the calculated and experimental curves allows us to conclude that the approximation of the slot currents by functions (4.4) is valid. Note that if the waveguide wall thickness is increased, the frequency response curves are narrowing.

When parameters of the radiating structures known as modulated leaky wave antennas [21, 31]): the slot length and distance between the slots vary according to the curves shown in Fig. 4.4, the RP of the slotted system is shown in Fig. 4.5, where the plots of the normalized current amplitude $\overline{J}_{0n} = |J_{0n}|/|J_{0n}|_{\max}$ along the array length are also shown. As can be seen, when the slot lengths are varied and $\Delta z_n = const = 12$ mm (Fig. 4.5a), the near and far side lobe levels (SLL) of the RP are decreasing, while the main lobe is widening. At the same time, when the distances between the slots is varied and $2L_n = const = 13$ mm (Fig. 4.5b), the width of the main lobe does not vary and the SSL decreases until the angle θ reaches about 90 degrees.

The RP of equidistant AA consisting of 100 slots with parameters: $2L_n = 11.5$ mm, $\Delta z_n = 10$ mm, $\lambda = 32$ mm, $d_n = 1.0$ mm, $h = 1.0$ mm is presented in

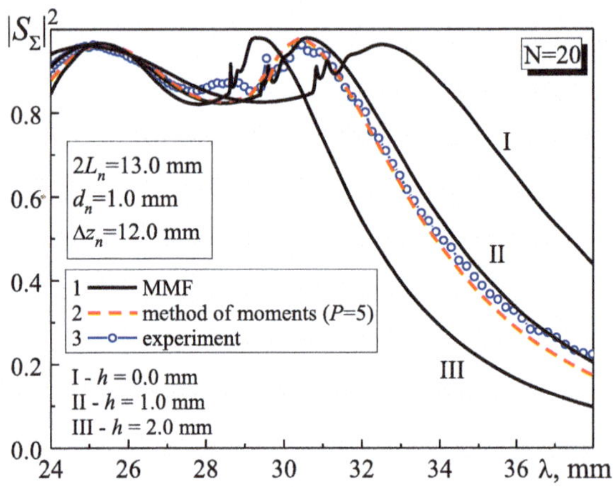

Fig. 4.3 Wavelength plots $|S_\Sigma|^2$ for the system consisting of 20 transverse slots of equal length ($2L_n/\lambda = 0.4$) and width ($d_n = 1$ mm) cut in the waveguide with the cross section $\{23 \times 10\}$ mm^2: 1—calculation by the formulas (4.10), (4.11)); 2—calculation by the method of moments ($J_n(s_n) = \sum_{p=1}^{P} J_{np} \sin \frac{p\pi(L_n+s_n)}{2L_n}$, $P = 5$); 3—experimental data

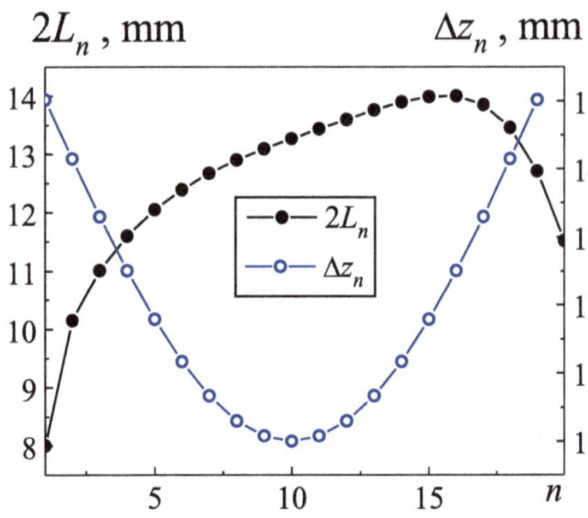

Fig. 4.4 Structure parameters $2L_n$ and Δz_n as functions of the slot number

Fig. 4.6 (curve 1). In this case, the power radiation coefficient $|S_\Sigma|^2$ is equal to 0.9134, while the distribution of current amplitude smoothly decreases along the array in the range from 1.0 to 0.3 (Fig. 4.6). If the number of slots in the structure is doubled and the distance between them is reduced ($N = 200$, $\Delta z_n = 5.0$ mm), the radiation coefficient $|S_\Sigma|^2$ significantly increases ($|S_\Sigma|^2 = 0.9972$), while the AA directivity noticeably decreases (curve 2). This can be explained by the fact that in this case, the slots closest to the generator are excited most efficiently, while the rest of them are weakly involved into formation of the radiation field. However, the distribution currents similar to that for the array with 100 slots can be obtained by appropriately choosing the AA parameters: $N = 200$, $\Delta z_n = 5.0$ mm , $2L_n = 9.98$ mm (curve 3). In this case, the radiation coefficient $|S_\Sigma|^2$ is also equal to 0.9134, while the far SLL of the RP are reducing.

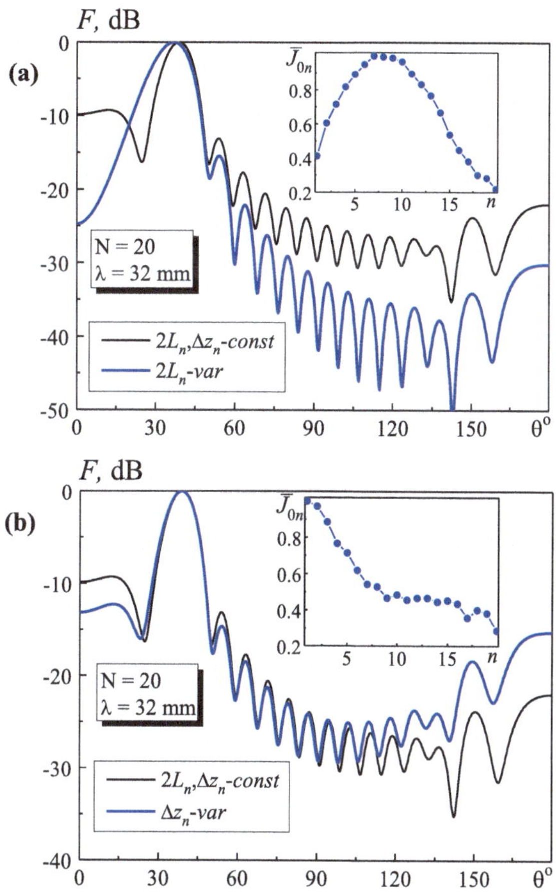

Fig. 4.5 Dependence of the slotted system RP on the structure parameters

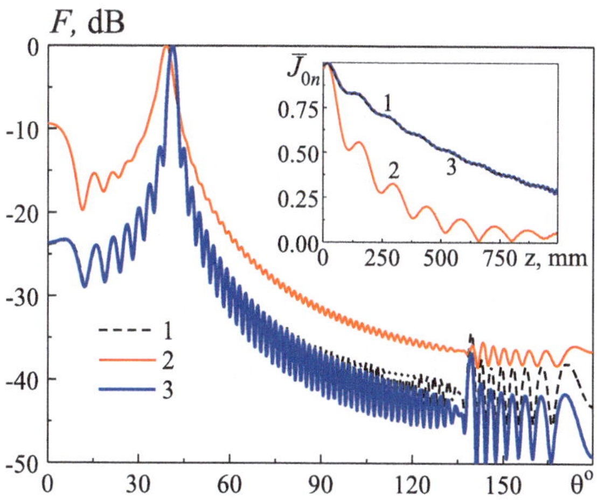

Fig. 4.6 Dependence of the antenna array RP on the structure parameters and slot number

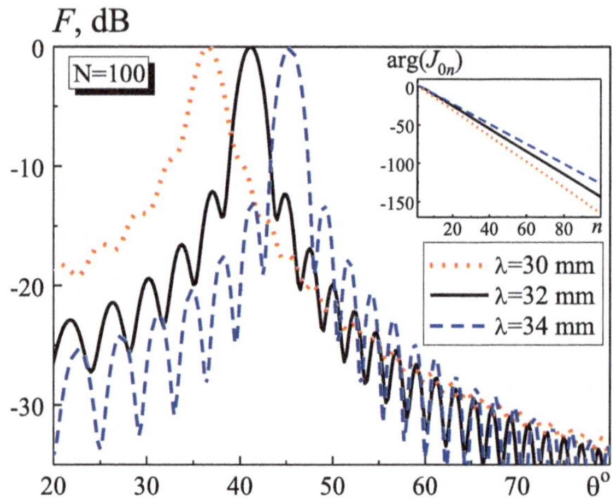

Fig. 4.7 Dependence of antenna array RPs on wavelength

In all cases considered above, the phase distribution of the AA current is determined mainly by the phase velocity of wave propagating in unperturbed transmission line (waveguide without slots) which approximately corresponds to that of traveling wave mode. Therefore, frequency scanning of AP main lobe in a sector of angles can be carried out by varying the operating wavelength as can be seen from the plots in Fig. 4.7, which were obtained by using the following parameters: $2L_n = 11.5\,\text{mm}$, $d_n = 1.0\,\text{mm}$, $\Delta z_n = 10.0\,\text{mm}$, $h = 1.0\,\text{mm}$. In this case, the leaky wave antenna characteristics, i.e. the direction and width of main lobe and SLL can be varied.

4.1.3 Near Electromagnetic Fields of Multi-slot Leaky-Wave Systems

In Sects. 4.1.1 and 4.1.2, the problem concerning the distribution of equivalent magnetic currents in the system of non-resonant transverse slots cut in a wide wall of rectangular waveguide was solved by the induced MMF method. Far-zone radiation characteristics of this structure with parameters: $(2L_n/\lambda) \neq 0.5$, $\Delta z_n \ll \lambda_g$, λ and λ_g, are the wavelengths in the free space and waveguide, respectively) were comprehensively analyzed. It was also shown that, the amplitude and phase distributions of slot magnetic currents in such systems can be controlled almost independently of each other, especially when number of slots is sufficiently large.

However, in a number of important practical applications, such as, for example, issues related to electromagnetic compatibility of radio electronic systems, medical diagnostics and hyperthermia, etc., distributions of electromagnetic fields in the near and intermediate zones should be known. This section is devoted to this problem solution.

4.1.3.1 Radiation into Free Space

As it was shown in Chap. 1, when extrinsic sources in the external half-space are absent, the radiation fields of the $n-$th slot in the coordinate system shown in Fig. 4.2, have the form:

$$\vec{E}_n(\vec{r}) = \frac{k}{\omega} rot \int_{S_n} \hat{G}_m(\vec{r}, \vec{r}_n')\vec{J}_n(\vec{r}_n')d\vec{r}_n' ,$$

$$\vec{H}_n(\vec{r}) = \frac{i}{\omega}(graddiv + k^2) \int_{S_n} \hat{G}_m(\vec{r}, \vec{r}_n')\vec{J}_n(\vec{r}_n')d\vec{r}_n' , \tag{4.13}$$

where $\vec{r}$ and $\vec{r}_n'$ are the radius-vectors of the observation point with coordinates x, y, z and source point, $\hat{G}_m(\vec{r}, \vec{r}_n')$ is the magnetic tensor Green's function for the half-space, and $\vec{J}_n(\vec{r}_n')$ is the surface density of the magnetic current at the aperture of the $n-$ th slot.

Let us further restrict ourselves to considering the magnetic field distribution in the planes $x = x_c = const$ Then, according to (4.13):

$$H_{xn}(x_c, y, z,) = \frac{2ik^2 J_{on}}{\omega} \int_{-L_n}^{L_n} f_n(s_n) \frac{e^{-ikR_n(x_c, y, z; s_n, z_n)}}{R_n(x_c, y, z; s_n, z_n)} ds_n ,$$

$$H_{yn}(x_c, y, z,) = H_{zn}(x_c, y, z) = 0 , \tag{4.14}$$

where $R_n(x_c, y, z; s_n, z_n) = \sqrt{(x_c - s_n)^2 + y^2 + (z - z_n)^2}$ and z_n is longitudinal axis coordinate of the $n-$ th slot. Then, the total magnetic field of the entire slot radiator system can be written as

$$H_x(x_c, y, z) = \sum_{n=1}^{N} H_{xn}(x_c, y, z) . \tag{4.15}$$

The curves of the normalized current amplitudes for the system consisting of 20 slots with parameters: $a \times b = 23.0 \times 10.0\,\text{mm}^2$, $\lambda = 32.0\,\text{mm}$, $d_n = h = 1.0\,\text{mm}$, $\Delta z_n = 12.0\,\text{mm}$, and total antenna length equals 230 mm are shown in Fig. 4.8 for two cases: $2L_n = const = 13.0\,\text{mm}$ (Fig. 4.8a) and $2L_n$ varies according to the curve in Fig. 4.8b.

The spatial distributions of near-fields for these arrays in the plane $x_c = 0$ are shown in Fig. 4.9. As can be seen, the high intensity of the near electromagnetic field in a predefined direction at a given wavelength can be achieved by selecting the system geometric parameters.

In this case, the radiation coefficient $|S_\Sigma|^2$ of the entire structure is rather high: for constant and variable slot lengths $2L_n$ it is equal to 0.852 and 0.912, respectively. The time required for computation $|S_\Sigma|^2$ by using the approximating function (4.4)

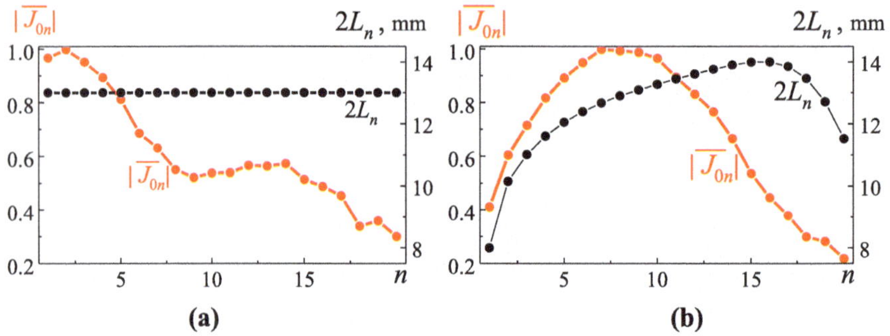

Fig. 4.8 Normalized amplitudes of the slot currents $|\overline{J}_{0n}|$

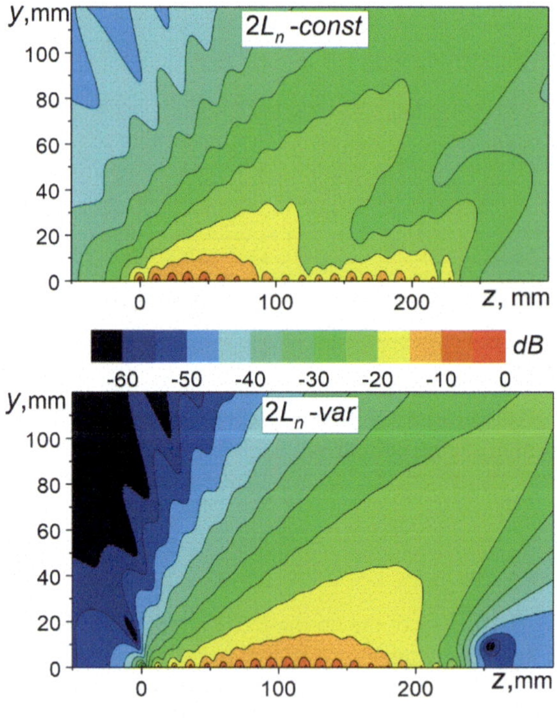

Fig. 4.9 Spatial distributions of the near-field in the plane $x_c = 0$ for the arrays with various structure parameters

which defines the current in each slot is significantly less as compared to that by applying the method of moments.

The distributions of current amplitudes and spatial distributions of the near field for the system consisting of 100 slots are plotted in Fig. 4.10 for two cases: (a) constant slot length $2L_n = 11.5$ mm,, (b) slot length is variable function of the slot number.

4.1.3.2 Radiation into Material Media

Let us consider a system consisting of closely spaced transverse slots cut in a wide wall of a rectangular waveguide which radiate into a lossy medium characterized by the complex permittivity $\varepsilon_1 = \varepsilon_1' - i\varepsilon_1''$ and permeability $\mu_1 = 1$. The problem was solved by the Galerkin method of moments. The trigonometric functions $J_n(s_n) = \sum_{p=1}^{Q} J_{nq} \sin \frac{q\pi(L_n+s_n)}{2L_n}$ were used as the distribution functions $f_{nq}(s_n)$ of equivalent magnetic currents in the slots. Then the expressions (4.13) can be transformed to the following form:

$$\vec{E}_n(\vec{r}) = \frac{k_1}{\omega} rot \int\limits_{S_n} \hat{G}_m(\vec{r}, \vec{r}_n')\vec{J}_n(\vec{r}_n')d\vec{r}_n',$$

$$\vec{H}_n(\vec{r}) = \frac{i}{\omega\mu_1}(graddiv + k_1^2) \int\limits_{S_n} \hat{G}_m(\vec{r}, \vec{r}_n')\vec{J}_n(\vec{r}_n')d\vec{r}_n', \tag{4.16}$$

where $\vec{J}_n(\vec{r}_n')$ is the surface density of the magnetic current at the aperture S_n of the $n-$ th slot as seen from the half-space above the waveguide, and $k_1 = k\sqrt{\varepsilon_1\mu_1}$.

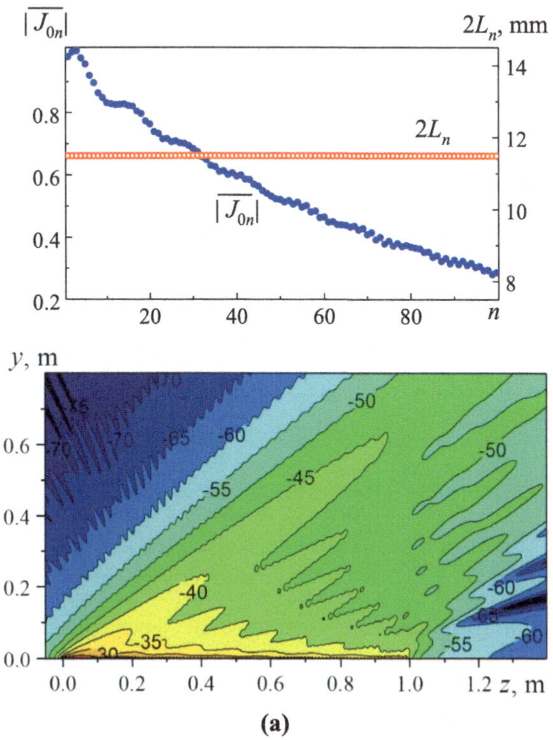

Fig. 4.10 Distributions of the amplitude of current and spatial near field for the system with parameters: $N = 100$, $a \times b = 23.0 \times 10.0\,\text{mm}^2$, $\lambda = 32\,\text{mm}$, $d_n = h = 1.0\,\text{mm}$, and $\Delta z_n = 12.0\,\text{mm}$:: **a** constant slot length equal to 11.5 mm, **b** slot length is variable function of the slot number

Fig. 4.10 (continued)

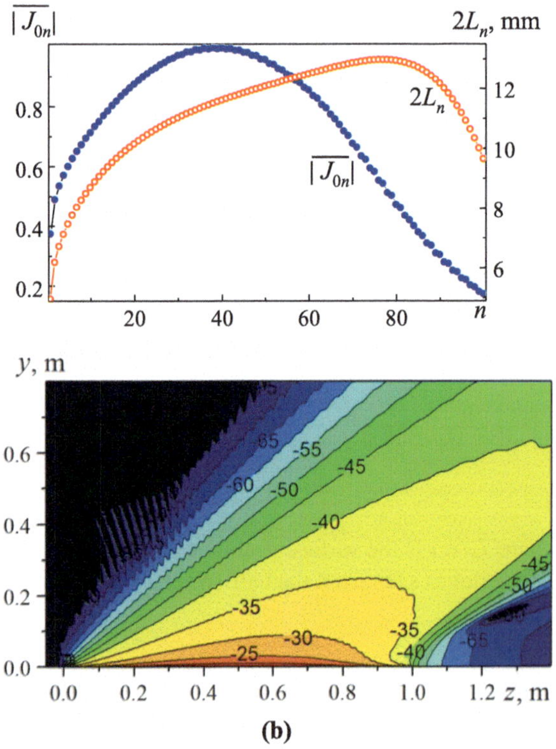

If the representation of the magnetic current in the slot is specified in the form $J_n(s_n) = \sum_{p=1}^{Q} J_{nq} f_{nq}(s_n)$, the expressions for the electric field components of $n-$th slot can be presented as:

$$E_{xn}(x, y, z) = 0,$$

$$E_{yn}(x, y, z) = -\frac{4ikk_1}{\omega} \sum_{q=1}^{Q} J_{nq} \int_{-L_n}^{L_n} f_{nq}(s_n) \frac{e^{-ik_1 R(s_n)}}{R^2(s_n)} \left[1 + \frac{1}{ik_1 R(s_n)}\right](z - z_n) \mathrm{d}s_n,$$

$$E_{zn}(x, y, z) = \frac{4ikk_1}{\omega} \sum_{q=1}^{Q} J_{nq} \int_{-L_n}^{L_n} f_{nq}(s_n) \frac{e^{-ik_1 R(s_n)}}{R^2(s_n)} \left[1 + \frac{1}{ik_1 R(s_n)}\right] y \mathrm{d}s_n, \tag{4.17}$$

where $R(s_n) = \sqrt{(x - s_n)^2 + y^2 + (z - z_n)^2}$ and z_n is the coordinate of the $n-$th slot longitudinal axis. Then the total electric field of the radiator system is equal to

$$\vec{E}(x, y, z) = \vec{y}_0 \sum_{n=1}^{N} E_{yn}(x, y, z) + \vec{z}_0 \sum_{n=1}^{N} E_{zn}(x, y, z), \tag{4.18}$$

where $\vec{y}_0$ and $\vec{z}_0$ are the unit vectors.

The distribution curves of the normalized current amplitudes $|J_n|/|J_n|_{\max}$ in the system consisting of 40 slots, whose length $2L_n$ defined by two different functions of the slot number marked by the indices 1 and 2 are presented in Fig. 4.11. The parameters of the system radiating into the medium with complex permittivity $\varepsilon_1 = 4.5 - i0.95$ which is equivalent to that of adipose tissue [32], were as follows: $a \times b = 23.0 \times 10.0\,\text{mm}^2$, $\lambda = 32.0\,\text{mm}$, $d_n = h = 1.0\,\text{mm}$, $\Delta z_n = 6.0\,\text{mm}$, and the total array length equal to 235 mm. In this case, the equivalent magnetic currents in the slots were approximated by the sum of trigonometric functions $J_n(s_n) = \sum_{p=1}^{3} J_{nq} \sin \frac{q\pi(L_n+s_n)}{2L_n}$. The spatial distributions of the total field $|\vec{E}|^2 = |E_y|^2 + |E_z|^2$ for the functions 1 and 2 in two mutually perpendicular planes $x = 0$ and $y = \lambda/2$ are shown in Fig. 4.12.

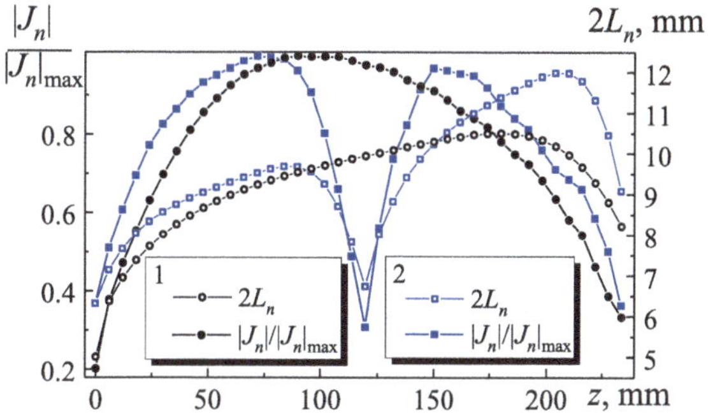

Fig. 4.11 Normalized distributions of current amplitude in the system consisted of 40 slots radiating into the medium with permittivity $\varepsilon_1 = 4.5 - i0.95$ when the slot length $2L_n$ is defined by two different functions of the slot number

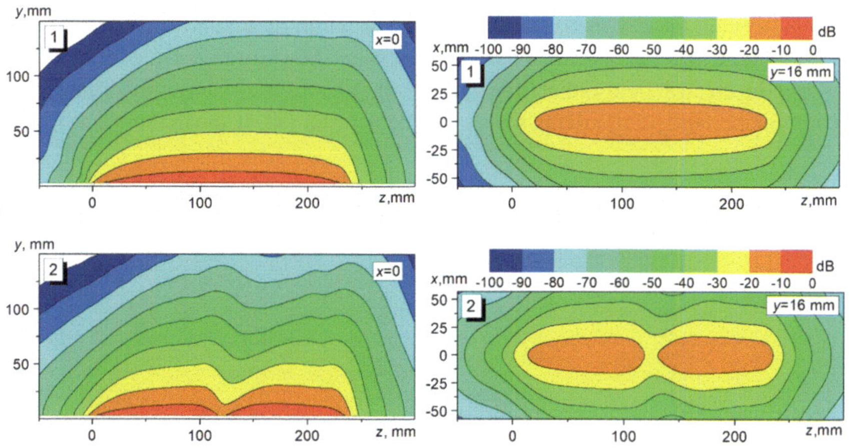

Fig. 4.12 Spatial distributions of the near-field in the system radiating into the adipose tissue with permittivity $\varepsilon_1 = 4.5 - i0.95$. The system parameters were selected as follows: $a \times b = 23.0 \times 10.0\,\text{mm}^2$, $\lambda = 32.0\,\text{mm}$, $d_n = h = 1.0\,\text{mm}$, $\Delta z_n = 6.0\,\text{mm}$

As can be seen, the required intensity of electromagnetic field $|\vec{E}|^2$ in one or more near regions can be achieved by selecting the geometric parameters of system radiators, in particular, the length of each slot. In this case, the radiation coefficients $|S_\Sigma|^2$ are high enough $\left(|S_\Sigma|^2 = 0.816 \text{ and } |S_\Sigma|^2 = 0.869\right)$ and for the both function.

4.2 Near Electromagnetic Fields Radiated by Combined Slotted Waveguide Arrays into Material Media

Slotted waveguide radiators and their systems have ample opportunities for formatting spatial distributions of near-zone fields under condition that electrophysical parameters of the external medium are known. These systems can be used to create medical applicators for microwave hyperthermia [33]. In our opinion, high efficiency of the microwave hyperthermia process can be achieved by using radiators with circular (elliptical) polarizations [22, 23, 34]. In this Section, the research subject concerns studying such systems of combined waveguide-slot radiators based on rectangular waveguides.

4.2.1 Problem Formulation and Solution

Consider a system of slotted radiators cut in a wide wall of a rectangular waveguide radiating into the lossy material medium. The medium is characterized by the relative complex permittivity $\varepsilon_1 = \varepsilon_1' - i\varepsilon_1''$ and permeability $\mu_1 = 1$. Let us numerate the slots and denote the n – th slot length by $2L_n$, its width by d_n, and slot centers by z_n and x_n in the Cartesian coordinate system (x, y, z) shown in Fig. 4.13.

Then the electric field radiated by the n – th slot into the material medium above the infinite perfectly conducting screen can be determined by the expression similar

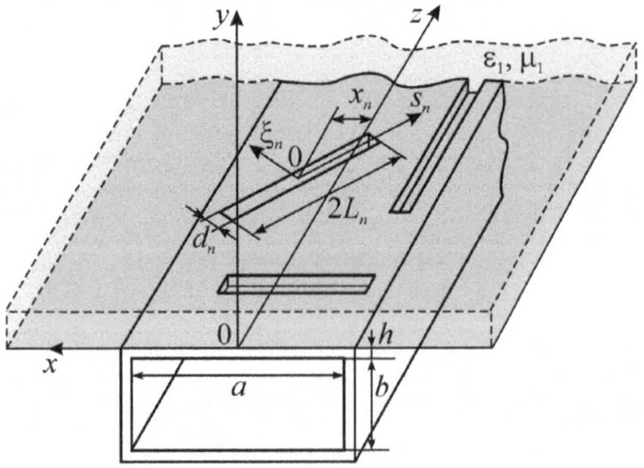

Fig. 4.13 The problem geometry and accepted notations

to (4.16)

$$\vec{E}_n(\vec{r}) = -\frac{1}{2\pi}\mathrm{rot}\int\limits_{\varsigma_n}\left[\vec{e}_{\varsigma_n}(\vec{r}')\ ,\vec{n}\right]\hat{G}^m(\vec{r},\vec{r}')\,d\varsigma_n, \tag{4.19}$$

where $\vec{r}_n'$ is radius vector of the source point, $\vec{r}$ is the radius vector of the observation point with coordinates x, y, z, $\hat{G}^m(\vec{r},\vec{r}_n')$ is the magnetic tensor Green's function of homogeneous or flat-layered half-space, $\vec{n}$ is the normal to the slot surface, and $\vec{e}_{\varsigma_n}(\vec{r}')$ is the electric field at the $n-$th slot aperture ς_n. Hence the field radiated by the system consisting of N slots can be written as $\vec{E}(\vec{r}) = \sum_{n=1}^{N}\vec{E}_n(\vec{r})$. Based on the conditions $d_n/(2L_n) << 1$, $d_n/\lambda << 1$, the electric field $\vec{e}_{\varsigma_n}(\vec{r}')$ can be presented as expansion over the trigonometric basis of the complete region considered in Chap. 2

$$\vec{e}_{\varsigma_n}(s_n) = \vec{\xi}_n^0\sum_{q=1}^{Q}V_{qn}f_{qn}(s_n), f_{qn}(s_n) = \sin\frac{q\pi(L_n+s_n)}{2L_n} \tag{4.20}$$

where s_n is the local longitudinal coordinate associated with the $n-$ th slot, $\vec{\xi}_n^0$ are unit vectors, V_{qn} are unknown coefficients determined by the Galerkin method of moments based on the continuity condition for the tangential components of total magnetic fields on the inner and outer surfaces of each slot. As a result, according to (4.19) and (4.20), the electric field components radiated by the longitudinal $n-$ th slot, whose axis $\{0s_n\}$ coincides with direction of the axis $\{0z\}$, can be presented by the double integrals

$$E_{xn}(x,y,z) = -\frac{ik}{\pi}\sum_{q=1}^{Q}V_{qn}\int\limits_{-d_n/2}^{d_n/2}\int\limits_{-L_n}^{L_n}f_{qn}(s_n)\frac{e^{-ik_1R(s_n,\xi_n)}}{R^2(s_n,\xi_n)}\left[1+\frac{1}{ik_1R(s_n,\xi_n)}\right]y\,ds_nd\xi_n,$$

$$E_{yn}(x,y,z) = \frac{ik}{\pi}\sum_{q=1}^{Q}V_{qn}\int\limits_{-d_n/2}^{d_n/2}\int\limits_{-L_n}^{L_n}f_{qn}(s_n)\frac{e^{-ik_1R(s_n,\xi_n)}}{R^2(s_n,\xi_n)}\left[1+\frac{1}{ik_1R(s_n,\xi_n)}\right](x-\xi_n-x_n)\,ds_nd\xi_n,$$

$$E_{zn}(x,y,z) = 0, \tag{4.21}$$

Where $R(s_n,\xi_n) = \sqrt{(x-\xi_n-x_n)^2+y^2+(z-s_n-z_n)^2}$. The formula (4.21) was obtained by expansion applicability of the condition $d_n/(2L_n) << 1$.

If the directions of the axes $\{0\xi_n\}$ and $\{0x\}$ coincide, the electric field components radiated by the transverse slot with number n, have the form:

$$E_{xn}(x,y,z) = 0,$$

$$E_{yn}(x,y,z) = \frac{ik}{\pi}\sum_{q=1}^{Q}V_{qn}\int\limits_{-d_n/2}^{d_n/2}\int\limits_{-L_n}^{L_n}f_{qn}(s_n)\frac{e^{-ik_1R(s_n,\xi_n)}}{R^2(s_n,\xi_n)}\left[1+\frac{1}{ik_1R(s_n,\xi_n)}\right]$$
$$(z-\xi_n-z_n)\,ds_nd\xi_n,$$

$$E_{zn}(x, y, z) = -\frac{ik}{\pi}\sum_{q=1}^{Q} V_{qn} \int_{-d_n/2}^{d_n/2}\int_{-L_n}^{L_n} f_{qn}(s_n)\frac{e^{-ik_1 R(s_n,\xi_n)}}{R^2(s_n, \xi_n)}\left[1 + \frac{1}{ik_1 R(s_n, \xi_n)}\right] y\, ds_n d\xi_n, \tag{4.22}$$

Where $R(s_n, \xi_n) = \sqrt{(x - s_n - x_n)^2 + y^2 + (z - \xi_n - z_n)^2}$. when a combination of N longitudinal and transverse slot radiators is used, the total electric field of the system can be determined by the expression

$$\vec{E}(x, y, z) = \vec{x}^0 \sum_{n=1}^{N} E_{xn}(x, y, z) + \vec{y}^0 \sum_{n=1}^{N} E_{yn}(x, y, z) + \vec{z}^0 \sum_{n=1}^{N} E_{zn}(x, y, z), \tag{4.23}$$

where $\vec{x}^0, \vec{y}^0, \vec{z}^0$ are unit vectors.

4.2.2 Numerical Results

The circular polarization of electromagnetic fields can be obtained by using two mutually orthogonal slots cut in a waveguide wide wall so that phase differences of electric fields excited in adjacent slots by the H_{10}-wave were equal to 90°, while field amplitudes radiated in the direction of the normal to the waveguide wide wall were equal. Such requirements can be satisfied by systems consisting of longitudinal and transverse slots cut in the wide wall of rectangular waveguide matched with a load (Fig. 4.14). In calculations below, the following parameters of radiating system were used: $a \times b = 86.36 \times 43.18$ mm^2 (WR-340), $h = 2.5$ mm, $f = 2.45$ GHz (frequency typically used for hyperthermia). The H_{10}-wave propagates in the positive direction of $\{0z\}$ axis. If the T-shaped slot arrangement is used (Fig. 4.14a), the slot lengths are close to the resonant values: $2L_1 = 59.0$ mm, $2L_2 = 57.5$ mm, $d_1 = d_2 = 3.0$ mm. The coordinates of the slot centers are: $x_1 = 41.0$ mm, $z_1 = 29.5$ mm, $x_2 = 10.0$ mm, $z_2 = 29.9$ mm. The geometrical dimensions of the slots and their location at the wide wall are selected in such a way that circularly polarized fields were formed in the direction normal to the slot plane. In this case, if the slotted system radiates into the free half-space ($\varepsilon_1 = \mu_1 = 1$), the radiation and reflection coefficients $|S_\Sigma|^2 = 0.9$ and $|S_{11}| = 0.004$. The cross-shaped radiators (Fig. 4.14b) with geometrical dimensions $2L_1 = 2L_2 = 38.0$ mm, $d_1 = d_2 = 3.0$ mm, $x_1 = x_2 = 23.5$ mm, and $z_1 = z_2$ can also generate a circularly polarized field in the normal direction to the slot plane.

The distributions of normalized near-field intensity $\left(|\vec{E}(x, y, z)|^2/|\vec{E}|^2_{\max}\right)$ in the planes $y = const$ radiated by the slotted systems at the frequency $f = 2.45$ GHz into material media with electrophysical parameters corresponding to some human tissues: skin $\left(\varepsilon_1' = 43.5\right)$, adipose tissue ($\varepsilon_1' = 5.56$, $\varepsilon_1'' = 0.86$) [32] are plotted in Figs. 4.15 and 4.16.

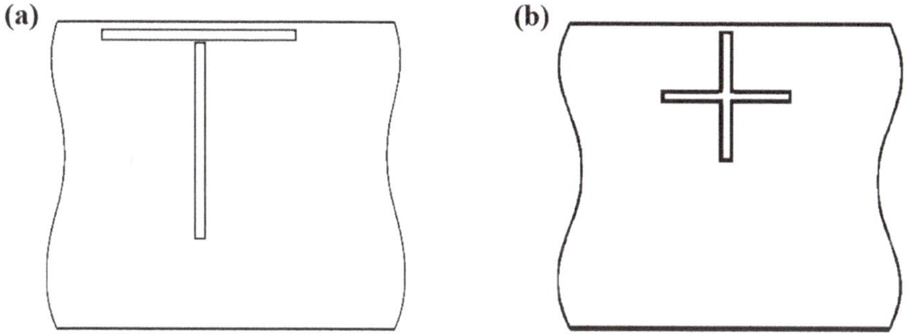

Fig. 4.14 The waveguide-slot radiators with the circular polarization: **a** T-shaped, **b** cross-shaped

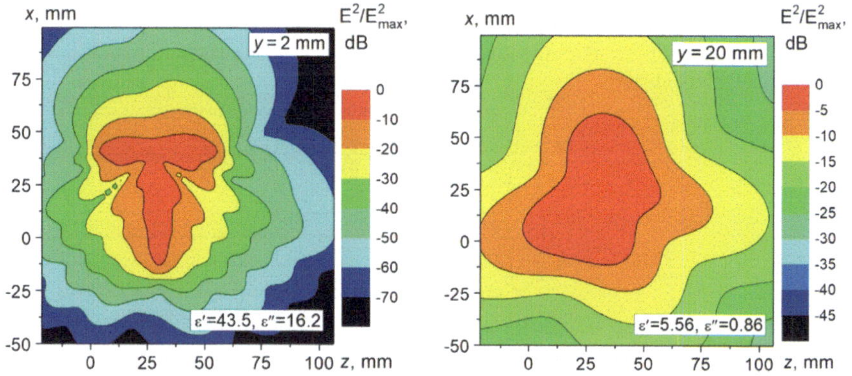

Fig. 4.15 Contour plots of the near field intensity for the T-shaped system (Fig. 4.14a) in material media with electrophysical parameters corresponding to human skin and adipose tissue

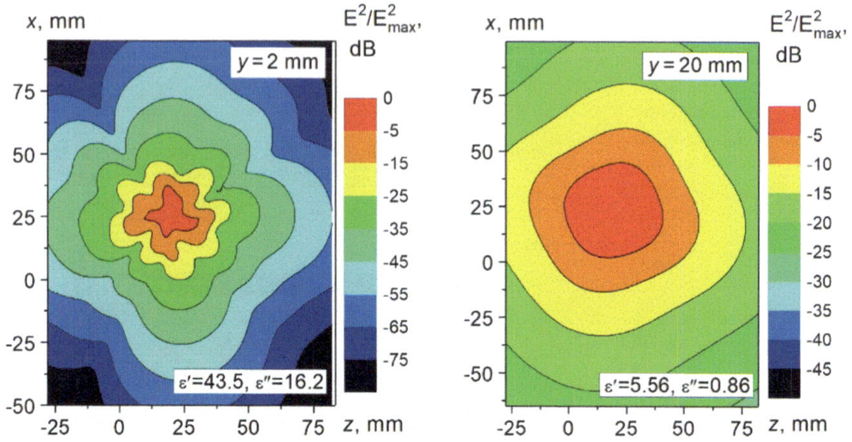

Fig. 4.16 Contour plots of the near field intensity for the cruciform slot (Fig. 4.14b) radiating in material media with electrophysical parameters corresponding to the skin and adipose tissue

Since the radiators generate circularly-polarized fields, the distribution of antenna near field intensities in the material medium at the distance $y = const$ are close to axisymmetric, therefore the fields are concentrated in a certain space region. Similar results can be obtained by using cruciform radiators formed by two slots cut in the wide wall of the rectangular waveguide at the angle 45° relative to the waveguide longitudinal axis.

If large areas of biological tissues should be heated by microwave hyperthermia, the array of waveguide-slot radiators shown in Fig. 4.17 can be used. In this case, the radiation coefficient and hence efficiency of such applicators is increasing. The contour plots of normalized near-field radiated by the multi-slot systems into the free half-space and material medium with electrophysical parameters corresponding to human skin in planes $y = const$ are shown in Figs. 4.18 and 4.19.

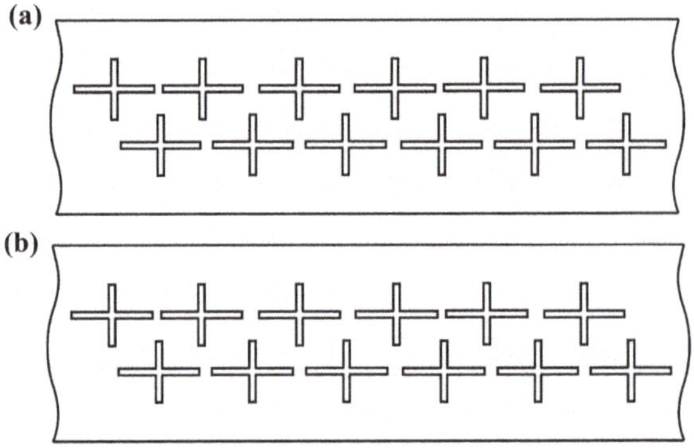

Fig. 4.17 Circularly polarized multi-slot radiator systems

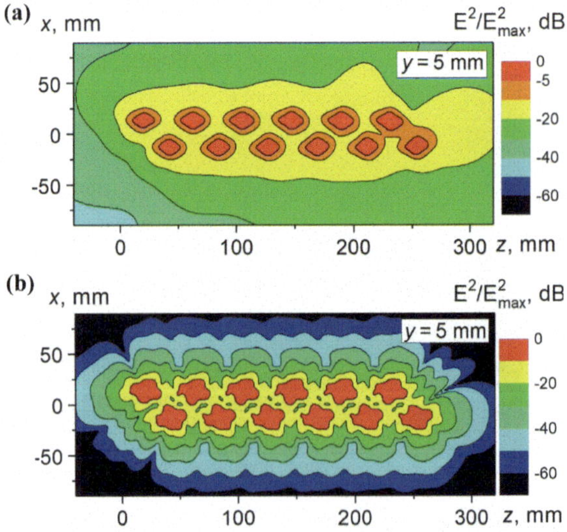

Fig. 4.18 Contour plots of the electric field intensity for the of cross-shaped slotted system shown in Fig. 4.17a radiating into: **a** the free half-space, **b** the material medium (human skin)

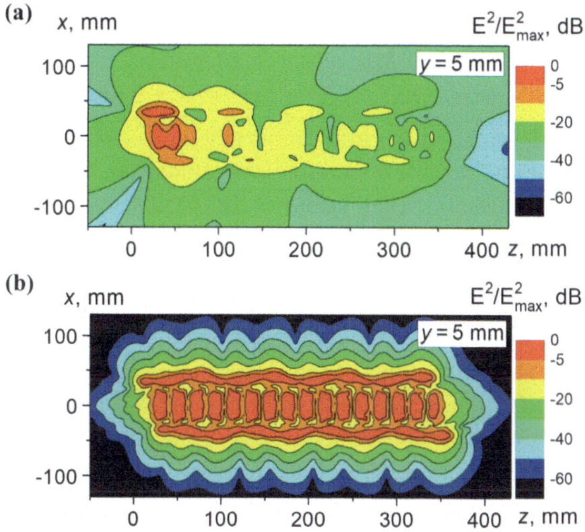

Fig. 4.19 Contour plots of the electric field intensity in the multi-slot system (Fig. 4.17b) radiating into: **a** the free half-space and **b** the material medium (human skin)

The contour plots of the field intensity for the system consisted of 12 cross-shaped slots (Fig. 4.17a) cut in the wide wall of rectangular waveguide at the distance between one another along the axis $\{0z\}$ equal to 21.0 mm are shown in Fig. 4.18. The radiator parameters were as follows: $2L_1 = 38.0\,\text{mm}$, $2L_2 = 28.0\,\text{mm}$, $d_1 = d_2 = 3.0\,\text{mm}$, $x_1 = x_2 = 13.0\,\text{mm}$ As can be seen from the plots, the field intensity distribution for the individual radiators radiating into the material media along the system length is uniform.

The distributions of the radiation field intensity in the structure shown in Fig. 4.17b are presented in Fig. 4.19. The system consists of 16 pairs of longitudinal and transverse slots with parameters: $2L_1 = 59.0\,\text{mm}$, $2L_2 = 56.0\,\text{mm}$, $d_1 = d_2 = 3.0\,\text{mm}$, $x_1 = 34\,\text{mm}$ and 41.0 mm, $x_2 = 0.0\,\text{mm}$, and the distance between the slots along the axis $\{0z\}$ is $\lambda_g/8 = 20.87\,\text{mm}$ (where λ_g is the wavelength in the waveguide). The substantially non-uniform distribution of the field intensity along the structure in the free half-space (Fig. 4.19a) becomes uniform when the system radiates into the biological media (Fig. 4.19b). The radiation and reflection coefficients for this slotted system radiating into the free half-space and skin, are $|S_\Sigma|^2 = 0.25$ and $|S_{11}| = 0.002$, respectively.

In conclusion, it should be noted that mass-dimensional parameters of these devices can be decreased by using low-profile waveguides, for example, with a cross section $a \times b = 86.36 \times 21.59\,\text{mm}^2$. In this case, the characteristics of the radiated field do not vary.

4.3 Multi-frequency Slotted Arrays on Multimode Waveguides

The development of multifrequency multi-channel antennas in the form of multiple antenna arrays placed in a single aperture, allows us to solve the problem of increasing the efficiency of both airborne and ground-based radiocomplexes. In a number of cases, it seems rational from the point of view of practical realization of needed characteristics of radiocomplexes to use multifunctional antennas in the form of multi-slot systems with different lengths of slot elements on the basis of a multi-mode waveguide. In the literature, there are practically no methods, calculation algorithms, and results of studies of the electrodynamic characteristics of slotted multi-element multi-frequency antennas on multimode waveguides, which certainly limits their application in antenna technology. The results obtained in this work allow us for the first time to apply the electrodynamic method of induced magnetomotive forces to calculate the radiation characteristics of multi-element waveguide-slotted arrays that include several (generally arbitrary type) subarrays with slotted elements of different lengths, operating at different frequencies and located in the walls of a single waveguide.

4.3.1 *Two-Frequency Slotted Array Based on Longitudinal Slots*

As an example, allowing on the basis of the results obtained in this work to conduct for the first time the computational studies of the characteristics and find out the physical peculiarities in the formation of the radiated field. Below the results of the study of the two-frequency in-phase waveguide-slot array, located in the wide wall of a rectangular waveguide are presented.

The radiating aperture is the two systems of slots located in a wide wall of thickness h of a rectangular waveguide (Fig. 4.20). The longitudinal slots with length l_1 and width d_1 are excited by dominant wave at frequency f_1, and the slots with length l_2 and width d_2 are excited by the H_{30} wave at frequency f_2. The first system operates in the long wavelength range and the second in the short wavelength range. We represent the electric field $\vec{E}_n$ (or magnetic current $\vec{J}_n$) in the slot with number n as a expansion in linearly independent vector-functions of slots $\vec{E}_{nq}$ with complex coefficients V_{nq}: $\vec{E}_n = \sum_q V_{nq}\vec{E}_{nq}$.

Then the field in the opening of the multifunctional antenna:

$$\vec{E} = \sum_r \sum_p V_{rp}^{(1)}\vec{E}_{rp}^{(1)} + \sum_\upsilon \sum_q V_{\upsilon q}^{(2)}\vec{E}_{\upsilon q}^{(2)}, \tag{4.24}$$

where v, r – slot numbers; p, q – numbers of harmonics on the slots. The first term is the field in the opening, created by the slots with length l_1, and the second term is the

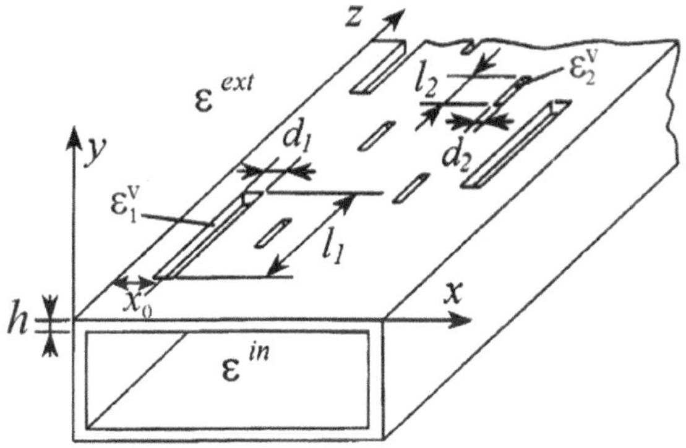

Fig. 4.20 Geometry of the dual-frequency waveguide-slotted array

field in the opening of short slots with length l_2. Having defined the electric field in the antenna's opening according to expression (4.24), we find energy characteristics and pattern of the multifrequency antenna.

The radiation pattern of the slotted system is defined by the expression:

$$F(\theta,\phi) = F_0(\theta,\phi)\sum_{n=1}^{N} d_n \frac{\sin\left(\frac{kd_n}{2}\cos\theta\sin\phi\right)}{\frac{kd_n}{2}\cos\theta\sin\phi} e^{ik(x_n\cos\theta\sin\phi + z_n\cos\theta\cos\phi)}$$
$$\times \int_{-L_n}^{L_n} J_n(s_n) e^{iks_n\cos\theta\cos\phi} ds_n. \tag{4.25}$$

Taking into account the representation of the equivalent magnetic current in the longitudinal slots in the form $J_n(s_n) = J^s_{0n} f^s_n(s_n) + J^a_{0n} f^a_n(s_n)$ and choosing as the functions $f^{s,a}_n(s_n)$ found by the averaging method (Sect. 2.3.3),

$$f^s_m(s_m) = \cos ks_m \cos k_c L_m - \cos kL_m \cos k_c s_m\,,$$
$$f^a_m(s_m) = \sin ks_m \sin k_c L_m - \sin kL_m \sin k_c s_m\,,$$

the expression for the integral in (4.25) is transformed to the form

$$\int_{-L_n}^{L_n} J_n(s_n) e^{iks_n \cos\theta\cos\phi} ds_n = 2J^s_{0n}\left[\frac{k\cos k_g L_n}{k^2-(k\cos\theta\cos\phi)^2}\Big(\sin kL_n \cos(kL_n\cos\theta\cos\phi)\right.$$
$$-\cos\theta\cos\phi\cos kL_n \sin(kL_n\cos\theta\cos\phi)\Big) - \frac{\cos kL_n}{k_g^2-(k\cos\theta\cos\phi)^2}$$
$$\left.\times\Big(k_g \sin k_g L_n \cos(kL_n\cos\theta\cos\phi) - k\cos\theta\cos\phi\cos k_g L_n \sin(kL_n\cos\theta\cos\phi)\Big)\right]$$
$$-2iJ^a_{0n}\left[\frac{k\sin k_g L_n}{k^2-(k\cos\theta\cos\phi)^2}\Big(\cos kL_n \sin(kL_n\cos\theta\cos\phi)\right.$$
$$-\cos\theta\cos\phi\sin kL_n \cos(kL_n\cos\theta\cos\phi)\Big) - \frac{\sin kL_n}{k_g^2-(k\cos\theta\cos\phi)^2}$$
$$\left.\times\Big(k_g \cos k_g L_n \sin(kL_n\cos\theta\cos\phi) - k\cos\theta\cos\phi\sin k_g L_n \cos(k_g L_n\cos\theta\cos\phi)\Big)\right]$$

In the case of slots with lengths greater than $(2.5 \div 2.75)\lambda$, the radiation patterns of the radiating systems in orthogonal planes when the electric field in the slots is approximated by functions of the form.

$$E_{np} = \frac{1}{d_n}\sin\frac{p\pi(L_n+s_n)}{2L_n},$$

can be calculated by the formula

$$F(\theta,\phi) = F_0(\theta,\phi)\sum_{n=1}^{N}\frac{\sin(\frac{kd_n}{2}\cos\theta\sin\phi)}{\frac{kd_n}{2}\cos\theta\sin\phi}e^{ik(x_n\cos\theta\sin\phi+z_n\cos\theta\cos\phi)}$$
$$\times\sum_{p=1}^{Q}V_{np}\frac{p\pi}{l_n}\frac{1-(-1)^p e^{ikl_n\cos\theta\cos\phi}}{\left(\frac{p\pi}{l_n}\right)^2-(k\cos\theta\cos\phi)^2}, \tag{4.26}$$

where $F_0(\theta, \phi)$ is the RP of the elementary magnetic dipole (note that in formulas (4.25) and (4.26) the slot lengths are designated as $2L$ and l, respectively. This is due to the convenience of representing the field distribution function in the slot when solving the integrodifferential equations by different methods).

In order to understand the physical features of the interaction of the slots in each subarray, as in the case of arrays with transverse slots (Sect. 4.1), it is advisable to give some results of studies of the radiation characteristics of a single longitudinal slot cut in the wall of a waveguide whose dimensions allow the propagation of higher wave types. The values of radiation and reflection coefficients of the longitudinal slot excited by wave H_{30} in the waveguide with a cross section of 28.5×12.6 мм^2 in the wavelength range are significantly less than the corresponding values when the $l \approx 0.5\lambda$ ($|S_\Sigma|^2 = 0.5$) of the main waveguide is excited by dominant mode H_{10} (Fig. 4.21).

The electrical length of the resonant slot l/λ_{res} (the length of the slot with the maximum radiation coefficient) depends on the dispersion properties of the

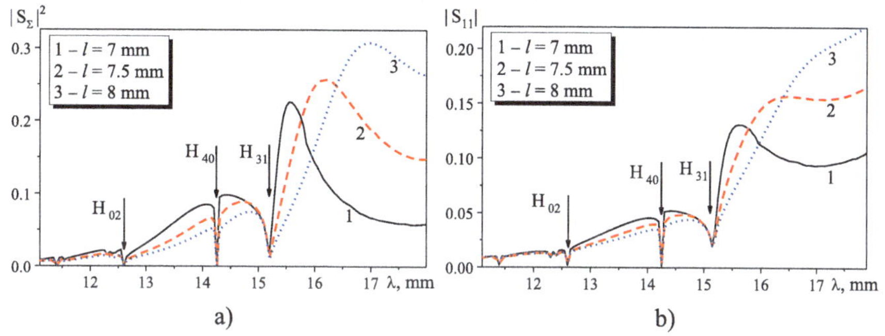

Fig. 4.21 Dependences of radiation (**a**) and reflection coefficients (**b**) on wavelength for slots ($d = 1$mm; $x_0 = 4.75$mm; $h = 1$mm) of different lengths excited by H_{30} wave

waveguide and varies with the ratio λ/λ_{cr}(λ_{cr} is the critical wavelength H_{30}, which excites the slot). As λ approaches λ_{cr}, the l/λ_{res} value tends to a value of 0.5, and reduces as the wavelength decreases. (For example, for the slot with $l = 7$mm at $\lambda = 15.6$mm($\lambda/\lambda_{cr} \approx 0.82$) the value of $l/\lambda_{res} \approx 0.449$). In the dependences $|S_{\sum}|^2 = f(\lambda)$ and $|S_{11}| = f(\lambda)$ the minimum values of the radiation coefficient appear at such λ at which higher order waves arise. Such a peculiarity of the dependences $S_{\sum}|^2 = f(\lambda)$ and $|S_{11}| = f(\lambda)$ must be taken into account when selecting the electrical dimensions of the slot radiators and the operating frequency band of the multi-element shortwave antenna.

Narrow slots are characterized by a very strong dependence of energy coefficients on the wall thickness h. The maximum radiation is observed at the value of h close to the value of d. By changing the ratio between d and h at a fixed x_0, it is possible to significantly change the slot radiation coefficient $|S_{\sum}|^2$ in multimode regime, which is especially important for the forming of the required amplitude-phase distribution in slotted array with a large number of elements (Figs. 4.22 and 4.23).

For the designing antennas with multi-parameter dependence of radiation characteristics, to estimate their efficiency it is necessary to take into account the transformation coefficients $\left|S_{30}^{10}\right|$ and $\left|S_{30}^{20}\right|$ $H_{30}-$ wave into H_{10} and H_{20} respectively for slots of different length (Fig. 4.24). (In the considered radiators (Fig. 4.24) the transformation coefficients are quite small and they can be disregarded for calculating the amplification of the slot radiator).

Taking into account the peculiarities in the formation of radiation characteristics by single slot radiators, here are some results of studies of two-frequency electrically short $L_A = (3 \div 8)\lambda$, ($N = 6 \div 11$) and electrically long $L_A = (30 \div 100)\lambda$, ($N = 60 \div 100$) arrays, whose complex structure has so far not allowed to apply electrodynamic methods for their calculation. Thus, Fig. 4.25 presents the results of calculation of amplitude distribution in the array of length$L_A = 134$mm, consisting of 6 slots of length l_1 and 11 shorter slots with $l_2 = 8$mm at their excitation by H_{30} wave with$\lambda = 16$mm. The centers of the slots with the size l_1 were located on the wall of the waveguide at points where the surface currents of the

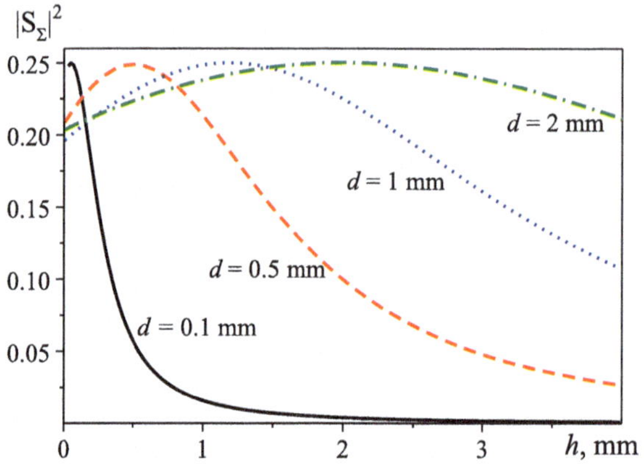

Fig. 4.22 Dependences of radiation coefficients on the wall thickness of the waveguide excited by the wave $H_{30}(\lambda = 16\,\text{mm})$ for slots $l = 7.5$ mm, $x_0 = 4.75$ mm with different width d

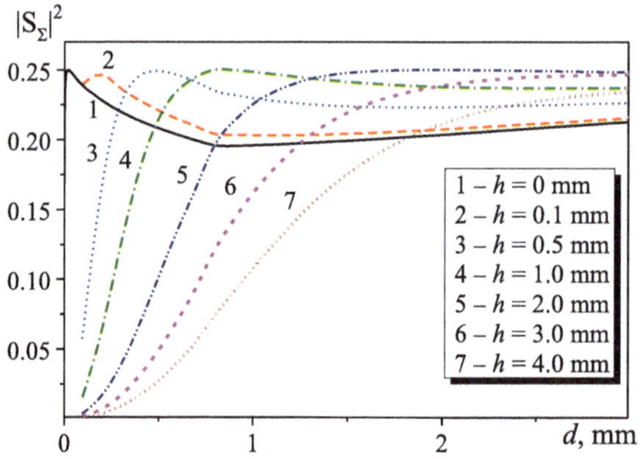

Fig. 4.23 Dependences of radiation coefficients on slot widths$l = 7.5$mm, $x_0 = 4.75$mm at different wall thickness h of the waveguide excited by wave H_{30} ($\lambda = 16$ mm)

H_{30} wave are equal to zero. Nevertheless, due to mutual influence these slots were excited. By selecting the sizes of the slots (value l_1 varied from 17 to 22 mm) and their positions it is possible to reduce the amplitude of the field induced by short slots with length l_2 in the slots with length l_1 and thus reduce the interaction of subarrays, and create the required amplitude-phase distribution in the antenna aperture at each of the working frequencies.

The patterns of the considered slotted array are shown in Fig. 4.26. Decreasing the level of side lobes to -20 dB in the pattern was achieved by creating a falling amplitude distribution in the antennas by changing the length of the slots and their location on the waveguide wall (Fig. 4.26, curves 1, 2). The influence of the long-wave antenna on the directivity diagram of the short-wave antenna was demonstrated

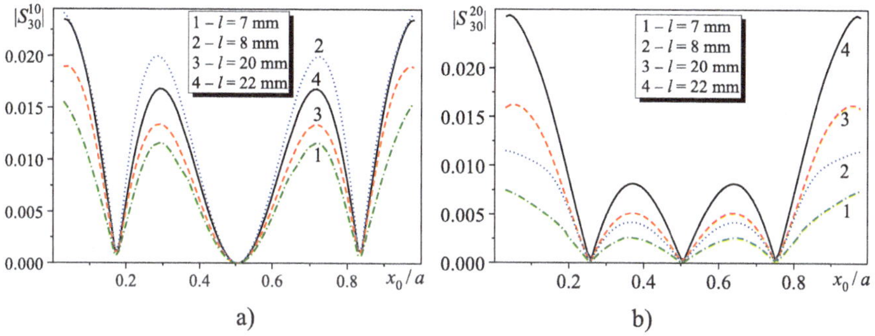

Fig. 4.24 Dependences of conversion coefficients of wave H_{30} to H_{10} (**a**) and H_{20} (**b**) from the parameter x_0 / a for slots of different lengths

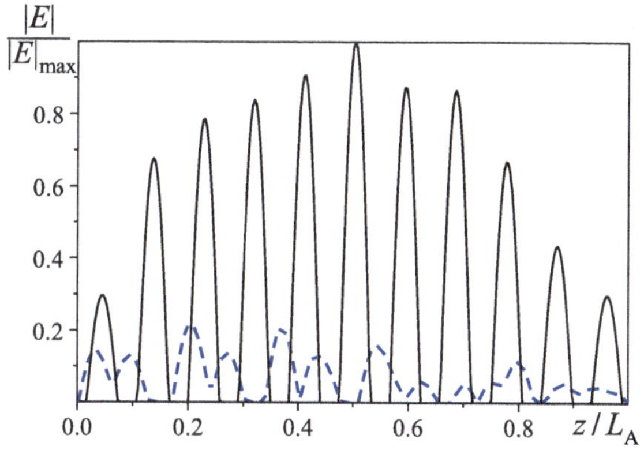

Fig. 4.25 Field distribution in two-frequency array («—» – $N = 11$, $l = 8$ mm; «——» – $N = 6$, $l = 20$ mm)

in the increase of the first side lobe to the value of −13.6 dB (Fig. 4.26, curve 3). It should be noted that when the frequency is changed within 8%, no increase in the side lobe above −13.6 dB is observed. For the given antenna at excitation by wave H_{30} with $\lambda = 16.5$mm the values of coefficients of radiation and reflection are equal correspondingly 0.949 and 0.012 (Fig. 4.27), and at excitation by wave H_{10} with $\lambda = 44$mm, $|S_{\sum}|^2 = 0.905$ and $|S_{11}| = 0.201$.

Using a modification of the induced magnetomotive force method developed by the authors [35], it is possible to calculate and optimize the characteristics of antenna arrays consisting of a large number of slot radiators of arbitrary length in general. So, for example, realization of the amplitude distribution falling to the edges in a linear array of length $L_A = 8.5\lambda$ by appropriate arrangement of the system of 100 short-wave slots in a wide wall of a waveguide with cross section $a \times b = 28.5 \times 12.6$ mm^2, $t = 1$ mm, excited by wave H_{30}, allows reducing the side lobes to the level -24 dB (Fig. 4.28). The width of the main lobe of the directional pattern of this antenna by the half-power level in the plane of vector $\vec{\mathrm{H}}$ is $\Delta\theta_{0.5}^{\mathrm{H}} = 1.5$, the directional coefficient is 144.6, the radiation and reflection coefficients are 0.997 and

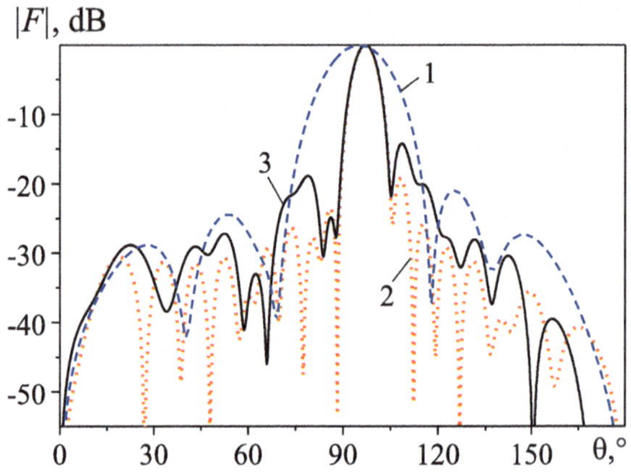

Fig. 4.26 Pattern of antenna arrays: 1—dual-frequency antenna, $\lambda = 44$mm; 2—single-frequency slot array with $l = 8$ mm, $\lambda = 16.5$ mm; 3—dual-frequency antenna, $\lambda = 16.5$ mm

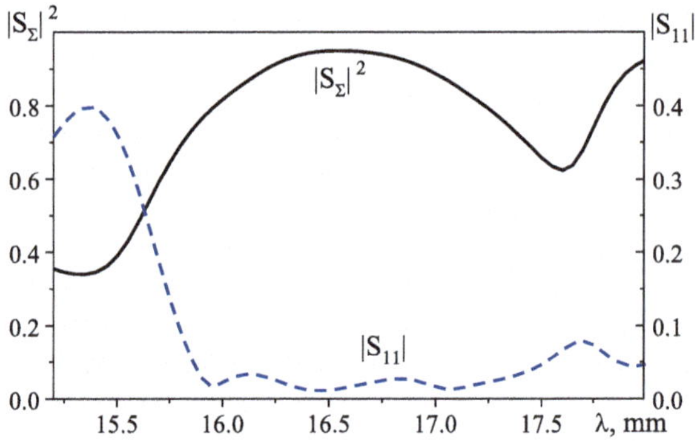

Fig. 4.27 Dependences of radiation and reflection coefficients of the two-frequency antenna on the wavelength by H_{30} wave excitation for $\lambda = 16.5$ mm

0.053, respectively. In the same aperture it is possible to place a system of 60 long-wave slots ($l_1 = 20$ mm) located so that their influence on the short-wave system is minimal. The influence of the long-wave antenna is appeared in the increase of the side lobes of the directional pattern to a level of –17.7 dB, which in most practical cases is acceptable (Fig. 4.29). At the same time the directional coefficient is 131.8.

Thus, the effectiveness of application of the method of induced magnetomotive forces using the obtained solution for determining the electrical field at the aperture of a multi-element slotted array to calculate the frequency-energy characteristics of waveguide-slotted arrays, including multi-element on the basis of multi-mode waveguides, whose characteristics cannot be studied by electrodynamic methods and are not available in the literature so far, is shown.

It is shown that the obtained solutions make it possible for the first time to apply the electrodynamic methods for calculation of multielement multi-frequency antennas and select by calculation the geometric parameters of subarray elements, which provide the best decoupling at operating frequencies while maintaining the main

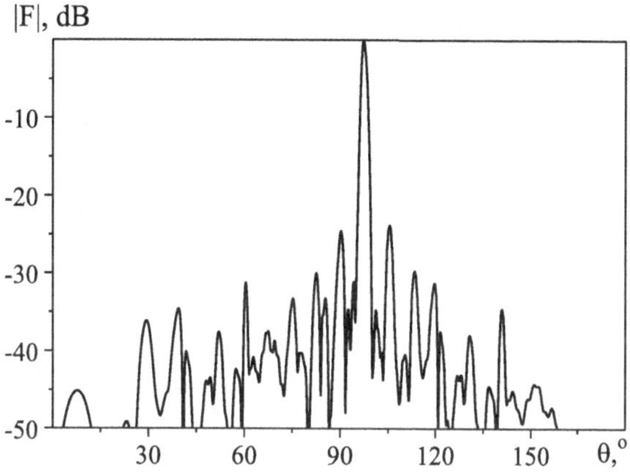

Fig. 4.28 Pattern of a single-frequency short-wave antenna array with $N = 100$, $l = 7.2$ mm, excited by H_{30} wave at $\lambda = 16.5$ mm

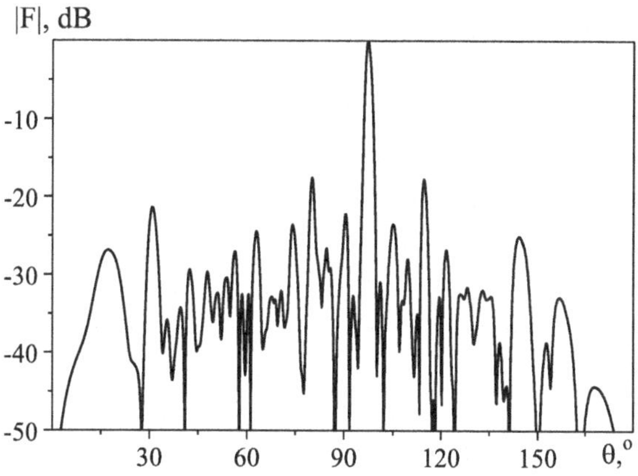

Fig. 4.29 Pattern of a dual-frequency antenna array with 60 slots of length $l_1 =$ 20 mm and 100 slots of length $l = 7.2$ mm, excited by wave H_{30} at $\lambda = 16.5$mm

characteristics of the antenna as a whole: the width of the radiation pattern, the directivity factor, gain, side lobes level, operating frequencies etc.

The constructed mathematical models and efficient calculation algorithms allow determining the selection criteria and conditions limiting the choice of electrical dimensions of the slot radiators and the operating frequency band for construction the waveguide-slotted array based on waveguides of big electrical sizes.

4.3.2 Radiation of Electromagnetic Waves by a System of Transverse and Longitudinal Slots in the Wide Wall of Rectangular Waveguide

Multi-mode waveguide-slotted arrays with combined aperture, subarrays of which radiate the fields of two orthogonal polarizations are promising for wide practical application. In this case, due to the orthogonality of the radiated fields, a minimum of mutual coupling between subarrays can be provided.

The array can be created on the basis of a system of subarrays of transverse and longitudinal slots in the wide wall of a rectangular multimode waveguide. The results obtained earlier in the monograph allow us to apply the electrodynamic method to calculate their characteristics and, for the first time, take into account the thickness of the waveguide wall and the complete interaction of the slots along the inner and outer spaces.

The radiating aperture represents two systems of slots located in a wide wall of thickness t of rectangular waveguide with cross-section $a \times b$ (Fig. 4.30). The longitudinal slots with lengths L_n and widths D_n are excited by the dominant mode H_{10} at frequency f_{10} (wavelength λ_{10}), and the transverse slots with lengths l_n and widths d_n are excited by H_{30} wave at frequency f_{30} (wavelengthλ_{30}). The first system operates in the long wavelength range and the second one works in the short wavelength band.

We represent the field in each slot as the expansion in linearly independent vector-functions of longitudinal and transverse slots $\vec{E}_{nq}$ and $\vec{e}_{n'q'}$ with complex coefficients V_{nq} and $W_{n'q'}$. Then the total electric field in the aperture of the multifunction antenna from the side of the waveguide (index "1") and the half-space above the infinite perfectly conducting screen (index "2") can be represented as follows:

$$\vec{E}^{(1)} = \sum_{n=1}^{N}\sum_{q=1}^{Q} V_{nq}^{(1)}\vec{E}_{nq}^{(1)} + \sum_{n'=1}^{N'}\sum_{q'=1}^{Q'} W_{n'q'}^{(1)}\vec{e}_{n'q'}^{(1)},$$

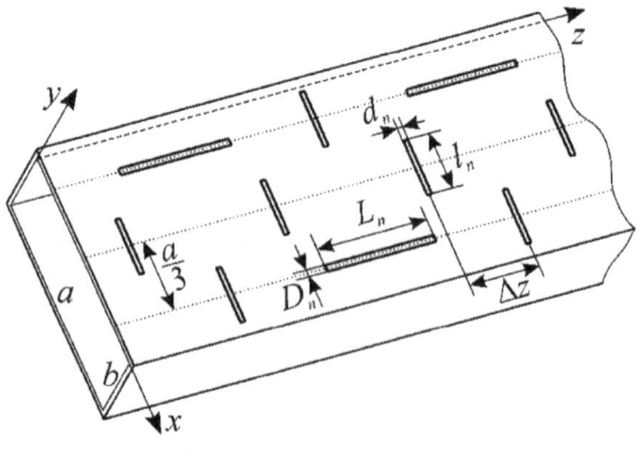

Fig. 4.30 The geometry of the system and the accepted notations

$$\vec{E}^{(2)} = \sum_{n=1}^{N}\sum_{q=1}^{Q} V_{nq}^{(2)}\vec{E}_{nq}^{(2)} + \sum_{n'=1}^{N'}\sum_{q'=1}^{Q'} W_{n'q'}^{(2)}\vec{e}_{n'q'}^{(2)}, \tag{4.27}$$

where $N,\ N'$ are the numbers of longitudinal and transverse slots, respectively, and $n,\ n'$ are their ordinal numbers. $Q,\ Q'$ are the number of field approximation functions on the surfaces of the longitudinal and transverse slots, $q,\ q'$ are the order numbers of vector-functions.

Substituting (4.27) into continuity equations for tangential components of magnetic field intensity vector on inner and outer surfaces of each slot (1.17) and using Galerkin method for solution of obtained integral equations, we obtain the system of linear algebraic equations regarding unknown quantities $V_{nq}^{(1),(2)}$ and $W_{n'q'}^{(1),(2)}$:

$$\begin{cases} \sum\limits_{n=1}^{N}\sum\limits_{q=1}^{Q} V_{nq}^{(1)}(Y_{mn,pq}^{\text{in}} + Y_{m,pq}^{\text{V}_{11}}\delta_{mn}\delta_{pq}) + \sum\limits_{n'=1}^{N'}\sum\limits_{q'=1}^{Q'} W_{n'q'}^{(1)} Y_{mn',pq'}^{\text{in}} \\ + \sum\limits_{q=1}^{Q} V_{mq}^{(2)} Y_{m,pq}^{\text{V}_{12}}\delta_{pq} = F_{mp}, \\ \sum\limits_{q=1}^{Q} V_{mq}^{(1)} Y_{m,pq}^{\text{V}_{21}}\delta_{pq} + \sum\limits_{n=1}^{N}\sum\limits_{q=1}^{Q} V_{nq}^{(2)}(Y_{mn,pq}^{\text{ext}} + Y_{m,pq}^{\text{V}_{22}}\delta_{mn}\delta_{pq}) \\ + \sum\limits_{n'=1}^{N'}\sum\limits_{q'=1}^{Q'} W_{n'q'}^{(2)} Y_{mn',pq'}^{\text{ext}} = 0, \\ m = 1, 2, \ldots N,\quad p = 1, 2, \ldots Q; \\ \sum\limits_{n=1}^{N}\sum\limits_{q=1}^{Q} V_{nq}^{(1)} Y_{m'n,p'q}^{\text{in}} + \sum\limits_{n'=1}^{N'}\sum\limits_{q'=1}^{Q'} W_{nq}^{(1)}(Y_{m'n,p'q}^{\text{in}} + Y_{m',p'q'}^{\text{V}_{11}}\delta_{m'n'}\delta_{p'q'}) \\ + \sum\limits_{q'=1}^{Q'} W_{m'q'}^{(2)} Y_{m',p'q'}^{\text{V}_{12}}\delta_{p'q'} = F_{m'p'}, \\ \sum\limits_{q'=1}^{Q'} W_{m'q'}^{(1)} Y_{m',p'q'}^{\text{V}_{21}}\delta_{p'q'} + \sum\limits_{n=1}^{N}\sum\limits_{q=1}^{Q} V_{nq}^{(2)} Y_{m'n,p'q}^{\text{ext}} \\ + \sum\limits_{n'=1}^{N'}\sum\limits_{q'=1}^{Q'} W_{n'q'}^{(2)}(Y_{m'n',pq}^{\text{ext}} + Y_{m',p'q'}^{\text{V}_{22}}\delta_{m'n'}\delta_{p'q'}) = 0, \\ m' = 1, 2, \ldots N',\quad p' = 1, 2, \ldots Q'. \end{cases}$$

The coefficients of the obtained SLAE are the intrinsic and mutual internal and external conductivities of the slots, as well as the conductivities of the slot volume:

$$Y_{mn,pq}^{\text{in,(ext)}} = \mp \int\limits_{\varsigma_m^{(1),(2)}} \left[\vec{E}_{mp}^{(1),(2)}, \vec{H}^{\text{in,(ext)}}(\vec{E}_{nq}^{(1),(2)})\right] \vec{n} d\varsigma,$$

$$Y_{m'n',p'q'}^{\text{in,(ext)}} = \mp \int\limits_{\varsigma_{m'}^{(1),(2)}} \left[\vec{e}_{m'p'}^{(1),(2)}, \vec{H}^{\text{in,(ext)}}(\vec{e}_{n'q'}^{(1),(2)})\right] \vec{n} d\varsigma,$$

$$Y_{m'n,p'q}^{\text{in,(ext)}} = \mp \int\limits_{\varsigma_{m'}^{(1),(2)}} \left[\vec{e}_{m'p'}^{(1),(2)}, \vec{H}^{\text{in,(ext)}}(\vec{E}_{nq}^{(1),(2)})\right] \vec{n} \mathrm{d}s, \tag{4.28}$$

$$Y_{m,pq}^{\text{v}_{11},(\text{v}_{22}),(\text{v}_{12})} = \int\limits_{\varsigma_m^{(1),(2),(1)}} \left[\vec{E}_{mp}^{(1),(2),(1)}, \vec{H}^{\text{v}}(\vec{E}_{mq}^{(1),(2),(2)})\right] \vec{n} d\varsigma,$$

$$Y_{m',p'q'}^{\text{v}_{11},(\text{v}_{22}),(\text{v}_{12})} = \int\limits_{\varsigma_{m'}^{(1),(2),(1)}} \left[\vec{e}_{m'p'}^{(1),(2),(1)}, \vec{H}^{\text{v}}(\vec{e}_{m'q'}^{(1),(2),(2)})\right] \vec{n} d\varsigma.$$

Here $F_{mp} = \int\limits_{\varsigma_m} \left[\vec{E}_{mp}, \vec{H}^0\right] \vec{n} d\varsigma$, $F_{m'p'} = \int\limits_{\varsigma_{m'}} \left[\vec{e}_{m'p'}, \vec{H}^0\right] \vec{n} \mathrm{d}s$, are partial magnetomotive forces; $\vec{H}^{\text{in,(ext),(v)}}(\vec{E}_{nq}^{(1),(2)})$, $\vec{H}^{\text{in,(ext),(v)}}(\vec{e}_{n'q'}^{(1),(2)})$, are magnetic fields in the "in" (inner region of the waveguide), "ext" (half-space above the slit) and "v" (inner region of the slot) regions, excited by electric fields $\vec{E}_{nq}^{(1),(2)}, \vec{e}_{n'q'}^{(1),(2)}$ on the slot surfaces; $\vec{H}^0$ is the magnetic field of external sources; ς_m is the surface area of the slot with number m; $\vec{n}$ is the normal vector to the slot surface, directed inside the waveguide; δ_{mn}, δ_{pq} are Kronecker symbols.

Let us consider the case, typical for practice, when the width of the slots is much less than their length and wavelength. Due to the electrodynamic boundary conditions, the electrical field in the slots is directed across them and does not depend on the transverse coordinate. Then as eigenvector functions of longitudinal and transverse slots we can take functions, respectively:

$$\vec{E}_{nq}^{(1),(2)} = \frac{\vec{x}^0}{D_n} \sin \frac{q\pi}{L_n}\left(z - \left(z_n - \frac{L_n}{2}\right)\right), \quad \vec{e}_{n'q'}^{(1),(2)} = \frac{\vec{z}^0}{d_{n'}} \sin \frac{q'\pi}{l_{n'}}\left(x - \left(x_{n'} - \frac{l_{n'}}{2}\right)\right), \tag{4.29}$$

where $\vec{z}^0, \vec{x}^0$ are the unit vectors directed, respectively, along the x and z axes; z_n, $x_{n'}$ are corresponding coordinates of the slot centers with numbers n and n'.

Let's obtain expressions for mutual conductivities of orthogonal slots located in the wide wall of rectangular waveguide (Fig. 4.31).

External mutual conductivity of orthogonal slots according to (4.28) taking into account (1.22) and (4.29):

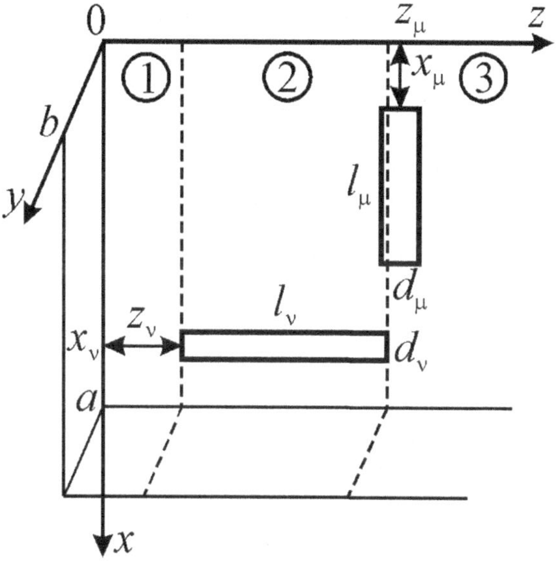

Fig. 4.31 Geometry of the system of two orthogonal slots

$$Y^{\text{ext}}_{\mu\nu,pq} = \frac{i}{2\pi k_0 \rho d_\mu d_\nu} \int\limits_{x_\mu}^{x_\mu+l_\mu} \int\limits_{z_\mu-\frac{d_\mu}{2}}^{z_\mu+\frac{d_\mu}{2}} \int\limits_{z_\nu}^{z_\nu+l_\nu} \int\limits_{x_\nu-\frac{d_\nu}{2}}^{x_\nu+\frac{d_\nu}{2}} \sin\frac{p\pi(x-x_\mu)}{l_\mu}\sin\frac{q\pi(z'-z_\nu)}{l_\nu}$$
$$\times \frac{\partial^2}{\partial x \partial z}\frac{e^{-ik\sqrt{(z-z')^2+(x-x')^2}}}{\sqrt{(z-z')^2+(x-x')^2}} \mathrm{d}x'\mathrm{d}z'\mathrm{d}z\,\mathrm{d}x. \tag{4.30}$$

By integrating a part of the integral in (4.30) over the parts, we obtain:

$$Y^{\text{ext}}_{\mu\nu,pq} = \frac{i}{2\pi k_0 \rho d_\mu d_\nu}\cdot\frac{p\pi}{l_\mu} \int\limits_{x_\mu}^{x_\mu+l_\mu} \int\limits_{z_\nu}^{z_\nu+l_\nu} \int\limits_{x_\nu-\frac{d_\nu}{2}}^{x_\nu+\frac{d_\nu}{2}} \sin\frac{p\pi(x-x_\mu)}{l_\mu}\sin\frac{q\pi(z'-z_\nu)}{l_\nu}$$
$$\times\left(\frac{e^{-ikr_1}}{r_1}-\frac{e^{-ikr_2}}{r_2}\right)\mathrm{d}x'\mathrm{d}z'\mathrm{d}x,$$

$$r_1=\sqrt{(z_\mu+\frac{d_\mu}{2}-z')^2+(x-x')^2},\ r_2=\sqrt{(z_\mu-\frac{d_\mu}{2}-z')^2+(x-x')^2},$$
$$r_2=\sqrt{(z_\mu-\frac{d_\mu}{2}-z')^2+(x-x')^2}.$$

According to expression (4.28), the internal mutual conductivity of orthogonal slots located in the wide wall of a rectangular waveguide (Fig. 4.31):

$$Y^{\text{in}}_{\mu\nu,pq} = -\int\limits_{x_\mu}^{x_\mu+l_\mu}\int\limits_{z_\mu-\frac{d_\mu}{2}}^{z_\mu+\frac{d_\mu}{2}} \frac{1}{d_\mu}\sin\frac{p\pi(x-x_\mu)}{l_\mu}\cdot H_x^{\text{in}}(\vec{e}_{\nu q})\text{dz}\,\text{dx}. \tag{4.31}$$

Here the x-th component of the magnetic field excited by the longitudinal slot according to (1.22) at $y = 0$ is defined by the expression

$$H_x^{\text{in}}(\vec{e}_{\nu q}) = \frac{-q\pi}{ik_0\rho abl_\nu d_\nu}\sum_{m,n}\varepsilon_m\varepsilon_n\sin\frac{m\pi x}{a}\cos\frac{m\pi x_\nu}{a}\sin\frac{m\pi d_\nu}{2a}\cdot\frac{I_{mn}(z)}{\gamma_{mn}^2-\left(\frac{q\pi}{l_\nu}\right)^2}, \tag{4.32}$$

where

$I_{mn}(z) = \left((-1)^q e^{-i\gamma_{mn}l_\nu} - 1\right)e^{i\gamma_{mn}(z-z_\nu)}$ for domain 1 in Fig. 4.31 ($z \le z_\nu$),
$I_{mn}(z) = \left(1-(-1)^q e^{i\gamma_{mn}l_\nu}\right)e^{-i\gamma_{mn}(z-z_\nu)}$ for domain 3 ($z \ge z_\nu + l_\nu$),
$I_{mn}(z) = (-1)^q e^{-i\gamma_{mn}l_\nu}e^{i\gamma_{mn}(z-z_\nu)} + e^{-i\gamma_{mn}(z-z_\nu)} - 2\cos\frac{q\pi(z-z_\nu)}{l_\nu}$ for domain 2

$$(z_\nu < z < z_\nu + l_\nu).$$

Substituting (4.32) into (4.31) and integrating over x and z (over the transverse slot area with number μ) gives the following expression for the transverse slot mutual conductivity:

$$Y^{\text{in}}_{\mu\nu,pq} = \frac{qp\pi^2}{ik_0\rho abl_\nu l_\mu d_\nu d_\mu}\sum_{m,n}\varepsilon_m\varepsilon_n\cos\frac{m\pi x_\nu}{a}\sin\frac{m\pi d_\nu}{2a}\cdot N_m\cdot\frac{F_{mn}}{\gamma_{mn}^2-\left(\frac{q\pi}{l_\nu}\right)^2}.$$

$$N_m = \frac{1}{\left(\frac{p\pi}{l_\mu}\right)^2-\left(\frac{m\pi}{a}\right)^2}\left[\sin\frac{m\pi x_\mu}{a}-(-1)^p\sin\frac{m\pi(x_\mu+l_\mu)}{a}\right],$$

$$F_{mn} = \int\limits_{z_\mu-\frac{d_\mu}{2}}^{z_\mu+\frac{d_\mu}{2}} I_{mn}(z)\,\text{dz},\; N_m = \frac{1}{\left(\frac{p\pi}{l_\mu}\right)^2-\left(\frac{m\pi}{a}\right)^2}\left[\sin\frac{m\pi x_\mu}{a}-(-1)^p\sin\frac{m\pi(x_\mu+l_\mu)}{a}\right],$$

$$F_{mn} = \int\limits_{z_\mu-\frac{d_\mu}{2}}^{z_\mu+\frac{d_\mu}{2}} I_{mn}(z)\,\text{dz},$$

The expressions for F_{mn} depending on the mutual arrangement of the slots will have different forms:

1. In the case where the slots do not intersect along the z-axis at (variation 1 in Fig. 4.32):

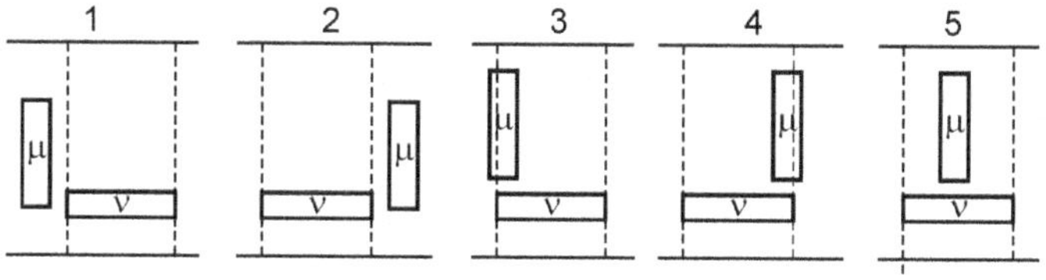

Fig. 4.32 Variations of mutual arrangement of orthogonal slots

$$F_{mn} = \frac{2}{\gamma_{mn}}\left((-1)^q e^{-i\gamma_{mn} l_v} - 1\right) e^{i\gamma_{mn}(z_\mu - z_v)} \sin\frac{\gamma_{mn}^d \mu}{2}$$

2. In the case where the slots do not intersect along the axis (variation 2 in Fig. 4.32):

$$F_{mn} = \frac{2}{\gamma_{mn}}\left(1 - (-1)^q e^{i\gamma_{mn} l_v}\right) e^{-i\gamma_{mn}(z_\mu - z_v)} \sin\frac{\gamma_{mn}^d \mu}{2}$$

3. For case 3 in Fig. 4.32

$$F_{mn} = \frac{2}{\gamma_{mn}}\left[e^{-i\gamma_{mn}\frac{d_\mu}{2}} \sin\left(\gamma_{mn}(z_\mu - z_v)\right) + (-1)^q e^{-i\gamma_{mn} l_v} \sin\frac{\gamma_{mn}^{d_\mu}}{2}\right] - \frac{2l_v}{q\pi}\sin\left(\frac{q\pi}{l_v}\left(z_\mu - z_v + \frac{d_\mu}{2}\right)\right)$$

4. For case 4 in Fig. 4.32

$$F_{mn} = \frac{2}{\gamma_{mn}}\left[e^{-i\gamma_{mn}(z_\mu - z_v)} \sin\frac{\gamma_{mn} d_\mu}{2} - (-1)^q e^{-i\gamma_{mn}\frac{d_\mu}{2}} \sin(\gamma_{mn}(z_\mu - z_v - l_v))\right] + \frac{2l_v}{q\pi}\sin\left(\frac{q\pi}{l_v}\left(z_\mu - z_v + \frac{d_\mu}{2}\right)\right).$$

5. For case 5 in Fig. 4.32

$$F_{mn} = \frac{2}{\gamma_{mn}\sin}\frac{\gamma_{mn} d_\mu}{2a}\left[(-1)^q e^{-i\gamma_{mn} 1_v e^{i\gamma_{mn}}}\left(z_\mu - z_v\right) - e^{-i\gamma_{mn}}\left(z_\mu - z_v\right)\right] - \frac{4l}{q\pi}\cos\frac{q\pi(z_\mu - z_v}{l_v}\sin\frac{q\pi d_\mu}{2l_v}.$$

By determining the electric field in the array aperture according to expression (4.27), we find the energy and directional characteristics of the multifrequency antenna. The radiation coefficient (the first summand is the radiation coefficient of longitudinal slots, the second summand is the radiation coefficient of transverse slots) will be equal to

$$|S_\Sigma|^2 = \left|S_\Sigma^{\text{long}}\right|^2 + \left|S_\Sigma^{\text{trans}}\right|^2 = \frac{1}{2}\text{Re}\sum_{n=1}^{N}\sum_{p=1}^{Q}\sum_{q=1}^{Q} V_{np}^{(2)} V_{nq}^{*(2)} Y_{nn,pq}^{\text{ext}}$$

$$+ \frac{1}{2}\text{Re}\sum_{n'=1}^{N'}\sum_{p'=1}^{Q'}\sum_{q'=1}^{Q'} W_{n'p'}^{(2)} W_{n'q'}^{*(2)} Y_{n'n',p'q'}^{\text{ext}},$$

and the reflection coefficient $|S_{11}| = -\frac{1}{4}\sum_{n=1}^{N}\sum_{p=1}^{Q} V_{np}^{(1)} F_{np} - \frac{1}{4}\sum_{n'=1}^{N'}\sum_{p'=1}^{Q'} W_{n'p'}^{(1)} F_{n'p'}$.

The pattern of the radiation system in the $x = 0$ plane is defined by two mutually perpendicular components of the electric field E_x and E_z:

$$F_{E_x}(\theta) = \sum_{n=1}^{N}\sum_{q=1}^{Q} V_{nq}^{(2)} \frac{q\pi}{L_n} \frac{1-(-1)^q e^{jkL_n\cos\theta}}{\left(\frac{q\pi}{L_n}\right)^2 - (k\cos\theta)^2} e^{jkz_n\cos\theta}\sin\theta,$$

$$F_{E_z}(\theta) = \sum_{n'=1}^{N'}\sum_{q'=1}^{Q'} W_{n'q'}^{(2)} \frac{\sin(\frac{kd_{n'}}{2}\cos\theta)}{\frac{kd_{n'}}{2}\cos\theta} \frac{l_{n'}}{q'\pi}\left(1-(-1)^{q'}\right)e^{jkz_{n'}\cos\theta},$$

where the angle θ is counted from the z-axis.

In order to understand the physical features of interaction of slots in each subarray, as in the case of systems with longitudinal slits (Sect. 4.1), it is advisable to give some results of the study of radiation characteristics of the single transverse slot cut in the wall of a waveguide, the sizes of which allow for propagation of higher types of waves.

Figure 4.33 shows the calculated dependences of radiation coefficients $|S_\sum|^2$ and reflection (for field strength) $|S_{11}|$ of transverse slot excited by wave H_{30} in the definite range of wavelengths. Curves 1, 2, 3 correspond to slots of different length: 7.5 mm, 8.5 mm, 9.5 mm at $x_1 = a / 6$. The dimensions of the waveguide cross section is $a \times b = 35 \times 15\text{mm}^2$, and –the slot width and wall thickness of the waveguide are 1 mm. From the graphs of Fig. 4.33 it is shown that the values of radiation and reflection coefficients in this case are significantly less than the corresponding values at excitation of slots by the dominant wave of the waveguideH_{10}. The most interesting peculiarity of the graphs in Fig. 4.33 is that in dependences $|S_\sum|^2 = f(\lambda)$ and $|S_{11}| = f(\lambda)$ there are minimum values of radiation and reflection coefficients at suchλ, at which higher order waves arise. Such a feature of the dependences $|S_\sum|^2 = f(\lambda)$ and $S_{11}| = f(\lambda)$ must be taken into account when selecting the electrical dimensions of the slot radiators and the operating frequency band of the multi-element shortwave antenna.

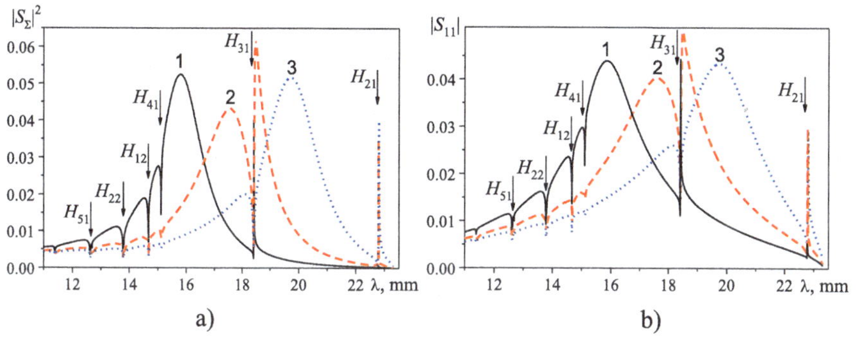

Fig. 4.33 Dependences of radiation coefficients (**a**) and reflection coefficients (**b**) for transverse slots of different wavelengths for H_{30} wave excitation (23.33 mm)

Figure 4.34a shows the dependences of the radiation and reflection coefficients on the wavelength, and in Fig. 4.34b the dependences of directional coefficients D and gain G on the wavelength of a two-frequency waveguide-slotted array consisting of 17 transverse (l_n = 8mm, d_n = mm, Δz = 16.25mm) and 8 longitudinal (L_n = 20mm, D_n = 1mm) slots for excitation by H_{30} wave. The transverse slots are located on the wall of the waveguide ($a \times b = 35 \times 15\text{mm}^2, t = 1.5\text{mm}$), where they are most excited by the H_{30} wave, and the centers of the longitudinal slits are located where the surface currents of the H_{30} wave crossing the slot are zero (Fig. 4.30). As can be seen from the graphs, the maximum directivity ($D = 55.7$) is achieved at $\lambda_{30} = 18.33$mm. At that $|S_\sum|^2 = 0.44$ and $|S_{11}| = 0.06$. Figure 4.35a shows the amplitude field distribution in the antenna aperture at a given wavelength, and Fig. 4.35b (curve 1) shows the corresponding pattern. The dotted curves in Fig. 4.35a represent the field distribution in the aperture of the longitudinal slots. Although the longitudinal slots are located at points where the surface currents of the H_{30} wave are equal to zero, nevertheless, these slots are excited due to their mutual influence. The long-wave antenna pattern in the vector $\vec{H}$ plane when excited by the H_{10} wave is represented by curve 2 in Fig. 4.35b. The distances between the slot radiators in the subarrays are chosen so as to exclude the effect of normal.

Let's note that for the considered array at excitation by H_{30} wave with λ_{30} = 18.33mm the values of radiation coefficients and directivity are$|S_\sum|^2 = 0.44$, D = 55.7, and at excitation by wave H_{10} with λ_{10} = 38mm are$|S_\sum|^2 = 0.85$,D = 23. The increase of the radiation coefficients and directivity can be achieved by increasing the number of radiators with a simultaneous change in the geometric sizes of the slots similar to what was done in Sect. 4.1.

The amplitudes of the fields induced in the longitudinal slots during the operation of the two-frequency array in the short-wave range, and hence the contribution of the longitudinal slots to its total radiation power, depend on the electric length of the longitudinal slots. This circumstance must be taken into account in the design of the two-frequency array. By changing the length of the longitudinal slots and their location, it is possible to reduce the amplitudes of the fields induced in the slots

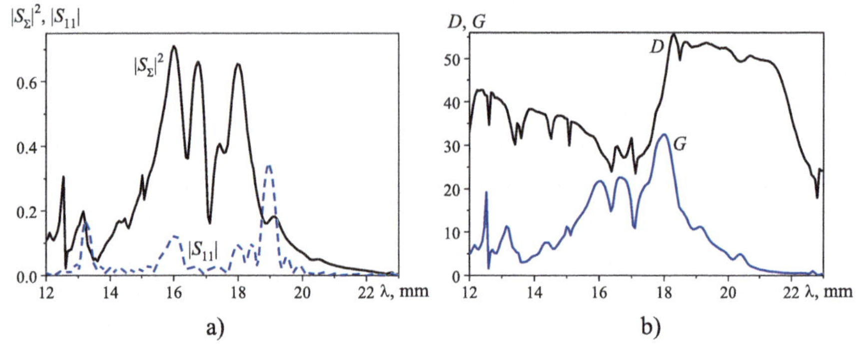

Fig. 4.34 Dependences of radiation and reflection coefficients (**a**), directivity and gain (**b**) of the two-frequency array on the wavelength by H_{30} wave excitation

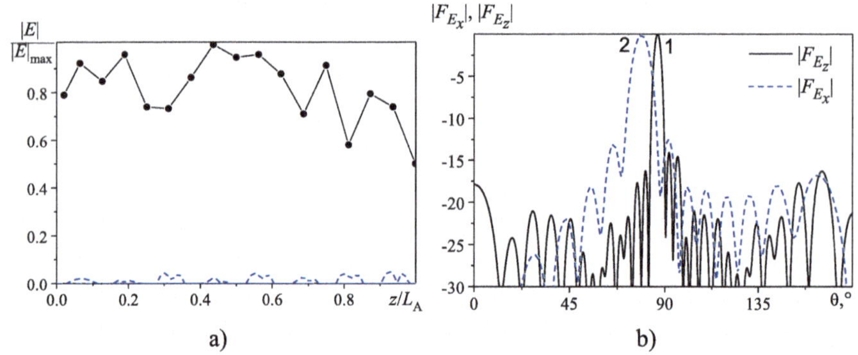

Fig. 4.35 Amplitude distribution of the electric field (**a**) in the aperture of the two-frequency array for H_{30} wave excitation with $\lambda_{30} = 18.33$ mm and the pattern (**b**) of the two-frequency array at $\lambda_{30} = 18.33$ mm(F_{E_z}) and $\lambda_{10} = 38$ mm(F_{E_x})

with lengths L_n and, thus, to reduce the influence of the long-wave array on the characteristics of the short-wave one. As an illustration for the structure discussed above, Fig. 4.36 shows the radiation power of the longitudinal slot system related to the radiation power of the transverse slot system as a function of the length of the longitudinal slots. The array is excited by the H_{30} wave with $\lambda_{30} = 18.33$mm.

When the system works in the long-wave range at frequency f_{10}, the transverse slots are practically not excited ($|S_\Sigma^{\text{trans}}| < 0.001$), since their electric length is much less than the resonance one, and they do not influence the radiation characteristics of the whole array.

It should be noted that in the subarrays decoupled by the polarization of the radiated field, the interaction between the high-frequency and low-frequency subarrays in the system under study appears itself to a much lesser degree compared to the array considered in Sect. 4.3.1.

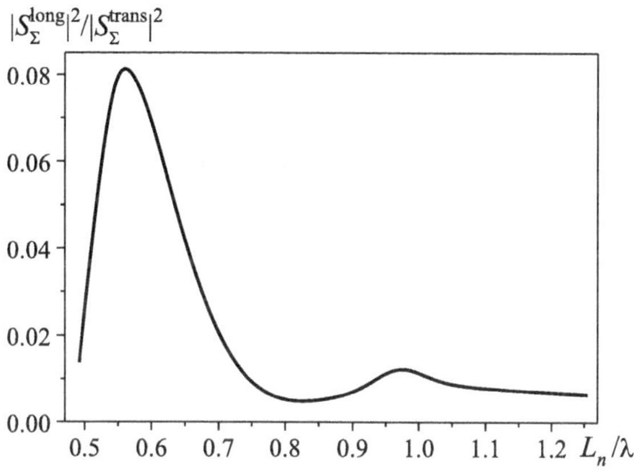

Fig. 4.36 Radiation power of the longitudinal slots system related to the radiation power of the transverse slots system, depending on the electric length of the longitudinal slots ($\lambda_{30} = 18.33$ mm)

4.4 Two-Dimensional Slotted Waveguide Arrays

The results obtained in Sects. 4.2 and 4.3, concerning the determination of the electrical field (equivalent magnetic current) in the slot with the application of basis functions in the method of magnetomotive forces, which most adequately describe the real current distribution in the slot, allow us to simplify significantly the SLAE solving in the problems of multi-element radiating systems and to extend significantly the possibilities of application of electrodynamic methods for calculation of two-dimensional radiating arrays.

Consider a two-dimensional resonant waveguide-slot array consisting of K linear arrays, i.e. short-circuited or with matched load at the end of rectangular waveguides with radiating slots on the wide wall (Fig. 4.37).

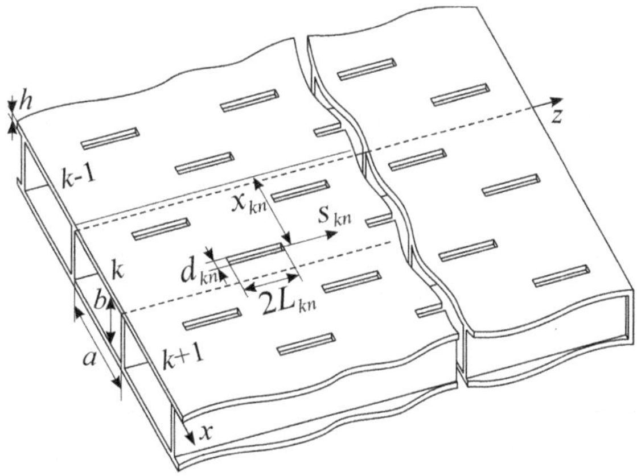

Fig. 4.37 The geometry of the system and the designations

We denote the numbers of slots in the radiating linear arrays by a pair of indices, where n is the number of slots along the linear array, k is the number of the linear array.

The unknown electrical fields on the slot surfaces with the number kn, bordering the inner region of the waveguide and the free half-space ($\vec{E}^{in}_{kn}$ and $\vec{E}^{e}_{kn}$, respectively) will be represented as linear combinations of coordinate basis functions:

$$\vec{E}^{in}_{kn}(s_{kn}) = \sum_{q=1}^{Q} \vec{V}^{in}_{knq}\vec{e}^{(1)}_{knq}(s_{kn}),\ \vec{E}^{e}_{kn}(s_{kn}) = \sum_{q=1}^{Q} \vec{V}^{e}_{knq}\vec{e}^{(2)}_{knq}(s_{kn}),$$

where V^{e}_{knq}, V^{in}_{knq} are unknown complex amplitude coefficients to be determined; $\vec{e}_{knq}$ are the linearly independent vector functions, which satisfy the boundary conditions on the slot contour; s_{kn} is a local coordinate associated with the kn-th slot; Q is the number of basis functions.

Based on the condition of narrowness of slots, the vector functions can be presented in the following form:

$$\vec{e}^{(1)}_{knq}(s_{kn}) = \vec{e}^{(2)}_{knq}(s_{kn}) = \vec{\xi}^{0}_{kn}\frac{1}{d_{kn}}\sin\frac{q\pi(L_{kn}+s_{kn})}{2L_{kn}}.$$

where $\vec{\xi}^{0}_{kn}$ is the unit vectors across the slots.

Coefficients V^{e}_{knq} and V^{in}_{knq} are calculated from conditions of continuity of tangential components of total magnetic fields on external and internal surfaces of slots with the use of Galerkin procedure. As a result, we obtain an SLAE with respect to the unknowns V^{e}_{knq} and V^{in}_{knq}

$$\begin{cases} \sum_{k=1}^{K}\sum_{n=1}^{N}\sum_{q=1}^{Q} V^{in}_{knq}\delta_{rk}\left(Y^{in}_{rm,kn,pq} + \delta_{mn}\delta_{pq}Y^{R_{11}}_{rm,kn,pq}\right) + \sum_{q=1}^{Q} V^{e}_{rmq}Y^{R_{12}}_{rm,rm,pq}\delta_{pq} = F_{rmp}, \\ \sum_{q=1}^{Q} V^{in}_{rmq}Y^{R_{21}}_{rm,rm,pq}\delta_{pq} + \sum_{k=1}^{K}\sum_{n=1}^{N}\sum_{q=1}^{Q} V^{e}_{knq}\left(Y^{e}_{rm,kn,pq} + \delta_{mn}\delta_{pq}\delta_{rk}Y^{R_{22}}_{rm,kn,pq}\right) = 0, \\ r = 1,2\ldots K;\ m = 1,2\ldots N;\ p = 1,2\ldots Q. \end{cases}$$

The coefficients of the resulting system of equations are the intrinsic and mutual internal, external conductivities of the slots and conductivities of the space formed by the slot cavity

$$Y^{in,(e)}_{rm,kn,pq} = \mp \int\limits_{S^{(1),(2)}_{rm}} \left[\vec{e}_{rmp}, \vec{H}^{in,(e)}(\vec{e}_{knq})\right]\vec{n}ds,\ Y^{R_{\nu\mu}}_{rm,kn,pq} = \int\limits_{S^{(\nu)}_{rm}} \left[\vec{e}^{(\nu)}_{rmp}, \vec{H}^{R}(\vec{e}^{(\mu)}_{knq})\right]\vec{n}ds,$$

$F_{rmp} = \int\limits_{S_{rm}} \left[\vec{e}_{rmp}, \vec{H}^{0}_{rm}\right]\vec{n}ds$ are partial magnetomotive forces.

By determining coefficients V^{e}_{knq} and V^{in}_{knq}, we find energetical and directional characteristics of the antenna, in particular, the radiation coefficient of the system

$$|S_{\Sigma}|^2 = \frac{2\mathrm{Re}\sum_{k=1}^{K}\sum_{n=1}^{N}\sum_{p=1}^{Q}\sum_{q=1}^{Q} V^{e}_{knp}(V^{e}_{knq})^{*}Y^{e}_{kn,kn,pq}}{\sum_{k=1}^{K}|A_k|^2 N_1},$$

where A_k is the amplitude of the wave propagating in the k-th waveguide, $N_1 = 120\pi kab(\pi/a)^2\gamma, k = 2\pi/\lambda$, γ is the wave propagation constant in the waveguide.

It is shown in Fig. 4.38a the patterns in two mutually perpendicular planes of the antenna system consisting of $K = 5$ waveguide-slotted linear arrays based on rectangular waveguides with the cross section$a \times b = 7.2 \times 4.8$ mm^2 and wall thickness $h = 1$mm excited by dominant type of wave with the same amplitudes at$\lambda = 8$mm. The number of slots in each of the waveguides $N = 40$, and their length and width $2L_{kn} = 3.5$mm and$d_{kn} = 0.5$mm. The distance between the slots in the waveguide-slotted linear array is $0.5\lambda_g$, where λ_g is the wavelength in the waveguide. The size of the radiating aperture along the z axis$L_A = 191.1$mm. The width of the pattern at the half-power level in the $\vec{H}$ and $\vec{E}$ vector planes and is $\Delta\theta_{0.5} = 2.1$ and $\Delta\varphi_{0.5} = 10.16$, respectively.

In contrast to such a "classical" waveguide-slotted array and its pattern, by selecting the length of each of the slots and the distance from the slot to the edge of the waveguide, we can create the desired amplitude field distribution in the antenna aperture, for example, the amplitude distribution, falling to the edges of the system, and form a radiation pattern with a much lower level of sidelobes. Thus, the patterns in Fig. 4.38b correspond to the array, the amplitude distribution of the electrical field in the aperture of which is shown in Fig. 4.39. The amplitude distribution falling to the edges of the array in the $\vec{H}$ vector plane is formed in the waveguide-slotted linear

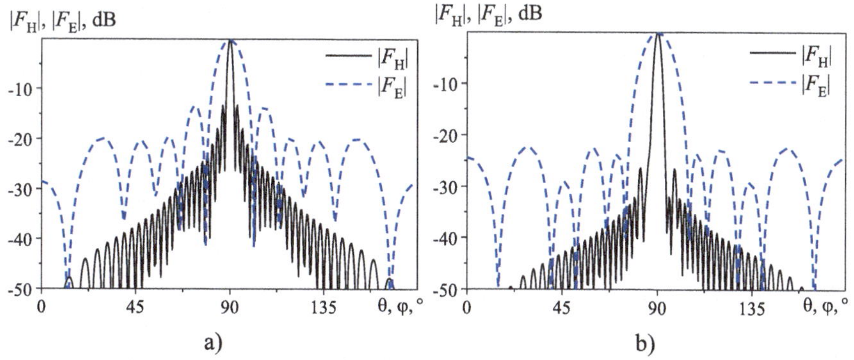

Fig. 4.38 Patterns of resonant antenna arrays

arrays, the slot lengths of which and their location relative to the edge of the waveguide vary in accordance with the values shown in Fig. 4.40. The amplitude distribution of the electrical field in the perpendicular plane (Fig. 4.39a) falling to the edges can be formed by excitation of the waveguide-slotted linear arrays by excitation of waves with different amplitudes. The patterns shown in Fig. 4.38b correspond to the case of $A_1 = A_5 = 0.5A_3$, $A_2 = A_4 = 0.8A_3$. The level of the side lobes of the pattern does not exceed −22dB in the $\vec{E}$ vector plane and −26dB in the $\vec{H}$ vector plane, and the width of the pattern in the corresponding planes $\Delta\varphi_{0.5} = 12.06°$ and $\Delta\theta_{0.5} = 2.85$. The distance from the center of the first slot in the waveguide-slotted linear array to the short circuit is chosen from the condition of the best matching with the tract and is equal to$L_j = 193.08$mm. The radiation coefficient of the array $|S_\Sigma|^2 = 0.98$.

A disadvantage of resonant WSA on the basis of semi-infinite waveguides, especially extended and with a large number of radiators, is the narrow operating band and strong sensitivity of the characteristics of the array to the inaccuracies in its technical implementation or to the temperature changes in the geometric sizes of the structure. Thus, a small change in L_j (e.g., due to changes in environmental temperature) leads to a significant increase in the reflection coefficient and, accordingly, to a decrease

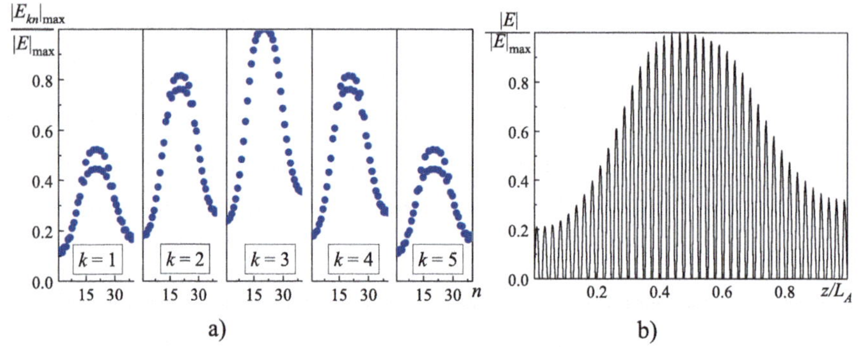

Fig. 4.39 Amplitude distribution of the electrical field in the array aperture

Fig. 4.40 Length of slots $2L_{kn}$ and their location x_{kn} depending on a slot number

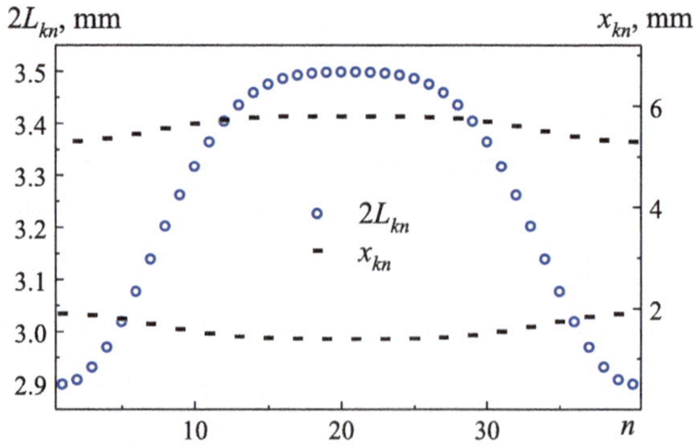

in the radiation coefficient. The antenna array consisting of waveguides with the matched load at the end is free from these drawbacks. A low reflection coefficient and a large radiation coefficient in this case can be provided by an appropriate choice of geometric sizes of the slots and their location in the wide wall of the waveguide.

The pattern of the antenna array consisting of $K = 8$ waveguide-slotted linear arrays based on rectangular waveguides $\left(a \times b = 6.2 \times 3.1\text{mm}^2, t = 0.5\text{mm}\right)$ with matched loads for excitation by the dominant type of wave with $\lambda = 8.45\text{mm}$ are shown in Fig. 4.41. The slot lengths in each of the linear arrays, depending on the slot number, change according to the graph shown in Fig. 4.42.

Due to this, an amplitude distribution decreasing towards the edges is formed along the z-axis in the radiating system (insertion in Fig. 4.42) and the pattern in the $\vec{H}$ vector plane has the low level of sidelobes(-40dB). The excitation of the waveguide-slot linear arrays by waves with different amplitudes $(A_1 = A_8 = 0.2A_5, A_2 = A_7 = 0.4A_5, A_3 = A_6 = 0.7A_5, A_4 = A_5)$ allows reducing the side lobes to -29dB in the perpendicular plane. To overcome the

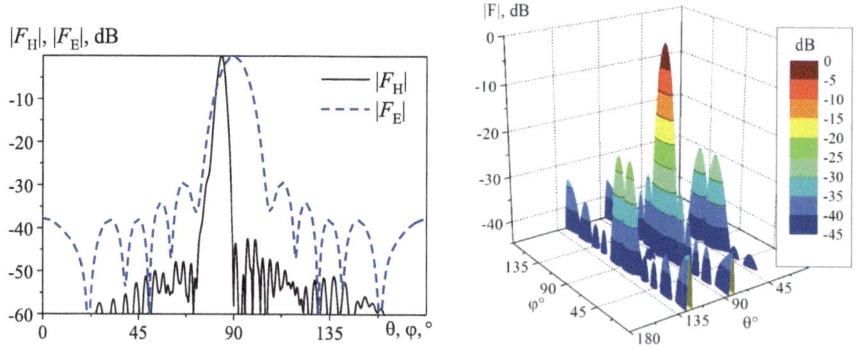

Fig. 4.41 Pattern of the nonresonant antenna array

Fig. 4.42 Dependences of slot lengths on the order number and the amplitude distribution of the electrical field in the aperture of the nonresonant array

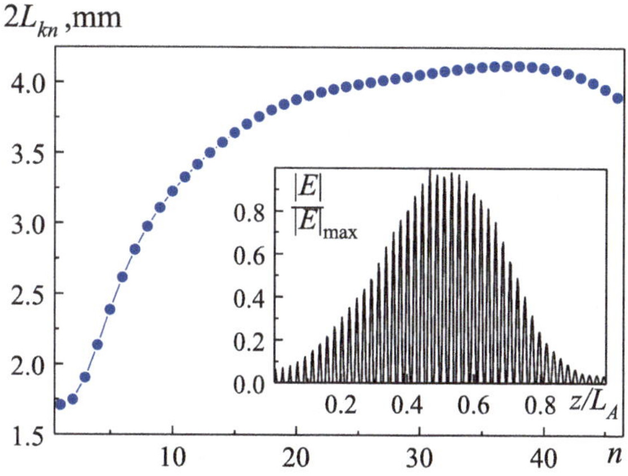

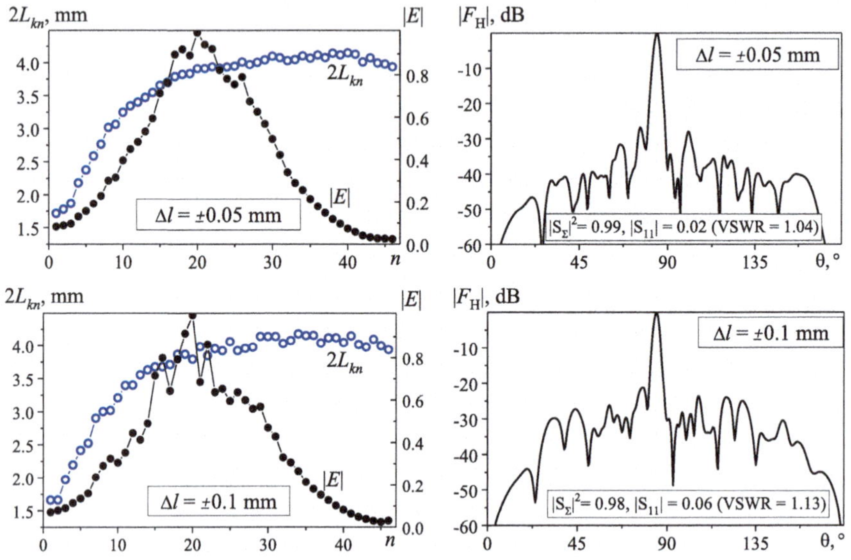

Fig. 4.43 Dependences of slot length and electrical field amplitude in the slot on its number in the array (**a**) and the corresponding pattern of the array (**b**)

so-called normal effect the distance between the slots in the arrays is different from $0.5\lambda_g$ and is equal to $0.546\lambda_g$. As a consequence, the main lobe of the pattern is deflected from the normal by an angle$\Delta\theta = 6$. The field reflection coefficients from the slot linear arrays do not exceed the value of0.004 (VSWR < 1.008), and the array radiation coefficient is $|S_\Sigma|^2 = 0.99$. It should be noted that the level of the array side lobes does not exceed -20dB even at a deviation of slot lengths from the values determined by the graph in Fig. 4.42, by ± 0.1mm (Fig. 4.43).

The materials of this Chapter are contained in the authors' publications [34, 36–45].

References

1. Hyneman RF (1959) Closely-spaced transverse slots in rectangular waveguide. IRE Trans Antennas Propag 7:335–342
2. Jones EMT, Shimizu JK (1960) A Wide-band transverse-slot flush-mounted array. IEEE Trans Antennas Propag 8:401–407
3. Elliott R, Kurtz L (1978) IEEE Trans Antennas Propag 26:214–219
4. Luebbers RJ, Munk BA (1978) Some effects of dielectric loading on periodic slot arrays. IEEE Trans Antennas Propag 26:536–542
5. Sangster AJ (1978) Slotted waveguide linear array with polarization control. Electronics Lett 14:616–618

6. Paoloni FJ (1980) Cavity-backed resonant slot array—theory and measurement. IEEE Trans Antennas Propag 28:259–262
7. Alexander MJ (1982) The improvement of sidelobe performance of slotted waveguide arrays. Marconi Review 45(226):165–188
8. Shavit R, Elliott RS (1983) Design of transverse slot arrays fed by a boxed stripline. IEEE Trans Antennas Propag 31:545–552
9. Elliott RS, O'Loughlin WR (1986) The design of slot arrays including internal mutual coupling. IEEE Trans Antennas Propag 34:1149–1154
10. Lo YT, Lee SW (1988) The design of waveguide-fed slot arrays (1988) Van Nostrand. Rheinhold, NY
11. Ohno T, Ogawa k, Teraoka T, Hirokawa J (2003) Proposal for a slot pair array having an invariant main beam direction with a cosecant radiation pattern using a post-wall waveguide. IEICE Trans Electron E86-C (2):176–183
12. Montisci G, Musa M, Mazzarella G (2004) Waveguide slot antennas for circularly polarized radiated field. IEEE Trans Antennas Propag 52:619–623
13. Manuilov MB, Lerer VA, Sinyavsky GP (2008) An efficient method for the electromagnetic analysis of slotted-waveguide antenna arrays. J Commun Technol Electron 53:1380–1390
14. Manuilov MB, Lerer AM, Sinyavsky GP (2011) Numerical-analytical implementation of Galerkin technique for analysis of waveguide and slotted waveguide antenna arrays. Proc. International Conference on Antenna Theory and Techniques 69–74
15. Mazzarella G, Montisci G (2011) Accurate modeling of coupling junctions in dielectric covered waveguide slot arrays. Prog Electromagn Res M 17:59–71
16. Liu J, Jackson DR, Long Y (2012) Substrate integrated waveguide (SIW) leaky-wave antenna with transverse slots. IEEE Trans Antennas Propag 60:20–29
17. Gatti RV, Rossi R (2016) A dual-polarization slotted waveguide array antenna with polarization tracking capability and reduced sidelobe level. IEEE Trans Antennas Propag 64:1567–1574
18. Camacho M, Boix RR, Medina F, Hibbins AP, Sambles JR (2020) Extraordinary transmission and radiation from finite by infinite arrays of slots. IEEE Trans Antennas Propag 68:581–586
19. Hou Y, Meng Z, Wang L, Li Y (2020) Endfire antenna array using microstrip-fed cavity-backed slot elements. 68:2448–2433
20. Zheng D, Lyu YL, Wu K (2020) Transversely slotted SIW leaky-wave antenna featuring rapid beam-scanning for millimeter-wave applications. IEEE Trans Antennas Propag 68:4172–4185
21. Siddiqui Z, Sonkki M, Tuhkala M, Myllymäki (2020) Periodically slotted coupled mode leaky coaxial cable with enhanced radiation performance. IEEE Trans Antennas Propag 68:7595–7600
22. Karmokar DK, Chen SL, Thslakotuna D, Qin PY, Bird TS (2020) Continuous backward-to-forward scanning 1-D slot-array leaky-wave antenna with improved gain. IEEE Antennas Wireless Propag Lett 19:89–93
23. Dimitrov KC, Lee Y, Min BW, Park J, Jeong J, Kim HJ (2020) Circularly polarized T-shaped slot waveguide array antenna for satellite communications. IEEE Antennas Wireless Propag Lett 19:317–321
24. Chen M, Fang XC, Wang W, Zhang HT, Huang GL (2020) Dual-band dual-polarized waveguide slot antenna for SAR applications. IEEE Antennas Wireless Propag Lett 19:1719–1723
25. Vilas Boas EC, Mittra R, Sodre AC (2020) A low-profile high-gain slotted waveguide antenna array with grooved structures. IEEE Antennas Wireless Propag Lett 19:2107–2111
26. Liu P, Pedersen GF, Zhang S (2022) Wideband Low-Sidelobe Slot Array Antenna With Compact Tapering Feeding Network for E-Band Wireless Communications. IEEE Trans Antennas Propag 70:2676–2685
27. Kan Y, Yang R, Zhang A, Lei Z, Jiao Y, Li J (2022) Meta-Surface Cavity-Based Waveguide Slot Array for Dual-Circularly Polarized Dual Beam. IEEE Trans Antennas Propag 70:3894–3898
28. Jin C, Chen J, Zhang B, Kong L, An S, He ZS, Liu J (2022) Low-Cost mm Wave Metallic Waveguide Based on Multilayer Integrated Vertical-EBG Structure and its Application to Slot Array Antenna Design. IEEE Trans Antennas Propag 70:2205–2213

29. Koli MNY, Afzal MU, Esselle KP (2022) Increasing the Gain of Beam-Tilted Circularly Polarized Radial Line Slot Array Antennas. IEEE Trans Antennas Propag 70:4392–4403
30. Ji S, Tomura T, Hirokawa J (2022) Fast analysis and bandwidth enhancement of the radiating subarray by method of moments for a parallel-plate slot array antenna with perpendicular-corporate feed. IEEE Trans Antennas Propag 70:2645–2655
31. Walter CH (1965) Traveling wave antennas. McGraw-Hill, NY
32. Berezovsky VA, Kolotilov NN (1990) Biophysical characteristics of human tissues. Naukova Dumka, Kiev
33. Kikuchi M (1995) Resent progress of electromagnetic techniques in hyperthermia treatment. IEICE Trans Electron E78-B: 799–808
34. Berdnik SL, Katrich VA, Nesterenko MV, Pshenichnaya SV (2010) Near-zone EM field of combined slotted waveguide radiators in a material medium. Telecommun Radio Eng 69(16):1491–1497
35. Nesterenko MV, Katrich VA, Penkin YuM, Berdnik SL (2008) Analytical and hybrid methods in theory of slot-hole coupling of electrodynamic volumes. Springer, New York
36. Berdnik SL, Katrich VA, Kiyko VI, Nesterenko MV (2005) Electrodynamic characteristics of the nonresonant system of transverse slots in the wide wall of rectangular waveguide. Radioelectron Commun Syst 48(1):52–57
37. Katrich VA, Lyashchenko VA, Nesterenko MV, Berdnik SL (2005) Slot antennas for various space and avionic applications. Proc Europ Microw Assoc 1(3):239–244
38. Berdnik SL (2010) Electromagnetic wave radiation by a set of longitudinal and transverse slots in a wide wall of a rectangular waveguide. Telecommun Radio Eng 9:763–771
39. Berdnik SL, Katrich VA, Lyashchenko VA (2002) Dual-frequency slot antennas on multi-mode waveguides. Bulletin of Karazin University. Radiophysics and electronics 570:18–21 (in Russian)
40. Berdnik SL, Katrich VA, Lyashchenko VA (2001) Double frequency slot antenna on a multi-mode waveguide. In: Proceedings of the 11th International Conference Microwave and Telecommunication Technology, Sevastopol, Ukraine, pp 387–388.
41. Berdnik SL, Katrich VA, Lyashchenko VA (2003) Closely spaced transverse slots in rectangular waveguide. In: Proceedings of the 4th International Conference on Antennas Theory and Techniques, Sevastopol, Ukraine, pp 273–275.
42. Berdnik SL, Katrich VA, Nesterenko MV (2005) Near field of non-resonant system of transverse slots in rectangular waveguide. In: Proceedings of the 15th International Crimean Conference Microwave & Telecommunication Technology, Sevastopol, Ukraine, pp 364–365
43. Berdnik SL, Katrich VA, Nesterenko MV (2007) Near Fields of Waveguide-Slot Arrays in Matter. In: Proceedings of the 17th international crimean conference - microwave & telecommunication technology, Sevastopol, Ukraine, pp 739–740.
44. Berdnik SL, Katrich VA (2007) Sidelobe reduction of two-dimensional waveguide-slot antenna arrays. In: Proceedings of the 6th International Conference on Antenna Theory and Techniques, Sevastopol, Ukraine, pp 461–463.
45. Berdnik SL, Katrich VA, Nesterenko MV, Pshenichnaya SV (2009) Near fields of combined waveguide-slot radiators in matter. In: Proceedings of the 19th international crimean conference microwave & telecommunication technology, Sevastopol, Ukraine, 2009, pp 442–443

Chapter 5
Combined Vibrator-Slot Antenna Arrays

5.1 Dual-Band Combined Vibrator-Slot Radiating Structures

In this Section the dual-frequency antennas based on vibrator-slot structures with similar RP in main polarization planes are suggested. Such antennas can be realized by diode switching the activity of slot and vibrator elements. The traditional combined radiators with passive vibrators are proposed to use at the main frequency. The work at alternative frequency is provided by active vibrators. The formation of radiation fields by the combined structures at the main and alternative frequencies is analyzed.

5.1.1 Switching Modes of Radiator Elements

The development of multi-frequency antennas is essential for creating modern antenna systems. Often, such antennas contain two types of aperture radiators operating in separate bands [1]. The integration of multi-frequency radiators in a single aperture results in a strong interaction between them and a distortions of antenna electrodynamic characteristics. The main goal in designing the combined antenna structures consists in reduction of such distortions. When the multi-frequency operating modes are separated in time, high-speed microwave switches based on semiconductor diodes can be used in the design of radiating elements [2]. For example, the characteristics of caseless p-i-n diodes satisfy modern requirements for switching microwave devices. The microwave switches can be effectively applied in complex combined structures, in which a part of a main radiator structure cut off by diode switching can be used as an alternative radiator.

One of known combined vibrator-slot radiators is the Clavin element consisting of a narrow radiating slot cut in the screen and two passive vibrators (monopoles) of

M. V. Nesterenko et al., *Vibrator and Slot Antenna Arrays for Modern Applications*,
Lecture Notes in Electrical Engineering 1244,
https://doi.org/10.1007/978-3-031-68346-6_5

a fixed length located on both sides of the slot at a specified distance from the slot center [3, 4]. The monopoles have electrical contact with the screen. Such antenna radiators with perfectly conducting monopoles are characterized by a close similarity of RP in the E- and H-planes [3, 4]. Note that the influence of electric lengths of the impedance monopoles and the distance between them on the Clavin element RP was analyzed in [5] under conditions that the relative level of lateral radiation in the E-plane and the difference of the RP width in the E- and H-planes at the level of -3 dB are taken into account. The combined radiator consisting of the slot cut in the screen and the horizontal vibrator is also characterized by similarity of the RP in the main polarization planes [6, 7]. However, studying such combined radiators in dual-frequency antennas has not been carried out, therefore the present work is aimed at this problem study.

Let us analyze two types of combined vibrator-slot radiating structures consisting of a slot and two vertical vibrators [3–5] or a horizontal vibrator [6, 7].

Consider the vibrator-slot structure of the first type, whose geometry is shown in Fig. 5.1a. The slot cut in an infinite perfectly conducting screen radiates into a half-space with the material parameters (ε_1, μ_1). The slot center is located at the beginning of the Cartezian coordinate system (x, y, z). The two asymmetric vibrators (monopoles) are located at the screen in the plane $\{y0z\}$. The slot length and width are $2L_3$ and d, the length and radii of the vibrators are L_1, L_2, r_1, and r_2 correspondingly. The vibrators are placed at distances y_{d1} and y_{d2} from the slot axis. All structural elements of the radiator satisfy the thin linear radiator approximations: $\frac{r_{(1,2)}}{L_{(1,2)}} \ll 1$, $\frac{r_{(1,2)}}{\lambda_1} \ll 1$, $\frac{d}{2L_3} \ll 1$, $\frac{d}{\lambda_1} \ll 1$, where λ_1 is the wavelength in the external medium. That is, all structural elements can be analyzed within the framework of the theory of thin linear radiators.

Let us consider using the first type of combined vibrator-slot structures as radiators for dual-frequency antennas. Suppose that, as in [3–5], the Clavin element with perfectly conducting vibrators have equal lengths $L_1 = L_2 = L_v$, radii $r_1 = r_2 = r_v$, and displacements $y_{d1} = y_{d2} = y_v$. The slot length and width are $2L_3 = 2L_{sl}$ and d. Let us also assume that the slot cut in the end wall of the rectangular waveguide adjacent to the screen is excited by the waveguide method from the lower half-space [8]. In this case, the excitation node allows resonant tuning of the combined radiator. It should be noted that the monopoles have electrical contact with the screen.

Several controlled frameless p-i-n diodes should be allocated in combined radiators of the double-frequency antenna array: two of them in bases of the vibrators and the rest distributed over the slot aperture including its central part as shown by red color in Fig. 5.1a. When the slot diodes are in the OFF state, the slot radiates electromagnetic waves into half-space ensuring the conventional operating mode of the combine radiator. The diodes in the ON state shunt the slot aperture providing the mode which can be called by conditional metallization. Correct operating of the radiator requires that commands ON for diodes in the slot and command OFF for excitation power supply to the waveguide section were synchronized.

In contrast to the traditional version, in the design of the combined radiator, in addition, we also provide for the possibility of supplying power to the bases of the

Fig. 5.1 The combined vibrator-slot radiators: **a** with two vertical monopoles; **b** with horizontal vibrator oriented perpendicular to the slot

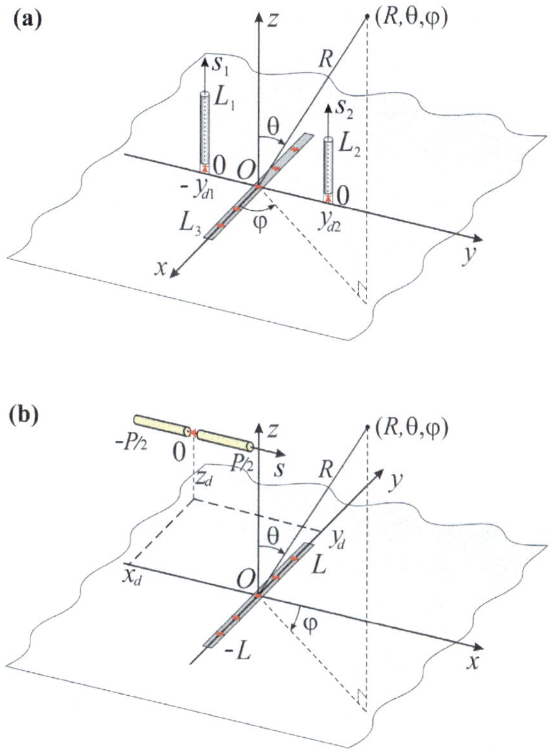

monopoles (which are a continuation of the central conductor of the coaxial line), dividing the excitation nodes and monopoly diode switches (diode 1—in the central conductor, diode 2—between monopole and screen). We believe that these diodes in the state diode 1 OFF and diode 2 ON shunt the monopoles with the screen, leaving them in the passive mode and ensuring the functioning of the combined emitter in the traditional form. On the contrary, in the state of diode 1 ON and diode 2 OFF, they connect the monopoles to the excitation generators and ensure their active mode of operation. Here, for technical correctness, we require synchronization of both commands ON and OFF both for diodes 1 in vibrators and for the operation of monopoles excitation generators. In this case, we will also require that the arrangement of the excitation generators ensure the tuning of the resonant radiation of monopoles, taking into account their mutual influence. Thus, in the ON state for all diode switches in the slot, and diodes 1 ON, and diodes 2 OFF in the vibrators, the two-frequency antenna is a pair of active vertical monopoles.

The slot (with length $2L = 2L_{sl}$ and width d) of the second type combined radiator (Fig. 5.1b) is also excited by the waveguide method with resonant radiator tuning. When the p-i-n diodes placed in the slot aperture are in the OFF state, the electromagnetic field radiation passes through the slot ensuring that the combined radiator operates in the conventional mode. When the diodes are in the ON state, they shunt the slot aperture and provide its conditional metallization. In the combined radiator of the second type, the power supply to the center of the tubular vibrator is fed to the vibrator center so that the excitation node and vibrator arms are separated by the diode switcher (Fig. 5.1b). The coordinates of the vibrator center are (x_d, y_d, z_d), the length and radius of the vibrator are P and r_v. When the switcher is in the ON state, it shunts the vibrator shoulders, and the vibrator is switched to the passive mode. The switch in the OFF state connects the vibrator to the excitation generator and ensures its active resonant operating mode. Thus, in the ON state for all diode switches in the slot and OFF in the vibrator, the two-frequency antenna of the second type is an active horizontal vibrator above the screen with symmetrical excitation.

5.1.2 *Radiation Fields of the Single Double -Frequency Antennas*

Let us introduce the spherical coordinate systems with polar axes directed along $\{0z\}$ and the angle φ measured from the axes $\{0x\}$, as shown in Fig. 5.1. The radiation fields of the antenna will be defined in the free half-space with the material parameters $(\varepsilon_1 = 1; \mu_1 = 1)$, therefore, the wavenumber $k = 2\pi/\lambda$, where λ is the wavelength in free space.

The vibrator-slot structure shown in Fig. 5.1a will be modelled by using analytical formulas defining the radiation fields obtained by the generalized method of electromagnetic motive forces (EMMF) [5, 7]. When time dependence of monochromatic field is specified as $e^{i\omega t}$ (t is time and ω is the circular frequency), the far zone electric field ($R \gg \lambda$) radiated by the combine radiator can be presented in the spherical coordinate system (R, θ, φ) in the following form

$$\vec{E}(\theta,\varphi) = \frac{ik^2 e^{-ikR}}{\omega R}\Big[\vec{\theta}^0 \sin\theta(\tilde{E}_1 e^{-iky_v \sin\theta\cos\varphi} + \tilde{E}_2 e^{iky_v \sin\theta\cos\varphi}) + \left(\vec{\varphi}^0 \cos\theta\cos\varphi + \vec{\theta}^0 \sin\varphi\right)2\tilde{E}_{sl}\Big], \tag{5.1}$$

where, $\vec{\theta}^0$ and $\vec{\varphi}^0$ are the unit vectors,

$$\tilde{E}_{1(2)} = \frac{2J_{1(2)}}{k(1-\cos\theta)^2}[\cos(kL_v\cos\theta)\sin(kL_v) - \sin(kL_v\cos\theta)\cos(kL_v)\cos\theta] - 2L_v\cos(kL_v)\frac{sin(kL_v\cos\theta)}{kL_v\cos\theta},$$

$$\tilde{E}_{sl} = \frac{2J_{sl}}{k - k(\sin\theta\cos\varphi)^2}[\cos(kL_{sl}\sin\theta\cos\varphi)\sin(kL_{sl}) - \sin(kL_{sl}\sin\theta\cos\varphi)\cos(kL_{sl})\sin\theta\cos\varphi] - \frac{2\cos(kL_{sl})\sin(kL_{sl}\sin\theta\cos\varphi)}{k\sin\theta\cos\varphi}.$$

The expression for the field $\tilde{E}_{sl}$ was written under the assumption of symmetric slot excitation. The complex amplitudes of electric currents on the monopoles $J_{1(2)}$ are determined from the rigorous problem solution taking into account the mutual influence of the vibrator-slot elements under the condition that magnetic current amplitude in the slot J_{sl} is given [5, 7].

The H-plane field ($\varphi = 0°$) of the conventional Clavin radiator at the main frequency has only the E_φ component which coincides with the field of the stand-alone slot, since currents are antiphase with equal amplitudes ($J_1 = -J_2$). In the E-plane ($\varphi = 90°$), the field has only the E_θ component which can be varied by the monopole currents. Hence, the fields in the H- and E-planes become closer. At the alternative frequency, when the slot is metalized and the vibrators are in the active mode, the expression for the radiator electric field is simplified to

$$\vec{E}(\theta, \varphi) = \frac{2ik^2 e^{-ikR} \sin\theta}{\omega R} \tilde{E}_{1(2)} \cos(ky_v \sin\theta \cos\varphi)\vec{\theta}^0, \qquad (5.2)$$

where the predefined vibrator currents are supposed to be equal ($J_1 = J_2$).

In the combined radiator of the second type, the magnetic current in the slot $\vec{J}^{\,m}_{sl}$ and electric current on the vibrator $\vec{J}^{\,e}_{v}$ can be approximated as (Chap. 2):

$$\vec{J}^{\,m}_{sl} = \vec{y}^0 \sum_{q=1}^{Q} \dot{V}_q \cos\left(\frac{q\pi y}{L_{sl}}\right), \quad q = 1, 2 \ldots Q,$$

$$\vec{J}^{\,e}_{v} = \vec{x}^0 \sum_{n=1}^{N} J^e_n \sin\left(\frac{n\pi(x - x_d + P/2)}{P}\right), \quad n = 1, 2 \ldots N, \qquad (5.3)$$

where $\vec{x}^0$ and $\vec{y}^0$ are the unit vectors, $\dot{V}_q$ and J^e_n are the unknown complex current amplitudes. Then, in the spherical coordinate system (R, θ, φ), the electric field components in the far zone can be written in the form:

$$E_\varphi(\theta, \varphi) = \frac{ike^{-ikR}}{4\pi R}(120\pi I_1 + I_2 \cos\theta) \sin\varphi,$$

$$E_\theta(\theta, \varphi) = -\frac{ike^{-ikR}}{4\pi R}(I_1 \cos\theta + I_2) \cos\varphi, \qquad (5.4)$$

where

$$I_1 = \sum_{n=1}^{N} J^e_n \left[\begin{array}{c} e^{ik(x_d \sin\theta\cos\varphi + y_d \sin\theta\sin\varphi + z_d \cos\theta)} \\ -e^{ik(x_d \sin\theta\cos\varphi + y_d \sin\theta\sin\varphi - z_d \cos\theta)} \end{array}\right] \frac{n\pi}{P}$$
$$\times \frac{(-1)^{n+1} e^{ik(x_d + P/2)\sin\theta\cos\varphi} + e^{ik(x_d - P/2)\sin\theta\cos\varphi}}{(n\pi/P)^2 - (k\sin\theta\cos\varphi)^2},$$

$$I_2 = -\sum_{q=1}^{Q} V_q \frac{q\pi}{2L_{sl}} \times \frac{(-1)^{q+1} e^{ikL_{sl}\sin\theta\cos\varphi} + e^{-ikL_{sl}\sin\theta\cos\varphi}}{(q\pi/2L_{sl})^2 - (k\sin\theta\cos\varphi)^2}.$$

According to the expression (5.4), the H-plane field at the main frequency has only the E_φ component, while in the E-plane, it has only the E_θ component. At the alternative frequency, when slot is metalized and the vibrator is in the active mode, the expression for the electric field is simplified to

$$\vec{E}(\theta,\varphi) = -\frac{ikI_1 e^{-ikR}}{4\pi R}\left(\vec{\theta}^0 \cos\theta\cos\varphi - \vec{\varphi}^0 120\pi \sin\varphi\right). \tag{5.5}$$

5.1.3 Numerical and Experimental Results

The combined structure consisting of two identical perfectly conducting monopoles symmetrically located relative to the half-wave slot (Fig. 5.1a) was studied in [3, 4]. The monopoles were placed at the distance $y_v = 0.086\lambda$. The length of the vibrators and slot were $L_v = 0.375\lambda$ and $2L_{sl} = 0.5\lambda$. These parameters allowing us to obtain the approximately equal RP in the main polarization planes, has been obtained based on the experimental studies by Clavin. Complex modeling carried out in [5, 7] has shown that the RP of the Clavin element in the main planes are similar when ($L_v = 0.3125\lambda$, $y_v = 0.086\lambda$) or ($L_v = 0.3\lambda$, $y_v = 0.086\lambda$) with the side radiation level of - 20 dB or -31 dB.

In the first case, the RP are close in width, while in the second case they have the lowest levels of lateral radiation. The normalized RP of Clavin radiators (5.1) with the parameters ($L_v = 0.3125\lambda, y_v = 0.086\lambda$) and ($L_v = 0.3\lambda, y_v = 0.086\lambda$) are shown in Fig. 5.2a. As can be seen from curve 1 in Fig. 5.2a, the RP in the H-plane coincide. The E-plane RP are shown by curves 2 and 3. As can be seen from Fig. 5.2a, the greatest similarity of RP in the main polarization planes is observed when ($L_v = 0.3125\lambda$, $y_v = 0.086\lambda$). The structure of this geometry will be considered as a radiator for the two-frequency antenna at the main frequency.

At the alternative frequency, the two identical active monopoles operate as the single radiator. The optimal operating wavelengths λ_a at which the monopoles are quarter-wave, i.e., their length is $L_v = 0.25\lambda_a = 0.3125\lambda$. Hence, the operating wavelength of the antenna at alternative frequency should be selected in the range $\lambda \le \lambda_a \le 1.25\lambda$. The results of numerical studies are presented at the alternative wavelength $\lambda_a = 1.1\lambda$ ($L_v = 0.284\lambda_a$ and $y_v = 0.078\lambda_a$). The normalized RP in the E-plane ($\varphi = 90°$) and H-plane ($\varphi = 90°$) for the two stand-alone monopoles (5.2) are shown in Fig. 5.2b by curves 1 and 2.

As can be seen from Fig. 5.2b, at the main frequency, the two in-phase monopoles form the maximum radiation along the screen plane ($\theta = \pm 90°$). Since the monopoles are displaced along the axis {0y} at $y_v = 0.078\lambda_a$, the radiation field

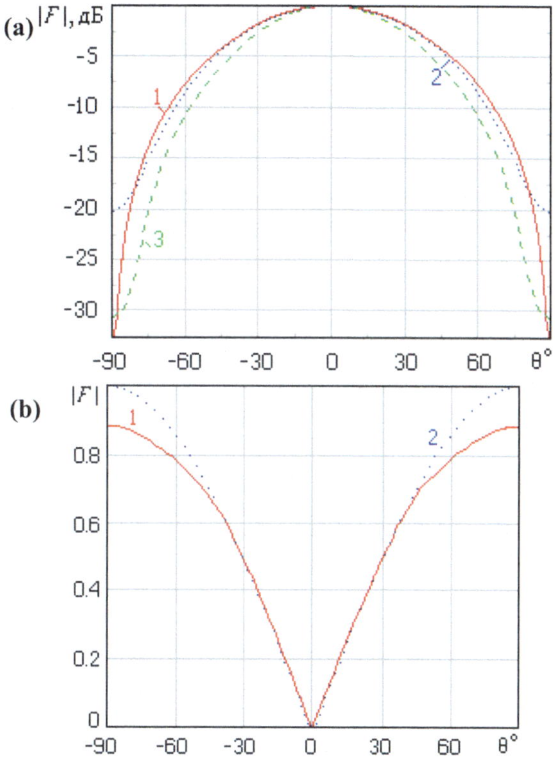

Fig. 5.2 The RP of the Clavin radiator: **a** 1—H-plane, 2,3—E-plane (2—$L_v = 0.3125\lambda$, $y_v = 0.086\lambda$), (3—$L_v = 0.3\lambda$, $y_v = 0.086\lambda$); **b** $\lambda_a = 1.1\lambda$, $L_v = 0.284\lambda_a$, $y_v = 0.078\lambda_a$, 1—E-plane, 2—H-plane

amplitude in the plane ($\theta = \pm 90°$) becomes uneven in the angle φ. The smallest amplitude observed in the H-plane is 12% lower than the field amplitude in the E-plane.

The modeling results for the structure with the slot and horizontal vibrator (Fig. 5.1b) can be presented in the same way. The combined structure with following parameters: $P = 0.407\lambda$, $2L_{sl} = 0.47\lambda$, $d = 0.07\lambda$ was studied in [6]. The vibrators were located symmetrically relative to the slot axis ($x_d = y_d = 0$) at the height $z_d = 0.105\lambda$. In the main planes, the normalized RP for this structure are almost coincide. Since the RP in both polarization planes are symmetric relative to $\theta = 0°$, the left-hand part of the curves in Fig. 5.3a, present E-plane ($\varphi = 0°$) RP, while the right-hand part corresponds to the H-plane ($\varphi = 90°$) RP. The calculated and experimental results are shown by curve 1 and 2 [6]. Note that the RP close in width can be obtained with another parameters P/λ, z_d/λ.

At an alternative wavelength λ_a, the standalone radiator acts as the active symmetrical horizontal vibrator. The optimal vibrator length is $P = 0.5\lambda_a$, $d = 0.05\lambda_a$, where the λ_a is the operating wavelength, thus, $\lambda_a = 0.804\lambda$. The normalized RP of the horizontal vibrator (5.5) in the E- and H-plane with these parameters are presented by the curves 1 and 2 in Fig. 5.3b.

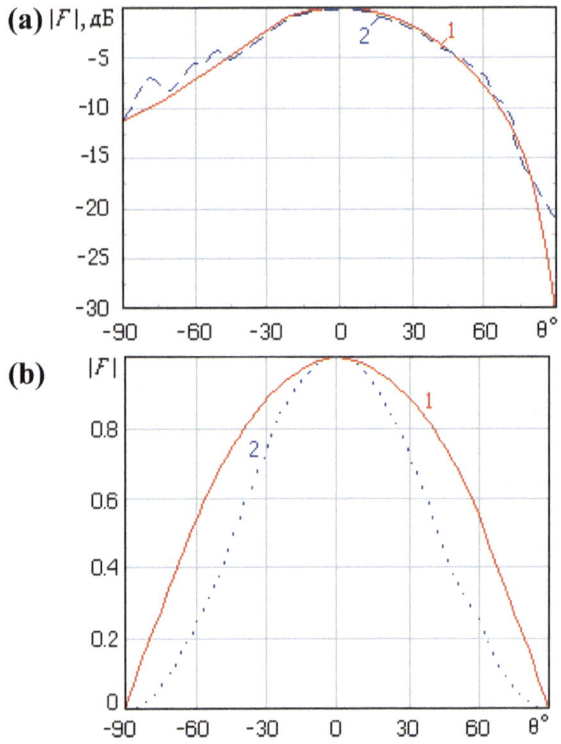

Fig. 5.3 The RP of the structure with the slot and horizontal vibrator:
a $P = 0.407\lambda$, $2L_{sl} = 0.47\lambda$, $d = 0.07\lambda$, 1—calculation (5.4), 2—experimental data [6] (кружками);
b $\lambda_a = 0.804\lambda$, $P = 0.5\lambda_a$, $d = 0.05\lambda_a$ 1—E-plane, 2—H-plane

5.2 Two-Frequency Operating Mode of Antenna Arrays with Switching Radiators of Clavin Type

In this Section the possibility of application Clavin type radiators in combined two-frequency antenna arrays with diode switching of vibrator and slot elements is validated. The conventional Clavin elements with passive monopoles operating at the main frequency and two active monopoles operating at the alternative frequency are used as the combined array radiator. Radiation fields of combined arrays at both frequencies are analyzed.

5.2.1 Problem Formulation and Solution

Let us consider the dual-frequency mode of operation of an antenna arrays (AA) made of Clavin type radiators (Fig. 5.1a). Knowing the expressions for the radiation field for single elements (5.1), (5.2) and assuming that all radiators at the nodes (x_n, y_m) of the planar arrays in Figs. 5.4 and 5.5 have the same geometry, the total radiation field $\vec{E}(r, \theta, \varphi)$ of the arrays can be defined as the sum of the radiation

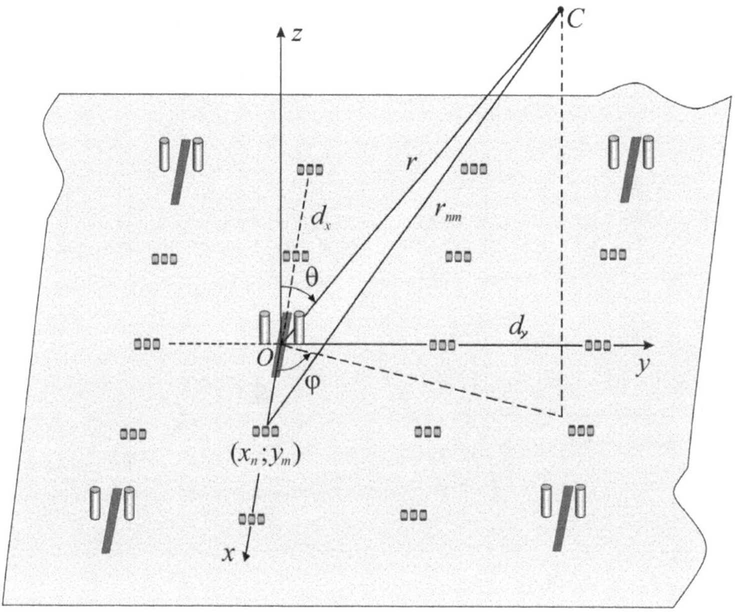

Fig. 5.4 The AA aperture operating at the main frequency

fields of each element, taking into account the amplitudes and phases with which these fields arrive at the observation point $C(r, \theta, \varphi)$:

$$\vec{E}(r, \theta, \varphi) = \sum_{n=1}^{N_x} \sum_{m=1}^{N_y} \frac{e^{-ikr_{nm}}}{r_{nm}} \vec{E}_{nm}(\theta, \varphi). \tag{5.6}$$

It should be recalled that here all combined radiators in the arrays are assumed to be tuned into resonance due to the selection of the values of their internal parameters and (or) the values of the grating parameters and , which make it possible to compensate for the mutual influence of the radiators. When radiators of periodic AA are characterized by equal amplitudes of currents, double summation in expression (5.6) can be replaced by a normalized lattice factor [9]:

$$\vec{E}(r, \theta, \varphi) = \frac{\vec{E}(\theta, \varphi)}{N_x N_y} \frac{\sin\left[\frac{N_x}{2}(kd_x \sin\theta\cos\varphi - \psi_x)\right]}{\sin\left[\frac{1}{2}(kd_x \sin\theta\cos\varphi - \psi_x)\right]} \frac{\sin\left[\frac{N_y}{2}\left(kd_y \sin\theta\sin\varphi - \psi_y\right)\right]}{\sin\left[\frac{1}{2}\left(kd_y \sin\theta\sin\varphi - \psi_y\right)\right]}, \tag{5.7}$$

where ψ_x is the phase shift between the currents of neighboring radiators in a row along the axis $\{0x\}$, and ψ_y is the phase shift between the currents of neighboring radiators in rows along the axis $\{0y\}$.

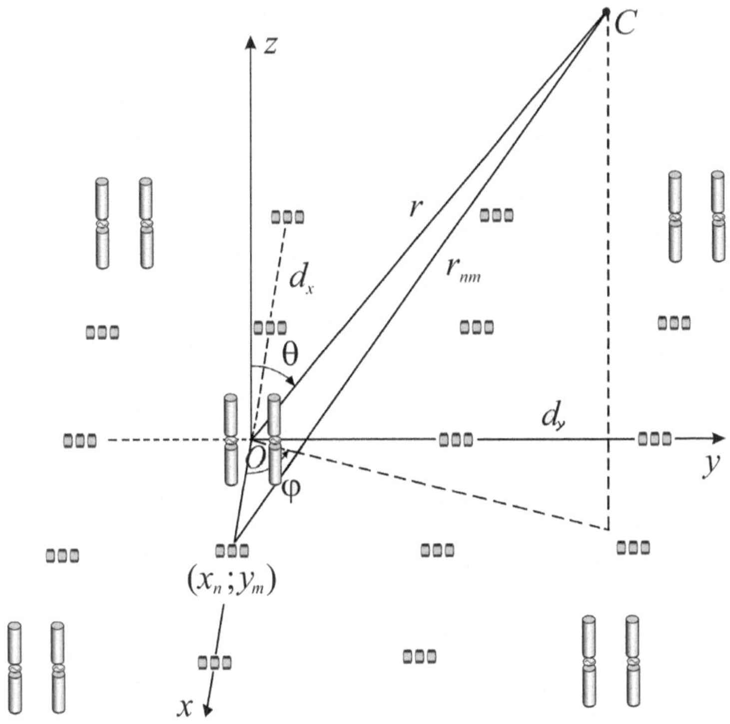

Fig. 5.5 Virtual AA aperture operating at the alternative frequency

At the main frequency, the maximum of RP of the single Clavin radiator (Fig. 5.1a) defined by the formula (5.1) is directed along the $\{0z\}$ axis. Since monopoles of the Clavin radiator are passive, the amplitudes and phases should be related to the magnetic current in the slot. As follows from physical considerations, the normal (transverse) radiation mode of the antenna array consisting of Clavin radiator (Fig. 5.4) can be achieved if $\psi_x = \psi_y = 0$, i.e., with the in-phase operation of the slotted elements. The form of array factor will be kept invariable in any plane ($\varphi = const$) passing through the $\{0z\}$ axis. Besides, the form of amplitude RP in the *H*- and *E*-planes do not depend upon the number of radiators in a row N_y and the number of rows N_y.

At the alternative frequency, the radiator of the antenna array (Fig. 5.5) can be presented by the two monopoles excited by the current of equal amplitudes. In this case, the phase shifts of the fields radiated by the adjacent radiators in the expression (5.7) should be took into account relative to phase centers monopole pairs. Moreover, from physical considerations it follows that the regime of axial radiation for the array operating at this mode can be realized under the conditions $\psi_x = kd_x$ and $\psi_y = kd_y$.

5.2.2 Numerical Results

Simulation of antenna array ($L_v = 0.3125\lambda$, $y_v = 0.086\lambda$) fields at the main frequency was carried out under conditions that the array periods $d_x = d_y = \lambda/2$, and phases $\psi_x = \psi_y = 0$. This selection minimizes the mutual influence between the antenna array elements [9]. The normalized RP in the H-plane obtained by using the expression (5.7) are presented in Fig. 5.6 for the following configuration: $N_x = N_y = 3$ (curve 1), $N_x = N_y = 5$ (curve 2), and $N_x = N_y = 7$ (curve 3). Simulation results have also shown that the array RP in the E-plane and H-plane coincide with good accuracy for arbitrary number of radiators in the array. As expected, we can state based on the well-known electrodynamic principles that the maximum of the RP is oriented normal to the screen plane, the main directional lobes of the RP become narrower, and the number of sides lobes increases if the number of radiators is increased.

At the alternative frequency (wavelength λ_a), the two identical active monopoles operate as a single radiator (Fig. 5.5), which physical location in the array stays unchanged. Since the electric lengths of the array periods d_x and d_y are changing due to transition to the alternative wavelength and differ from $\lambda_a/2$, the resulting phase shifts should be compensated by some or other method. When such compensation is achieved, the directions of the RP maxima for the single radiator and array coincide. The specified compensation can be only achieved by using the phase shifters in the vibrator transmission lines. Since technical possibilities for implementing such phase shifters are quite limited, the main and alternative frequencies should be fairly close to each other. On the other hand, the optimal operating wavelength λ_a of the vibrator radiator is achieved if the monopole length is $L_v = 0.25,\ \lambda_a = 0.3125\lambda$. Therefore, the alternative frequency should be selected from the wavelength band $\lambda \leq \lambda_a \leq 1.25\lambda$.

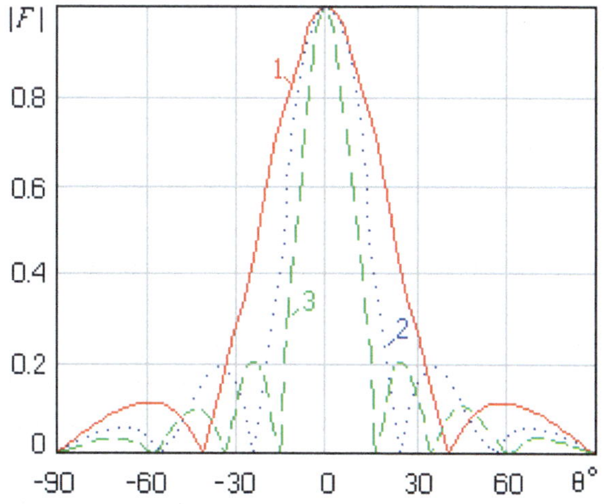

Fig. 5.6 The RP of radiating structures at the main frequency: ($L_v = 0.3125\lambda$, $y_v = 0.086\lambda$, $d_x = d_y = \lambda/2$, $\psi_x = \psi_y = 0$) 1—$N_x = N_y = 3$, 2 -$N_x = N_y = 5$, 3 -$N_x = N_y = 7$

If the simulation of antenna array fields based on the formula (5.7) is carried out at the alternative frequency, the array periods are $d_x = d_y = \lambda/2 = \lambda_a/2.2$ and the parameters are $\psi_x = \psi_y = 0.91\pi$. The simulated H-plane RP of square antenna arrays are shown in Fig. 5.7a for the arrays with $N_x = N_y = 3$ (curve 1), $N_x = N_y = 5$ (curve 2), and $N_x = N_y = 7$ (curve 3). It should be noted, the E-plane RP are similar to the H-plane RP shown in Fig. 5.7a for identical number of array radiators. Besides, the field amplitudes decrease in the directions $\theta = \pm 90°$. The RP in the both planes are asymmetric relative to the direction $\theta = 0°$ due to the phase mismatching between the radiators. If the number of radiators is increased, and hence, phase errors in the observation point are accumulated, the skew of the RP increases. The phase compensations, ensuring equality $\psi_x = \psi_y = \pi$ eliminates the RP asymmetry (Fig. 5.7b). The maxima of all RP shown in Fig. 5.7 are oriented along the screen plane, the width of main lobes decreases and the number of side lobes increases if the number of radiators is increased.

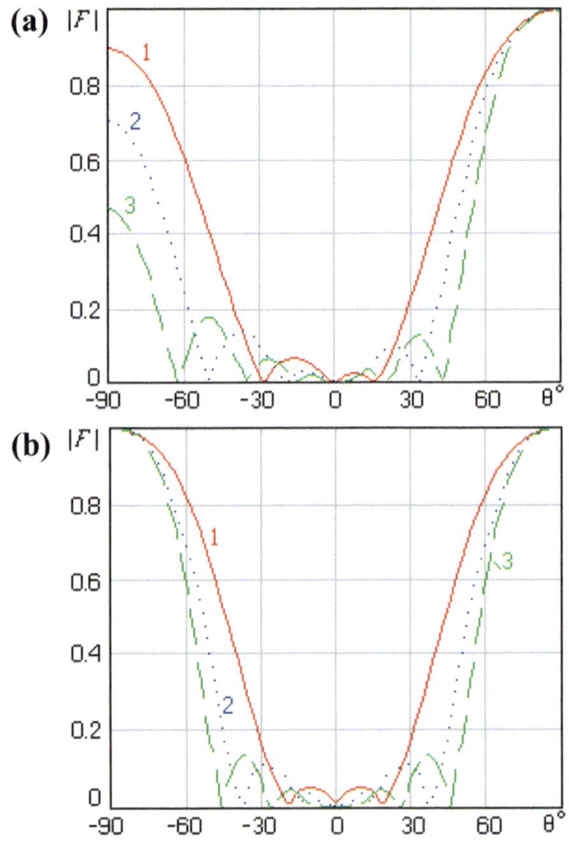

Fig. 5.7 The RP of the antenna arrays with different number of radiators at the alternative frequency ($d_x = d_y = \lambda/2 = \lambda_a/2.2$): 1—$N_x = N_y = 3$, 2—$N_x = N_y = 5$, 3—$N_x = N_y = 7$, **a** without phase compensation ($\psi_x = \psi_y = 0.91\pi$); **b** with phase compensation ($\psi_x = \psi_y = \pi$)

The simulation results carried out for other ratios λ_a/λ from the interval $\lambda \leq \lambda_a \leq 1.25\lambda$ have confirmed the validity of the approach for obtaining the field characteristics of the AA with the Calvin type radiators.

5.3 Yagi-Uda Combined Vibrator-Slot Radiating Structures

In this Section an electrodynamically rigorous mathematical model of a combined vibrator-slot structure consisting of a narrow radiating slot cut in a rectangular waveguide end wall and several thin impedance vibrators placed over the infinite screen is presented. Numerical results concerning internal and external electrodynamic characteristics of the antennas with optimized structural parameters have confirmed the possibility of constructing the Yagi-Uda combined radiating structures in the microwave and extremely high frequency (EHF) bands.

5.3.1 Formulation and Solution of the Diffraction Problem

Radiating arrays known as Yagi-Uda (wave-channel) antennas with linear vibrators examined in detail in Chap. 3. The structural diversity of the Yagi-Uda antennas is not limited by using only electric elements. The principle of the wave-channel antenna was applied to magnetic type elements. For example, an antenna consisting of loop elements, i.e., electrodynamic equivalents of magnetic dipoles, was considered in [10]. This loop antenna consists of the reflector, active element, and vibrator directors coaxially placed in parallel on the common rod so that they are dielectrically isolated from one another. All loop elements are open-ended with a minimal clearance, and their planes are. A more complex design of the loop wave channel was proposed in [11], where the symmetric relative to the longitudinal antenna axis antenna elements are made in the form of grouped pairs of loop radiators. The Yagi-Uda antennas with the vibrator or loop elements are usually used in the meter and decimeter wavelength ranges. First of all, this is related to the physical principles of radiation by the active elements and the power supply implemented by using coaxial cables. It should also be noted that to minimize the design dimensions of the Yagi-Uda antennas in the decimeter wavelength range, microstrip devices based on low-profile structures have been proposed [12, 13].

The wave-channel radiating structures can be used in the microwave and EHF bands under two conditions: the power should be supplied to the antenna by the waveguides and the magnetic elements must be applied. These conditions were realized in antenna design developed in [14, 15]: apertures of rectangular waveguides were proposed to use as antenna structure elements [14], and slot in microstrip metal patches were excited by a plane dielectric waveguide [15]. However, this type antennas operating in the centimeter wave band are characterized by insufficient isolation of the cross-polarization components, required for the remote sensing and wave polarimetry. To minimize this drawback, electrically thin radiating elements should be used in these antennas.

The microwave antennas with linear elements with good polarization isolation, can be built based on combined vibrator-slot radiating structures [16–19]. Electrodynamic characteristics of these radiating structures with waveguide excitation can be easily obtained by a generalized method of induced EMMF [7]. Unfortunately, such approach for this problem solution was never discussed earlier. The article is aimed at developing based on the mathematical model a Yagi-Uda combined vibrator-slot radiating structure (YUCVSRS) operating in the centimeter and millimeter wave ranges. The antenna characteristics will be modelled by generalized method of induced EMMF, since modelling conducted by existing commercial programs are resource-intensive and low-efficient.

Let the H_{10}-wave propagates in a semi-infinite rectangular waveguide (marked by index "Wg") from the direction $z = -\infty$. A narrow longitudinal slot cut in the waveguide end wall symmetrically with respect to the side walls and oriented along the axis $\{0x\}$ radiates into the half-space above the infinite perfectly conducting plane (marked by index "Hs"). A system of N thin impedance vibrators each of them is placed at fixed distances z_n from the plane as shown in Fig. 5.8. The cross section of the waveguide is $\{a \times b\}$, the material parameters of the medium filling the waveguide are ε_1, μ_1, the slot length and width are $2L_0$ and d. The length and radius of the vibrator with the number $n \in [1, N]$ are $2L_n$ and r_n.

Let us assume that dimensions of the slot and vibrators satisfy the following inequalities $\frac{r_n}{2L_n} \ll 1$, $\frac{d}{2L_0} \ll 1$, $\frac{d}{\lambda_{1(2)}} \ll 1$ ($\lambda_{1(2)}$ are the wavelengths in the media with parameters $\varepsilon_{1(2)}, \mu_{1(2)}$). Since the electric currents $J_n(s_n)$ on the vibrators and the equivalent magnetic current in the slot $J_0(s_0)$ should satisfy the boundary conditions $J_n(\pm L_n) = 0$, $J_0(\pm L_0) = 0$, the initial system of integral equations relative to unknown currents [7, 17] can be written as.

$$\left(\frac{\mathrm{d}^2}{\mathrm{d}s_n^2} + k_2^2\right)\left\{\begin{aligned}&\int_{-L_1}^{L_1} \mathrm{J}_1(s_1')\mathrm{G}_{s_1}^{HsE}(s_n, s_1')\mathrm{d}s_1' + \\ &\int_{-L_2}^{L_2} \mathrm{J}_2(s_2')\mathrm{G}_{s_2}^{HsE}(s_n, s_2')\mathrm{d}s_2' + \cdots\end{aligned} + \int_{-L_N}^{L_N} \mathrm{J}_N(s_N')\mathrm{G}_{s_N}^{HsE}(s_n, s_N')\mathrm{d}s_N'\right\}$$

$$- ik\vec{e}_{s_n}\mathrm{rot}\int_{-L_0}^{L_0} \mathrm{J}_0(s_0')\mathrm{G}_{s_0}^{HsM}(s_n, s_0')\mathrm{d}s_0' = i\omega\varepsilon_2 z_{in}(s_n)\mathrm{J}_n(s_n), \quad n = 1, 2, \ldots N;$$

$$\frac{1}{\mu_1}\left(\frac{\mathrm{d}^2}{\mathrm{d}s_0^2} + \mathrm{k}_1^2\right)\int_{-L_0}^{L_0} \mathrm{J}_0(s_0')\mathrm{G}_{s_0}^{WgM}(s_0, s_0')\mathrm{d}s_0'$$

$$+\frac{1}{\mu_2}\left(\frac{d^2}{ds_0^2}+k_2^2\right)\int_{-L_0}^{L_0} J_0(s_0')G_{s_0}^{HsM}(s_0,s_0')ds_0'$$

$$+ik\vec{e}_{s_0}\text{rot}\sum_{n=1}^{N}\int_{-L_n}^{L_n} J_n(s_n')E_{s_n}^{HsE}(s_0,s_n')ds_n' = -i\omega H_{0s_0}(s_0), \tag{5.8}$$

where $z_{in}(s_n)$ are the vibrator internal linear impedances, measured in [Ohm/m], $H_{0s_0}(s_0)$ is the projection of the external source field on the slot axis, $G_{s_m}^{HsE}(s_n, s_m')$ and $G_{s_m}^{HsM,WgM}(s_n, s_m')$ are components of the electric and magnetic tensor Green's functions [7] (marked by indices E and M) for the vector potentials of coupling electrodynamic volumes (half-space above an infinite ideally conducting plane and a semi-infinite rectangular waveguide), $k = 2\pi/\lambda$, λ is the wavelength in free space, $\vec{e}_{s_0}$, $\vec{e}_{s_n}$ are unit vectors directed along the slot and vibrator axes, and s_0, s_n are local coordinates associated with the longitudinal axes of the slot and vibrators. The magnetic

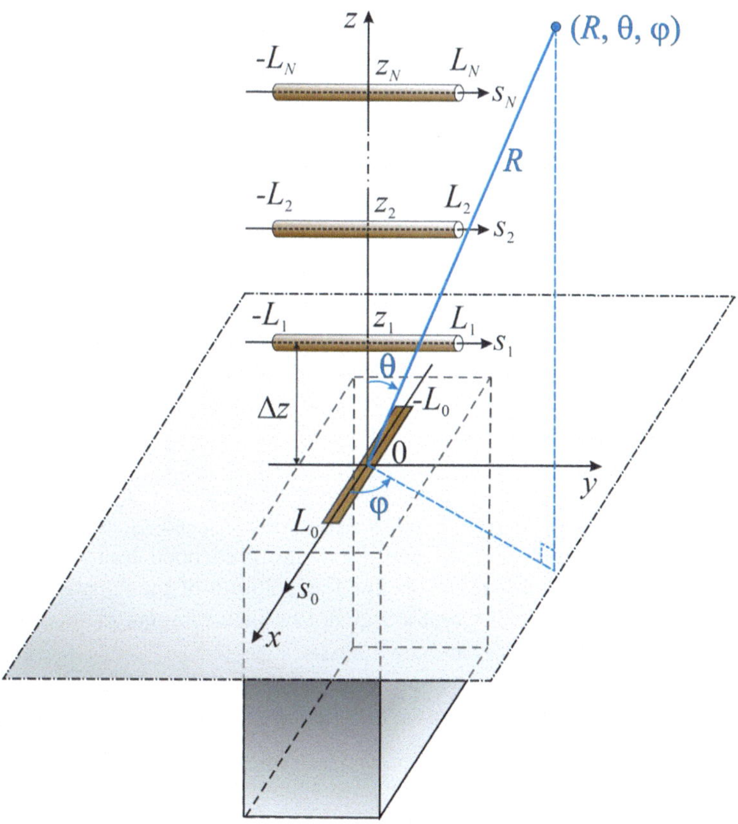

Fig. 5.8 The YUCVSRS geometry and accepted notations

Green's function in "*Hs*" can be presented as $G_{s_0}^{HsM}(s_0, s_0') = \frac{2e^{-ik\sqrt{(s_0-s_0')^2+(d_e/4)^2}}}{\sqrt{(s_0-s_0')^2+(d_e/4)^2}}$ where $d_e = de^{-\frac{\pi h}{2d}}$ is the effective slot width allowing to take into account the actual thickness of the waveguide wall h (Chap. 2).

Since the projection of the field of external sources on the slot axis is $H_{0s_0}(s_0) = H_0 \cos\frac{\pi s_0}{a} = H_0^s(s_0)$, the magnetic current in the slot can be described by a symmetric function $J_0(s_0) = J_0^s(s_0)$ relative to the slot center. Then, Eq. (5.8) under condition $\varepsilon_{1(2)} = \mu_{1(2)} = 1$ can be transformed to

$$\left(\frac{d^2}{ds_n^2}+k^2\right)\left\{\begin{array}{l}\int\limits_{-L_1}^{L_1} J_1(s_1')G_{s_1}^{HsE}(s_n, s_1')ds_1' \\ +\int\limits_{-L_2}^{L_2} J_2(s_2')G_{s_2}^{HsE}(s_n, s_2')ds_2' + \ldots\end{array} + \int\limits_{-L_N}^{L_N} J_N(s_N')G_{s_N}^{HsE}(s_n, s_N')ds_N'\right\}$$

$$+ik\int\limits_{-L_0}^{L_0} J_0^s(s_0')\tilde{G}_{s_0}^{HsM}(s_n, s_0')ds_0' = i\omega z_{in}(s_n)J_n(s_n), \quad n = 1, 2, \ldots N;$$

$$\left(\frac{d^2}{ds_0^2}+k_1^2\right)\int\limits_{-L_0}^{L_0} J_0^s(s_0')\left[G_{s_0}^{WgM}(s_0, s_0') + G_{s_0}^{HsM}(s_0, s_0')\right]ds_0'$$

$$-ik\sum_{n=1}^{N}\int\limits_{-L_n}^{L_n} J_n(s_n')\tilde{G}_{s_n}^{HsE}(s_0, s_n')ds_n' = -i\omega H_0^s(s_0). \tag{5.9}$$

where $\tilde{G}_{s_n}^{HsE}(s_0, s_n') = -\frac{\partial}{\partial z}G_{s_n}^{HsE}(s_0, 0, z; 0, s_n', z_n)\big|_{z=0}$, $\tilde{G}_{s_0}^{HsM}(s_n, s_0') = \frac{\partial}{dz}G_{s_0}^{HsM}(0, s_n, z; s_0', 0, 0)\big|_{z=z_n}$. In the above expressions, the fixed values of the variable z are substituted after differentiation.

The equations system (5.9) can be solved by the generalized method of induced EMMF, using the functions $J_n(s_n) = J_{0n}f_n(s_n)$ and $J_0^s(s_0) = J_{00}^s f_0^s(s_0)$ as approximating expressions for the currents. In this expression, J_{0n} and J_{00}^s are the unknown complex current amplitudes, $f_n(s_n)$ and $f_0^s(s_0)$ are predefined basic functions of current distributions, which can be obtained as solution of the equation defining the currents on the standalone vibrator and slot by the averaging method [8]. The expressions for the YUCVSRS can be written as

$$f_n(s_n) = \cos\tilde{k}_n s_n - \cos\tilde{k}_n L_n, f_0^s(s_0) = \cos ks_0 - \cos kL_0 \tag{5.10}$$

where $\tilde{k}_n = k + \frac{i\alpha_n 2z_{in}^{av}}{60\,\text{Ohm}}$, $\alpha_n = \frac{1}{2\ln[r_n/(2L_n)]}$ are natural small parameters, and $z_{in}^{av} = \frac{1}{2L_n}\int\limits_{-L_n}^{L_n} z_{in}(s_n)ds_n$ are internal impedances averaged over the vibrators [7].

Let us apply the generalized method of the induced EMMF to the system of Eq.(5.9). If the waveguide is excited by the H_{10}-wave with amplitude H_0, the system of linear algebraic equations (SLAE) relative to unknown current amplitudes J_{0n} and J_{00}^s can be presented as

$$\begin{cases} J_{00}^s(Z_{00}^{sWg} + Z_{00}^{sHs}) + J_{01}Z_{01} + J_{02}Z_{02} + \ldots + J_{0N}Z_{0N} = -\dfrac{i\omega}{2k}H_0^s, \\ J_{00}^s Z_{10} + J_{01}(Z_{11} + F_1^{\overline{Z}}) + J_{02}Z_{12} + \ldots + J_{0N}Z_{1N} = 0, \\ J_{00}^s Z_{20} + J_{01}Z_{21} + J_{02}(Z_{22} + F_2^{\overline{Z}}) + \ldots + J_{0N}Z_{2N} = 0, \\ \ldots \\ J_{00}^s Z_{N0} + J_{01}Z_{N1} + \ldots + J_{0(N-1)}Z_{N(N-1)} + J_{0N}(Z_{NN} + F_N^{\overline{Z}}) = 0, \end{cases} \tag{5.11}$$

where $H_0^s = H_0 \int\limits_{-L_0}^{L_0} \cos\frac{\pi s_0}{a} f_0^s(s_0)ds_0$, Z_{mn} and $F_n^{\overline{Z}}$ are dimensionless coefficients. For example, the expression for the Z_{mn} coefficients can be written as

$$Z_{mn} = \int\limits_{-L_m}^{L_m} f_m(s_m)\left(\frac{d^2}{ds_m^2} + k^2\right) \int\limits_{-L_n}^{L_n} f_n(s_n')G_{s_n}^{HsE}(s_m, s_n')ds_n'ds_m,$$

$$n = 1, 2, \ldots N; \quad m = 1, 2, \ldots N, \tag{5.12}$$

which can be interpreted as the vibrator own or mutual effective resistances if $n = m$ or $n \neq m$.

The electrodynamic characteristics of the vibrator-slot antenna can be found by solving SLAE (5.11) relative to the currents amplitudes J_{0n} and J_{00}^s. For example, the field reflection coefficient S_{11} defined by the slot inhomogeneity in the semi-infinite rectangular waveguide operating in the single-mode regime can be presented as

$$S_{11} = \left\{1 - \frac{16\pi\gamma J_{00}^s}{abk\omega H_0} \cdot \frac{\sin kL_0 \cos\frac{\pi L_0}{a} - \frac{ka}{\pi}\cos kL_0 \sin\frac{\pi L_0}{a}}{1 - [\pi/(ka)]^2}\right\} e^{2i\gamma z}, \tag{5.13}$$

where $\gamma = \sqrt{k^2 - (\pi/a)^2}$ is the propagation constant. The power reflection coefficient $|S_\Sigma|^2$ of the slot into the half-space can be obtained by using the energy balance equation

$$|S_{11}|^2 + |S_\Sigma|^2 = 1\,. \tag{5.14}$$

Let us introduce a system of spherical coordinates shown in Fig. 5.8. The total wave zone field ($R \gg \lambda$) reradiated by the combined vibrator director array equal to the sum of the secondary radiation fields of each vibrator taking into account the phases of fields arriving to the observation point $C(R, \theta, \varphi)$ can be written as

$$\vec{E}(R,\theta,\varphi) = \frac{2k^2}{\omega}\frac{e^{-ikR}}{R}\left[i\left(\vec{\theta}^0\sin\varphi + \vec{\varphi}^0\cos\theta\cos\varphi\right)J_{00}^{s}F_{C0}\right.$$
$$\left. + \left(\vec{\theta}^0\cos\theta\sin\varphi + \vec{\varphi}^0\cos\varphi\right)\sum_{n=1}^{N}J_{0n}F_{Cn}\sin(kz_n\cos\theta)\right], \qquad (5.15)$$

where and are the unit vectors, , and . If the relations (5.10) is taken into account, the function F_{C0} can be written as

$$F_{C0} = \frac{2}{k^2-(k\sin\theta\cos\varphi)^2}[k\cos(kL_0\sin\theta\cos\varphi)\sin(kL_0)$$
$$-k\sin(kL_0\sin\theta\cos\varphi)\cos(kL_0)\sin\theta\cos\varphi]$$
$$-2L_0\cos(kL_0)\frac{\sin(kL_0\sin\theta\cos\varphi)}{kL_0\sin\theta\cos\varphi}, \qquad (5.16a)$$

$$\mathrm{F}_{Cn} = \frac{2}{\tilde{\mathrm{k}}_n^2-(\mathrm{k}\sin\theta\sin\varphi)^2}\left[\tilde{\mathrm{k}}_n\cos(\mathrm{kL}_n\sin\theta\sin\varphi)\sin\left(\tilde{\mathrm{k}}_n\mathrm{L}_n\right)\right.$$
$$\left.-\mathrm{k}\sin(\mathrm{kL}_n\sin\theta\sin\varphi)\cos\left(\tilde{\mathrm{k}}_n\mathrm{L}_n\right)\sin\theta\sin\varphi\right]$$
$$-2\mathrm{L}_n\cos\left(\tilde{\mathrm{k}}_n\mathrm{L}_n\right)\frac{sin(\mathrm{kL}_n\sin\theta\sin\varphi)}{\mathrm{kL}_n\sin\theta\sin\varphi}. \qquad (5.16b)$$

As can be seen from the expressions (5.15) and (5.16), the E-plane ($\varphi = 90°$) and H-plane ($\varphi = 0°$) components of the electric field are $E_\varphi = 0$ and $E_\theta = 0$. The longitudinal component of the field in the far zone is $E_r = 0$.

5.3.2 Numerical Results

The electrodynamic characteristics were simulated for the YUCVSRS consisting of the rectangular waveguide and N vibrator directors. The waveguide was excited by the H_{10}-wave with amplitude $H_0 = 1$ at the frequency. $f = 9.2$ GHz ($\lambda = 32.6$ mm). The waveguide cross-section, wall thickness, slot width, and the radii of the perfectly conducting vibrators ($z_{in}^{av} = 0$) where $a \times b = 23 \times 10$ mm^2, $h = 1.0$ mm, $d = 1.5$ mm, and $r_n = 0.4$ mm. The remaining parameters of the YUCVSRS: the slot length, vibrator length, and distances between structural elements were selected by the numerical optimization to ensure the maximum directivity D in the direction of $\{0z\}$ axis under condition of satisfactory waveguide matching (the voltage standing wave ratio VSWR < 1.1). The directivity D and VSWR were calculated by the following relationships:

$$\text{VSWR} = \frac{1 + |S_{11}|}{1 - |S_{11}|}, \tag{5.17}$$

$$D = \frac{4\pi}{\int_0^{2\pi} \int_0^{\pi/2} [F(\theta, \varphi)]^{2s} \sin\theta d\theta d\varphi} \tag{5.18}$$

where $F(\theta, \varphi) = \frac{|\vec{E}(R,\theta,\varphi)|}{|\vec{E}(R,\theta_m,\varphi_m)|}$ is the RP normalized at the electric field modulus in the radiation maximum direction.

The YUCVSRS optimization was carried out under condition that all its vibrator directors are positions at equal distances $\Delta z = z_1 = z_{n+1} - z_n$ from each other. The optimal parameters of the YUCVSRS: $2L_0$, $2L_n$ and Δz for different number of vibrator directors are shown in Table 5.1.

The calculated RP isn the main polarization planes for the optimized YUCVSRS, the angular plots $|F| = |E_\theta(\theta, \varphi = 0°)|$ and $|F| = \left|E_\varphi(\theta, \varphi = 90°)\right|$, are presented in Fig. 5.9. All plots are normalized to the maximum of the electric field modulus for a particular antenna structure. As can be seen from Fig. 5.9 the main lobe of the RP becomes narrower if the number of vibrator elements is increased. At the same time, the widths of the RP main lobes (at the level $|F| = 0.5$) in the orthogonal polarization planes tend to approach to one other. As was shown in [6], the two-element vibrator-slot system with the vibrator director, whose length was $2L_1 = 0.407\lambda$, and distance between the resonant slot and the vibrators was $\Delta z = 0.105\lambda$ is characterized by almost identical RPs in E- and H- planes. This effect was experimentally proved in [6]. The simulation results obtained for the two-element YUCVSRS have confirmed the similarity of the RP in the two polarization planes. Thus, this prove out the mathematical model validity and its physical adequacy.

Properties of the side radiation for the optimized YUCVSRS can be analyzed by comparing the RP for $N = 2$, $N = 3$, $N = 4$, and $N = 5$ presented in Fig. 5.10. As can be see, if the number of radiating elements in the YUCVSRS is increased, the number of the RP side lobes also increases in accordance with electrodynamics principles. It was also established that the power levels for the side lobes do not exceed 10% of that for the main lobe.

Table 5.1 Parameters of the YUCVSRS

N	D	$2L_0$	$2L_n$	Δz
1	6.89	0.521λ	0.398λ	0.269λ
2	11.90	0.536λ	0.405λ	0.294λ
3	16.13	0.54λ	0.406λ	0.321λ
4	19.73	0.528λ	$0,402\lambda$	0.349λ
5	22.97	0.518λ	0.399λ	0.367λ

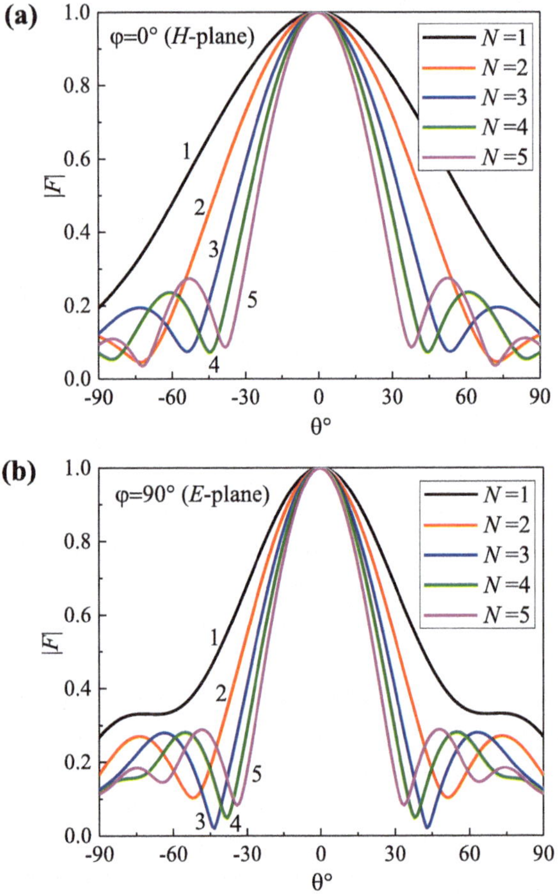

Fig. 5.9 The RP of the YUCVSRS with various vibrator numbers: **a** $|F| = |E_\theta(\theta, \varphi = 0°)|$, **b** $|F| = |E_\varphi(\theta, \varphi = 90°)|$

The internal and external energy characteristics of the YUCVSRS with the parameters listed in Table 5.1 are presented in Fig. 5.11. The frequency plots of the coefficients $|S_{11}|$, $|S_\Sigma|^2$, and VSWR defined by the formulas (5.13), (5.14), and (5.17) for different N are shown in Fig. 5.11a. The frequency plots of the directivity D[dB] (5.18) and the antenna gain G defined by the formula

$$\mathrm{G} = \mathrm{D} \cdot |\mathrm{S}_\Sigma|^2 \tag{5.19}$$

are presented in Fig. 5.11b.

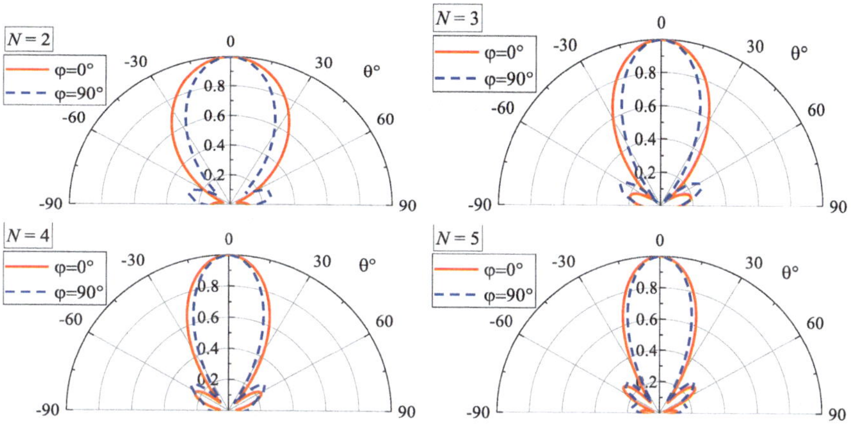

Fig. 5.10 The RP of the YUCVSRS with optimized parameters

The plots shown in Fig. 5.11 were calculated in the frequency range $f \in [7.5; 10.5]$ GHz belonging to the single-mode waveguide regime. As can be seen, the curves $|S_{11}|$, VSWR have minima, and the curves $|S_\Sigma|^2$, D and G have maxima in the vicinity of the frequency $f = 9.2$ GHz. It can be also stated that if the number of vibrators is increased, the operating frequency band becomes narrower.

The simulation results have confirmed the possibility of reducing the main lobe width and increasing their directivity and gain by using the system of several passive vibrator directors arranged in the Yagi-Uda configuration. It is shown that the YUCVSRS with optimized parameters can provide the maximum directivity and good matching of the waveguide transmission line with VSWR < 1.1.

The materials of this Chapter are contained in the authors' publications [20–22].

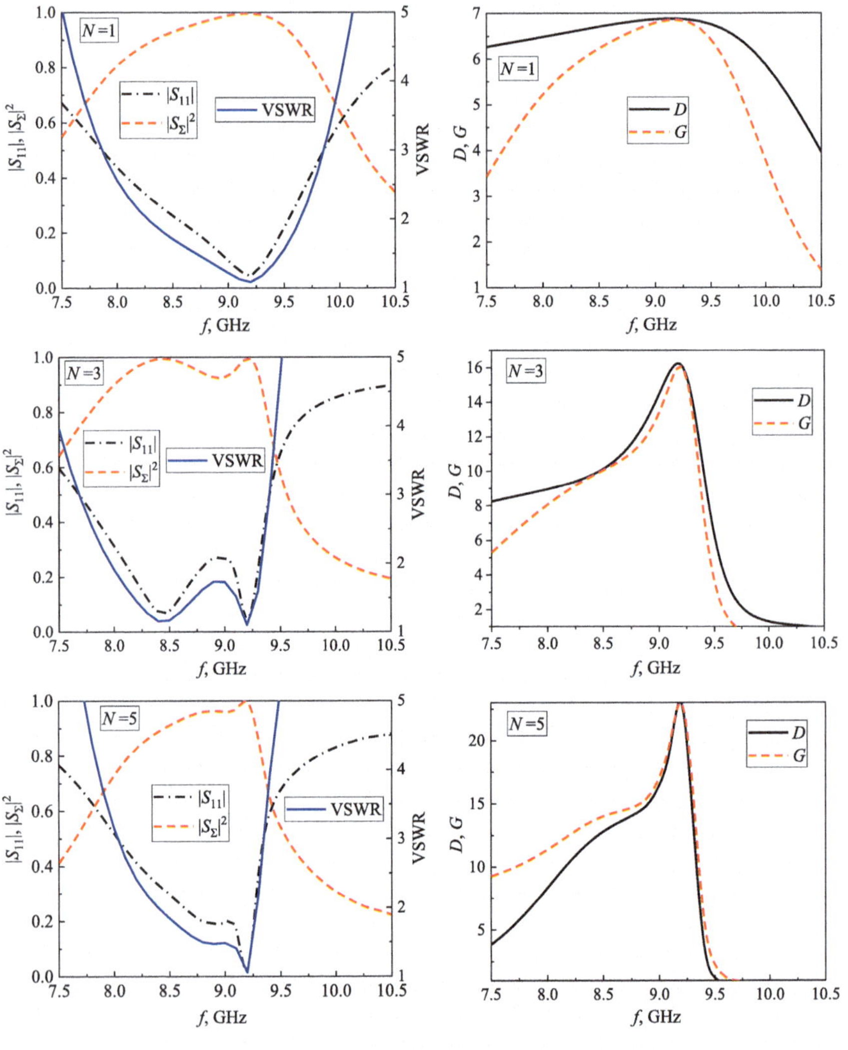

Fig. 5.11 The frequency plots of the energy characteristics for the YUCVSRS with various *N*

References

1. Ponomarev LI, Stepanenko VI (2009) Scanning multi-frequency combined antenna arrays. Radiotekhnika, Moscow (in Russian)
2. Mortenson KE (1964) Microwave semiconductor control devices. Microwave J 7:49–57
3. Clavin A, Huebner DA, Kilburg FJ (1974) An improved element for use in array antennas. IEEE Trans Antennas Propag 22:521–526
4. Clavin A (1975) A multimode antenna having equal E- and H-planes. IEEE Trans Antennas Propag 23:735–737

5. Berdnik SL, Katrich VA, Nesterenko MV, Penkin YuM (2020) Waveguide radiation of the combined vibrator-slot structures. Prog Electromagn Res B 87:151–170
6. Lee Y-H, Hong D-H, Ra J-W (1983) Waveguide slot antenna with a coupled dipole above the slot. Electron Lett 19(8):1–10
7. Nesterenko MV, Katrich VA, Penkin YuM, Berdnik SL, Dumin OM (2020) Combined vibrator-slot structures: theory and applications. Springer Nature Swizerland AG, Cham, Swizerland
8. Nesterenko MV, Katrich VA, Penkin YuM, Berdnik SL (2008) Analytical and hybrid methods in theory of slot-hole coupling of electrodynamic volumes. Springer Science+Business Media, New York
9. Amitay N, Galindo V, Wu CP (1972) Theory and analysis of phased array antennas. John Wiley & Sons Inc., New York
10. Campbell RW (1970) Endfire antenna array having loop directors. U.S. Patent 3,491,361, issued January 20
11. Petlya II, Panchenko VA, (2001) Loop antenna. RF Patent 2174272, issued September 27
12. Liu H, Gao S, Loh T-H (2013) Small director array for low-profile smart antennas achieving higher gain. IEEE Trans Antennas Propag 61:162–168
13. Wang Z, Liu XL, Yin Y-Z, Wang JH, Li Z (2014) A novel design of folded dipole for broadband printed Yagi-Uda antenna. Prog Electromagn Res C 46:23–30
14. Altshuler EE (1962) A monopole array driven from a rectangular waveguide. IRE Trans Antennas Propag 10:558–560
15. Zhang Z, Cao X-Y, Gao J, Li S-J, Liu X (2015) Compact microstrip magnetic Yagi antenna and array with vertical polarization based on substrate integrated waveguide. Prog Electromagn Research C 59:135–141
16. Berdnik SL, Katrich VA, Nesterenko MV, Penkin YuM, Penkin DYu (2015) Radiation and scattering of electromagnetic waves by a multielement vibrator–slot structure in a rectangular waveguide. IEEE Trans Antennas Propag 63(9):4256–4259
17. Penkin DYu, Berdnik SL, Katrich VA, Nesterenko MV, Kijko VI (2013) Electromagnetic fields excitation by a multielement vibrator-slot structures in coupled electrodynamics volumes. Prog Electromagn Res B 49:235–252
18. Hoseini M, Forooraghi K, Abdolali F (2022) Mutual impedance computation of a waveguide slot-fed arbitrary patch using combined conventional moment method and equivalent electric and magnetic dipole method. Prog Electromagn Res C 123:61–73
19. Garufo A, Carluccio G (2022) Active admittance and impedance of hexagonally sampled dipole and slot connected arrays. IEEE Antennas Wireless Propag Lett 21:341–345
20. Penkin Yu, Katrich V, Nesterenko M, Berdnik S, Dumin O (2020) Dual-band combined vibrator-slot radiating structures. In: Proceedings of the 2020 IEEE Ukrainian Microwave Week. Kharkiv, Ukraine pp 149–153
21. Penkin YuM, Katrich VA, Nesterenko MV (2019) Two-frequency operating mode of antenna arrays with radiators of Clavin type and switching vibrator and slot elements. Prog Electromagn Res M 87:171–178
22. Berdnik SL, Katrich VA, Nesterenko MV, Penkin YuM, Dumin OM (2020) Yagi-Uda combined radiating structures of centimeter and millimeter wave bands. Prog Electromagn Res M 93:89–97

Chapter 6
Impedance Synthesis for Vibrator Antenna Arrays

6.1 Impedance Synthesis for Linear Radiating Vibrators Array

In this Section a new method of impedance synthesis of antenna array radiation fields based on a single methodological conception is presented. At first, an approximate solution for the current in the thin vibrator with variable impedance was obtained using the partial averaging operation of the integral–differential equation. The variable impedance of the vibrator was taken into account in the form of an integral coefficient averaged along the vibrator length. The approach turns out to be common for radiators with impedance coatings of different configurations and/or different distributions of lumped impedances. It was established that the shape of the vibrator RP does not depend on the form of the impedance distribution function, and it is determined only by the averaged value of the impedance distribution along the vibrator axis. The solution shows that the impedance coating of a symmetrical thin vibrator excited at the center by the voltage δ-generator affects the shape of the RP in the wave zone, and the effect is directly proportional to the small natural parameter of the problem. The synthesis problem of the radiator impedances for the spatial scanning of the RP was solved for the linear vibrator array. The analytical solution of the problem was obtained for the equidistant array of symmetric vibrators with equal excitation currents. The possibility of changing the RP shape over a wide range by varying the intrinsic complex impedances of the vibrators is demonstrated for an equidistant linear array consisting of 5 half-wave vibrators located at a distance of one eighth wavelength from each other in the free space.

M. V. Nesterenko et al., *Vibrator and Slot Antenna Arrays for Modern Applications*, Lecture Notes in Electrical Engineering 1244,
https://doi.org/10.1007/978-3-031-68346-6_6

6.1.1 Analytical Current Presentation on a Vibrator with Variable Surface Impedance

The problem of AA synthesis consists in finding a form, dimensions and amplitude-phase distribution (APD) of currents in the array elements by using a RP in the wave zone [1]. Since the shape of the antenna array is supposed to be known in most cases, the problem solution is reduced to determining numbers of the radiators, distances between them and complex current amplitudes in them. At present, the problems of current APD synthesis is fairly well understood theoretically, but, unfortunately, cannot help practical implementation of a particular antenna array design. Alternative problems of an AA constructive synthesis, which were apparently first considered in [2], are still poorly covered in the literature, since they are much more complex than the problems in a conventional formulation. Taking into account constraints arising in practice, these problems are usually inverse boundary value problems, which can often be reduced to non-linear and multiexternal problems. Therefore, the problem can be solved by sophisticated numerical methods, whose results cannot be directly applied as a basis for array control algorithms required, e.g., for spatial scanning of the RP in the wave zone. Therefore, the solutions of the constructive synthesis problems in a spatial frequency representation can be obtained by using analytical or numerical analytic methods. In the latter case, the current practice is to obtain the satisfactory results by using analytical or numerical and analytical methods.

On the other hand, the problems of constructive synthesis are actual for issues of electromagnetic compatibility of electronic equipment. In that case, field profiles in the wave and near-field radiation zones are studied (see for example, [3–6]). Complex structures containing several integrated circuits can often be modelled based on the use of dipole arrays. The electromagnetic radiation in the near zone can be defined by using direct numerical methods developed for frequency-time domain. However, extrapolation of these methods to simulation of fields in the wave zone may encounter considerable difficulties [4].

The conception of impedance synthesis for linear vibrator array has aroused from the analysis of impedance vibrator arrays [7, 8], and it can be considered as improved method for synthesis of antenna arrays with reactive loads [9, 10].

Consider a problem of electromagnetic wave excitation by a thin impedance vibrator in infinite medium with material parameters (ε_1, μ_1). The vibrator intrinsic impedance is supposed to be an arbitrary complex-valued function of the vibrator length. The initial integral equation of the problem satisfying a boundary condition on the vibrator surface S can be written as (Chap. 2)

$$\frac{1}{i\omega\varepsilon_1}(\text{grad div} + k_1^2) \int\limits_S \hat{G}(\vec{r}, \vec{r}')\vec{J}(\vec{r}')d\vec{r}' = -\vec{E}_0(\vec{r}) + z_i(\vec{r})\vec{J}(\vec{r}) \tag{6.1}$$

where $z_i(\vec{r})$ is intrinsic linear impedance ([Ohm/m]), $\vec{E}_0(\vec{r})$ is field of extraneous sources, $\hat{G}(\vec{r},\vec{r}\,')$ is tensor Green's function of a free space for electric vector potential $k_1 = k\sqrt{\varepsilon_1\mu_1}$, $k = 2\pi/\lambda$, λ is wavelength in a free space.

In the thin wire approximations, Eq. (6.1) can be reduced to an integral equation with a quasi-one-dimensional kernel (2.16)

$$\left(\frac{d^2}{ds^2} + k_1^2\right)\int\limits_{-L}^{L} J(s')\frac{e^{-ik_1R(s,s')}}{R(s,s')}ds' = -i\omega\varepsilon_1 E_{0s}(s) + i\omega\varepsilon_1 z_i(s)J(s), \tag{6.2}$$

where L is the vibrator half-length, $R(s,s') = \sqrt{(s-s')^2 + r^2}$, r is the vibrator radius, and $E_{0s}(s)$ is a projection of extraneous source fields at the vibrator axis. We will solve Eq. (6.2) by a small parameter method [11]. Let us isolate a logarithmic singularity of Eq. (6.2) kernel with the help of an artificial technique (Chap. 2, remind the reader)

$$\int\limits_{-L}^{L} J(s')\frac{e^{-ik_1R(s,s')}}{R(s,s')}ds' = \Omega(s)J(s) + \int\limits_{-L}^{L}\frac{J(s')e^{-ik_1R(s,s')} - J(s)}{R(s,s')}ds' \tag{6.3}$$

Here

$$\Omega(s) = \int\limits_{-L}^{L}\frac{ds'}{\sqrt{(s-s')^2 + r^2}} = \Omega + \gamma(s) \tag{6.4}$$

where $\gamma(s) = \ln\frac{[(L+s)+\sqrt{(L+s)^2+r^2}\,][(L-s)+\sqrt{(L-s)^2+r^2}\,]}{4L^2}$ is a function which is equal to zero at the vibrator center and reaches maximal values at the vibrator end The current at the vibrator ends is zero since the boundary conditions $J(\pm L) = 0$ are fulfilled, and $\Omega = 2\ln\frac{2L}{r}$ is the large parameter. Then, taking into account (6.3), Eq. (6.2) can be converted to the integral–differential equation for the vibrator electric current

$$\begin{aligned}&\frac{d^2J(s)}{ds^2} + k_1^2J(s)\\&= \alpha\{i\omega\varepsilon_1 E_{0s}(s) + F[s, J(s)] - i\omega\varepsilon_1 z_i(s)J(s)\}\end{aligned} \tag{6.5}$$

where $\alpha = \frac{1}{2\ln[r/(2L)]}$ is the natural small parameter of the problem ($|\alpha| \ll 1$). A functional

$$F[s, J(s)] = -\frac{dJ(s')}{ds'} \frac{e^{-ik_1 R(s,s')}}{R(s, s')}\Bigg|_{-L}^{L} + \left[\frac{d^2 J(s)}{ds^2} + k_1^2 J(s)\right]\gamma(s)$$
$$+ \int_{-L}^{L} \frac{\left[\frac{d^2 J(s')}{ds'^2} + k_1^2 J(s')\right] e^{-ik_1 R(s,s')} - \left[\frac{d^2 J(s)}{ds^2} + k_1^2 J(s)\right]}{R(s, s')} ds' \qquad (6.6)$$

defines the vibrator self-field in the spatial domain. Let us introduce a denomination

$$\begin{aligned}\tilde{k}^2(s) &= k_1^2\left[1 + i\alpha\omega\varepsilon_1 z_i(s)/k_1^2\right] \\ &= k_1^2\left[1 + i2\alpha\overline{Z}_S(s)/(\mu_1 kr)\right],\end{aligned} \qquad (6.7)$$

where

$$\overline{Z}_S(s) = 2\pi r z_i(s)/Z_0 \qquad (6.8)$$

is distribution of surface impedance, normalized to the characteristic impedance of the medium $Z_0 = \sqrt{\mu_1/\varepsilon_1}$ [Ohm]. Then, Eq. (6.5) can be written as

$$\frac{d^2 J(s)}{ds^2} + \tilde{k}^2(s) J(s) = \alpha\{i\omega\varepsilon_1 E_{0s}(s) + F[s, J(s)]\}. \qquad (6.9)$$

The integral–differential Eq. (6.9) has a variable parameter $\tilde{k}(s)$, therefore it cannot be solve by methods proposed for vibrators with a constant impedance [8, 11]. However, an approximate analytical solution of Eq. (6.9) can be obtained since the parameter $\tilde{k}(s)$ in the expression (6.7) has two terms and the second term is proportional to a small parameter α. However, an approximate analytical solution of Eq. (6.9) can be obtained, since the parameter $\tilde{k}(s)$ in the expression (6.7) has two terms and the second term is proportional to the small parameter α. Therefore, it would be appropriate to approximately represent the vibrator impedance $\overline{Z}_S(s)$ by its mean value along the vibrator length. Then Eq. (6.9) can be written as

$$\frac{d^2 J(s)}{ds^2} + \tilde{k}_m^2 J(s) = \alpha\{i\omega\varepsilon_1 E_{0s}(s) + F[s, J(s)]\} \qquad (6.10)$$

where $\tilde{k}_m^2 = k_1^2\left(1 + \frac{i2\alpha}{\mu_1 kr}\frac{1}{2L}\int_{-L}^{L}\overline{Z}_S(s)ds\right)$ is mean value of the coefficient $\tilde{k}(s)$. The solution of Eq. (6.10) can be called the first approximation to the solution of Eq. (6.9) [12, 13], Appendix B.

The integral–differential Eq. (6.10), can be solved by the series expansion in the small parameter α

$$\begin{aligned}
&\frac{d^2 J_1(s)}{ds^2} + \tilde{k}_m^2 J_1(s) = i\omega\varepsilon_1 E_{0s}(s)\,,\\
&\frac{d^2 J_2(s)}{ds^2} + \tilde{k}_m^2 J_2(s) = F[s, J_1(s)]\,,\\
&\dots\\
&\frac{d^2 J_n(s)}{ds^2} + \tilde{k}_m^2 J_n(s) = F[s, J_{n-1}(s)]\,.
\end{aligned} \tag{6.11}$$

The solution of integral–differential equations at each step should be determined by taking into account the boundary conditions for the currents $J_1(\pm L) = J_2(\pm L) = \dots = J_n(\pm L) = 0$. The general solution for the current can be obtained as an expansion in powers of the small parameter α

$$J(s) = \alpha J_1(s) + \alpha^2 J_2(s) + \dots + \alpha^n J_n(s) \tag{6.12}$$

If the system (6.11) is formally supplemented by homogeneous equation for the zero-approximation of the current J_0, it would have a solution $J_0(s) = C_1 \cos \tilde{k}_m s + C_2 \sin \tilde{k}_m s$, that does not depend upon the exciting field $E_{0s}(s)$.

If any losses are present in the medium or vibrator, the trigonometric functions in the solution become complex and, hence, cannot be equal to zero for any arguments. Then to satisfy the boundary conditions $J_0(\pm L) = 0$, the constants C_1 and C_2 should be equal to zeroes. Therefore, identities $J_0 \equiv 0$ and $F[s, J_0(s)] \equiv 0$ hold for any vibrator length, the first approximation of the vibrator current becomes equal to

$$J(s) \gg \alpha J_1(s) = -\alpha \frac{i\omega\varepsilon_1 / \tilde{k}_m}{\sin 2\tilde{k}_m L} \times \begin{cases} \sin \tilde{k}_m (L - s) \int\limits_{-L}^{s} E_{0s}(s') \sin \tilde{k}_m (L + s') ds' \\ + \sin \tilde{k}_m (L + s) \int\limits_{s}^{L} E_{0s}(s') \sin \tilde{k}_m (L - s') ds', \end{cases} \tag{6.13}$$

and one can see that it does not depend upon the vibrator eigenfield (6.6).

If the vibrator is excited in the middle by a δ-generator with the voltage amplitude V_0, i.e., $E_{0s}(s') = V_0 \delta(s - s')$, the formula (6.13) can be reduces to

$$\begin{aligned}
J(s) \gg \alpha J_1(s) &= -\frac{i\omega\varepsilon_1 \alpha V_0}{2\tilde{k}_m \cos \tilde{k}_m L} \sin \tilde{k}_m (L - |s|)\\
&= J_0 \sin \tilde{k}_m (L - |s|) = J_0 f(s)
\end{aligned} \tag{6.14}$$

As can be seen, the approximate solution for the current on the vibrator with the variable surface impedance (6.14) automatically ensures the vibrator current continuity is preserved even for piecewise-constant impedance distributions. A characteristic property of the solution consists in that the variable impedance is taken into account as averaged integral coefficient $\frac{1}{2L}\int_{-L}^{L}\overline{Z}_S(s)ds$. Unlike the current distribution functions known from the literature [14, 15], the function $f(s)$ in Eq. (6.14) contains the information about the impedance distribution along vibrator $\overline{Z}_S(s)$. Moreover, this approach is valid for vibrators with one or several local inclusions of lumped impedance loads.

6.1.2 Radiation Pattern of a Symmetric Vibrator with Variable Surface Impedance

When the vibrator current and RP are known, the vibrator radiation field, i.e., the dependence of the radiated field intensity upon the direction can be obtained by summation of fields produced by all vibrator sections. The field in the wave zone at a distance R ($R \gg \lambda_1$, λ_1 is the wavelength in the medium) induced by the current element $J(s')\mathrm{d}s'$ can be calculated by neglecting terms, which are decreasing with distance faster than $1/R$. Under this condition, the field induced in the wave zone by a symmetrical vibrator whereby the current (6.14) flows can be represented, as

$$E = E_\theta = \frac{60\pi i}{\lambda_1}\int\limits_{-L}^{L}\sin\theta_S\frac{e^{-ik_1R_S}}{R_S}J(s')\,\mathrm{d}s'. \tag{6.15}$$

The formula (6.15) is written in a spherical coordinate system whose axis coincides with the vibrator axis (Fig. 6.1). Since the distance R_S can be expanded in powers of the variable s' so that $R_S = R - s'\cos\theta + \frac{(s')^2\sin^2\theta}{R} - \ldots$, we can assume that $R_S \approx R$ and $R_S \approx R - s'\cos\theta$ in the denominator and exponent of the formula (6.15), respectively. If the distance to the observation point is large, we can assume that $\sin\theta_S = \sin\theta$. Taking into account the above approximation, we can rewrite (6.15) as

$$E_\theta = \frac{60\pi i}{\lambda_1}\cdot\frac{e^{-ik_1R}}{R}\sin\theta\int\limits_{-L}^{L}J(s')e^{ik_1s'\cos\theta}\,\mathrm{d}s'. \tag{6.16}$$

Analysis of the expression (6.16) shows that small deviations of the current distribution $J(s')$ from the exact shape are averaged during the integration and do not significantly affect the spatial shape of the RP. Therefore, we can use the approximate current distribution (6.14). After substituted (6.14) into (6.16), we obtained the following formula

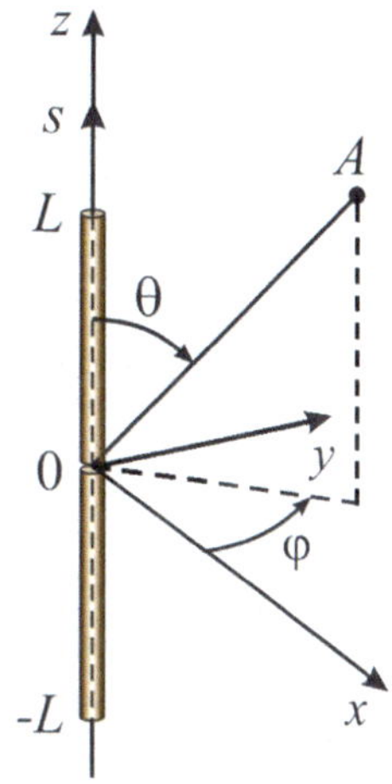

Fig. 6.1 The problem geometry and accepted notations

$$E_\theta = \frac{120\pi i}{\lambda_1} \cdot \frac{e^{-ik_1 R}}{R} J_0 \frac{\cos(k_1 L\cos\theta) - \cos\left(\tilde{k}_m L\right)}{\tilde{k}_m^2 - (k_1\cos\theta)^2} \tilde{k}_m \sin\theta. \tag{6.17}$$

Before proceeding to the synthesis problem, we perform some identical transformations in (6.17). Let us introduce the notation $\overline{\beta} = \frac{1}{2L\mu_1 kr}\int_{-L}^{L} \overline{Z}_S(s)ds$. Then, taking into account the relations (6.7), the expression $\tilde{k}_m^2 - (k_1\cos\theta)^2$ in the denominator of (6.17) can be written as $k_1^2\left(1 + 2i\alpha\overline{\beta}\right) - k_1^2\cos^2\theta = k_1^2\left(\sin^2\theta + 2i\alpha\overline{\beta}\right)$. Since the parameter α is small, we can write $\tilde{k}_m^2 = k_1^2\left(1 + 2i\alpha\overline{\beta}\right) \approx k_1^2\left(1 + 2i\alpha\overline{\beta} - \alpha^2\overline{\beta}^2\right) = k_1^2\left(1 + i\alpha\overline{\beta}\right)^2$ and, hence, $\tilde{k}_m = k_1\left(1 + i\alpha\overline{\beta}\right)$. Thus, the multiplier $\frac{\cos(k_1 L\cos\theta) - \cos\left(\tilde{k}_m L\right)}{\tilde{k}_m^2 - (k_1\cos\theta)^2}\tilde{k}_m\sin\theta$ in (6.17) can be represented as

$$\begin{aligned} &\frac{\cos(k_1 L\cos\theta) - \cos\left(\tilde{k}_m L\right)}{\tilde{k}_m^2 - (k_1\cos\theta)^2}\tilde{k}_m\sin\theta \\ &= \frac{\left(1 + i\alpha\overline{\beta}\right)\sin\theta}{k_1\left(\sin^2\theta + 2i\alpha\overline{\beta}\right)}\left[\cos(k_1 L\cos\theta) - \cos\left(k_1 L + i\alpha\overline{\beta}k_1 L\right)\right]. \end{aligned} \tag{6.18}$$

Taking into account the series expansions of trigonometric functions $\sin x = x - {x^3}/{3!} + {x^5}/{5!} \mp \ldots$ and $\cos x = 1 - {x^2}/{2!} + {x^4}/{4!} \mp \ldots$, we can write the expression (6.18) with an accuracy up to the parameter α as

$$\begin{aligned} &\frac{\left(1 + i\alpha\overline{\beta}\right)\sin\theta}{k_1\left(\sin^2\theta + 2i\alpha\overline{\beta}\right)}\left[\cos(k_1 L\cos\theta) - \cos(k_1 L) + i\alpha\overline{\beta}k_1 L\sin(k_1 L)\right] \\ &= \frac{\left[Fc(\theta) + i\alpha\overline{\beta}(Fc(\theta) + k_1 L\sin(k_1 L))\right]\sin\theta}{k_1\left(\sin^2\theta + 2i\alpha\overline{\beta}\right)} \end{aligned}$$

$$
\begin{aligned}
&= \frac{1}{k_1 \sin^3\theta}\left[Fc(\theta) + i\alpha\overline{\beta}(Fc(\theta) + k_1 L \sin(k_1 L))\right]\left(\sin^2\theta - 2i\alpha\overline{\beta}\right)\\
&= \frac{1}{k_1 \sin\theta}\left[Fc(\theta) - i\alpha\overline{\beta}\left(Fc(\theta)\frac{1+\cos^2\theta}{\sin^2\theta} - k_1 L \sin(k_1 L)\right)\right], \qquad (6.19)
\end{aligned}
$$

where $Fc(\theta) = \cos(k_1 L \cos\theta) - \cos(k_1 L)$. The final expression for the electric field E_θ can be presented as

$$
\begin{aligned}
E_\theta &= \frac{120\pi}{\lambda_1}\cdot\frac{e^{-ik_1 R}}{R}\cdot\frac{J_0}{k_1 \sin\theta}\left[iFc(\theta) + \alpha\overline{\beta}\right.\\
&\left.\left(Fc(\theta)\frac{1+\cos^2\theta}{\sin^2\theta} - k_1 L \sin(k_1 L)\right)\right]\\
&= \frac{e^{-ik_1 R}}{R}\cdot\frac{60 J_0}{\sqrt{\varepsilon_1\mu_1}\sin\theta}\left[iFc(\theta) + \alpha\overline{\beta}\right.\\
&\left.\left(Fc(\theta)\frac{1+\cos^2\theta}{\sin^2\theta} - k_1 L \sin(k_1 L)\right)\right]. \qquad (6.20)
\end{aligned}
$$

The first term in square brackets of the expression (6.20) determines the RP of a perfectly conducting vibrator, and the second term defines contribution of the impedance vibrator coating to the radiation field. If $\overline{Z}_S(s) = 0$, the expression (6.20) reduces to the well-known formula for the RP of a symmetric perfectly conducting vibrator in the wave zone (see, for example, [8]).This fact confirms a correctness of the approximation.

Formula (6.20) directly allows us to make two important conclusions. Firstly, the impedance coating of a thin symmetrical vibrator excited in the center by a voltage generator influences the RP in the wave zone. The effect is directly proportional to the product of the parameter $\alpha = \frac{1}{2\ln[r/(2L)]}$ and the mean value of the variable impedance $\frac{1}{2L}\int_{-L}^{L}\overline{Z}_S(s)ds$, and it is inversely proportional to the coefficient $(\mu_1 kr)$. Consequently, the shape of the vibrator RP cannot be varied in wide limits by applying the impedance coating to the vibrator. Secondly, the form of the RP does not depend upon the function $\overline{Z}_S(s)$ if the mean value of the impedance is $\frac{1}{2L}\int_{-L}^{L}\overline{Z}_S(s)ds$.

6.1.3 Problem of Impedance Synthesis of a Linear Vibrators Array

Let us consider a linear array of N symmetrical impedance vibrators (Fig. 6.2). The array spacing is d, all vibrators are of length $2L$ and are characterized by the variable impedances $\overline{Z}_{Sn}(s)$. If the vibrators are resonantly tuned by selecting intrinsic resistors of the δ-generators, mutual influences between them can be neglected, and the total field radiated by the array becomes equal to the sum of the field radiated by all vibrators, taking into account the phases of waves arriving to an observation

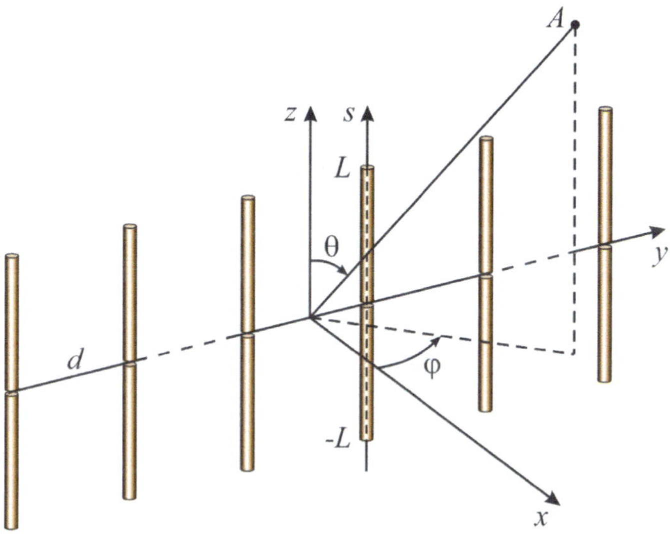

Fig. 6.2 A linear vibrator array

point A. The difference of propagation paths between neighboring vibrators and the observation point in the wave zone is equal to $d \sin\theta \cos\varphi$.

Let us introduce a notation $u = kd \sin\theta \cos\varphi$. Then, taking into account (6.20), we can write the expression for the resulting field in the free space as [7, 10]

$$E_\theta = 60\frac{e^{-ikR}}{R}e^{-i(N-1)u/2}$$
$$\left\{\begin{array}{l} \frac{J_1}{\sin\theta}\left[iFc(\theta) + \alpha\overline{\beta}_1\left(Fc(\theta)\frac{1+\cos^2\theta}{\sin^2\theta} - kL\sin(kL)\right)\right] \\ +\frac{J_2 e^{iu}}{\sin\theta}\left[iFc(\theta) + \alpha\overline{\beta}_2\left(Fc(\theta)\frac{1+\cos^2\theta}{\sin^2\theta} - kL\sin(kL)\right)\right] \\ +\frac{J_3 e^{i2u}}{\sin\theta}\left[iFc(\theta) + \alpha\overline{\beta}_3\left(Fc(\theta)\frac{1+\cos^2\theta}{\sin^2\theta} - kL\sin(kL)\right)\right]\ldots \\ +\frac{J_N e^{i(N-1)u}}{\sin\theta}\left[iFc(\theta) + \alpha\overline{\beta}_N\left(Fc(\theta)\frac{1+\cos^2\theta}{\sin^2\theta} - kL\sin(kL)\right)\right] \end{array}\right\}, \qquad (6.21)$$

where $\overline{\beta}_n = \frac{1}{2Lkr}\int_{-L}^{L}\overline{Z}_{Sn}(s)ds$ and $\{J_n\}$ is the current of the n-th vibrator ($n = 1, 2 \ldots N$).

If the vibrators are perfectly conducting and the vibrator currents are equal, i.e., $J_1 = J_2 = \cdots = J_N = J_0$, the expression (6.21) can be reduced to

$$E_\theta = 60iJ_0 \frac{e^{-ikR}}{R} \frac{Fc(\theta)}{\sin\theta} e^{-i(N-1)u/2} \\ \left\{1 + e^{iu} + e^{i2u} + \cdots + e^{i(N-1)u}\right\} \tag{6.22}$$

According to formula (6.22), the maximum of the array RP is reached in direction $\left(\theta = \frac{\pi}{2};\ \varphi = \frac{\pi}{2}\right)$ when $u = 0$. On the other hand, it is known [7, 10] that if the phase shift between neighboring vibrators is equal Δu, the maximum of the array RP is shifted in the direction $(\theta_{\max};\ \varphi_{\max})$ determined from the relation $\sin\theta_{\max}\cos\varphi_{\max} = \Delta u\big/kd$ [3, 6]. In this case, we can write

$$E_\theta = 60iJ_0 \frac{e^{-ikR}}{R} \frac{Fc(\theta)}{\sin\theta} e^{-i(N-1)(u-\Delta u)/2} \\ \left\{1 + e^{i(u-\Delta u)} + e^{i2(u-\Delta u)} + \cdots + e^{i(N-1)(u-\Delta u)}\right\}. \tag{6.23}$$

Since amplitudes of the vibrator currents are arbitrary, we rename $J_0 e^{i(N-1)\Delta u/2}$ by J_0, perform several identical transformations in (6.21), and obtain

$$E_\theta = 60iJ_0 \frac{e^{-ikR}}{R} \frac{Fc(\theta)}{\sin\theta} \\ e^{-i(N-1)(u-\Delta u)/2} \left\{ \begin{array}{l} \left[1 - i\alpha\overline{\beta}_1 \left(\frac{1+\cos^2\theta}{\sin^2\theta} - \frac{kL\sin(kL)}{Fc(\theta)}\right)\right] \\ +e^{iu}\left[1 - i\alpha\overline{\beta}_2 \left(\frac{1+\cos^2\theta}{\sin^2\theta} - \frac{kL\sin(kL)}{Fc(\theta)}\right)\right] \\ +e^{i2u}\left[1 - i\alpha\overline{\beta}_3 \left(\frac{1+\cos^2\theta}{\sin^2\theta} - \frac{kL\sin(kL)}{Fc(\theta)}\right)\right] \ldots \\ +e^{i(N-1)u}\left[1 - i\alpha\overline{\beta}_N \left(\frac{1+\cos^2\theta}{\sin^2\theta} - \frac{kL\sin(kL)}{Fc(\theta)}\right)\right] \end{array} \right\}. \tag{6.24}$$

The expressions (6.23) and (6.24) become identical if

$$e^{-i(n-1)kd\sin\theta\cos\varphi} = \left[1 - i\alpha\overline{\beta}_n \left(\frac{1+\cos^2\theta}{\sin^2\theta} - \frac{kL\sin(kL)}{Fc(\theta)}\right)\right]\Bigg|_{\theta=\theta_{\max};\ \varphi=\varphi_{\max}} \tag{6.25}$$

for every $n = 1, 2 \ldots N$. The relations (6.25) for the predefined angles $(\theta_{\max};\ \varphi_{\max})$ can be used to uniquely find a solution vector for unknowns $\left\{\overline{\beta}_n\right\}$. Thus, we can affirm that the maximum of the antenna RP in the wave zone can be redirected to predefined position by synthesizing certain impedances on the array vibrators.

6.1.4 Numerical Results

Let us consider the vibrator array consisting of five ($N = 5$) symmetric half-wave $(2L = \lambda/2,\ kL = \pi)$ impedance vibrators with equal radii $(r = L/25 = \lambda/100)$. The problem can be formulated as searching such impedances $\overline{Z}_{Sn}(s)$ that the maximum of the RP in the main plane $\theta_{\max} = \pi/2$ would be in the direction of the predefined angle $\varphi_{\max}$. In this case, it is not difficult to verify that the relations (6.25) can be simplified and reduced to

$$e^{-i(n-1)kd\cos\varphi_{\max}} = 1 - i\alpha\overline{\beta}_n \tag{6.26}$$

Without losing the approach commonality, we assume that the complex vibrator impedances $\overline{Z}_{Sn}(s) = \overline{R}_{Sn} + i\overline{X}_{Sn}$ do not depend upon the variable s. Then, taking into account that $\overline{\beta}_n = \frac{\overline{R}_{Sn}+i\overline{X}_{Sn}}{kr}$, the relations (6.26) can be represented in the following form

$$\cos(kd(n-1)\cos\varphi_{\max}) - i\sin(kd(n-1)\cos\varphi_{\max}) \\ = 1 - i\alpha\frac{\overline{R}_{Sn} + i\overline{X}_{Sn}}{kr} \tag{6.27}$$

Equating the real and imaginary components of the left and right sides of Eq. (6.27), we finally obtain the formulas ready for the numeric computing

$$\begin{aligned} \overline{R}_{Sn} &= 2kr\ln[r/(2L)]\sin[kd(n-1)\cos\varphi_{\max}], \\ \overline{X}_{Sn} &= -2kr\ln[r/(2L)]\{1 - \cos[kd(n-1)\cos\varphi_{\max}]\}. \end{aligned} \tag{6.28}$$

The formulas (6.25) and (6.28) are valid for any number of vibrators in the array and for arbitrary distances between the vibrators. However, in general, we cannot guarantee that the impedances computed in such a way and defined here as effective physical quantities have the positive real parts, $\overline{R}_{Sn} \geq 0$. Namely, this condition, put forward from the energy considerations determines the possibility of practical realization of the impedance $\overline{Z}_{Sn} = \overline{R}_{Sn} + i\overline{X}_{Sn}$ as the vibrator intrinsic impedance. From physical considerations, we can assume that the conditions $\overline{R}_{Sn} \geq 0$ cab be satisfied for arrays with a single directional lobe. The study of more exact requirements for the array dimensions are beyond the scope of this article. As is known, the antenna array can have a single label RP oriented in the direction $\varphi_{\max}$ in the range of real angles under condition $\frac{d}{\lambda} < \frac{1}{1+|\cos\varphi_{\max}|}$ [10, 16]. Therefore, we selected the distance between the vibrators equal to $\lambda/8$, $(kd = \pi/4)$. As follows from formula (6.28), if the direction of the RP main lobe is defined by the angle $\varphi_{\max} = 90°$, the impedances $\overline{Z}_{Sn}$ should be equal to zeros. Estimated value of the impedance $\overline{Z}_{Sn}$ for $\varphi_{\max}$ equal to 100°, 110°, 120° and 130°are shown in Table 6.1.

The Table 6.1 shows that to direct the main lobe maximum at the angle other than 90°, the modulus of the impedances $|\overline{Z}_{Sn}|$ should be increased. Note, the impedance

Table 6.1 The estimated values of the impedances $\overline{Z}_{Sn}$ for various angles of radiation maxima

n	$\varphi_{\max} = 100°$		$\varphi_{\max} = 110°$		$\varphi_{\max} = 120°$		$\varphi_{\max} = 130°$	
	$\overline{R}_{Sn}$	$\overline{X}_{Sn}$	$\overline{R}_{Sn}$	$\overline{X}_{Sn}$	$\overline{R}_{Sn}$	$\overline{X}_{Sn}$	$\overline{R}_{Sn}$	$\overline{X}_{Sn}$
1	0.0	0.0	0.0	0.0	0.0	0.0	0.0	0.0
2	0.0081	0.089	0.031	0.172	0.066	0.246	0.107	0.306
3	0.032	0.175	0.121	0.323	0.246	0.426	0.382	0.479
4	0.072	0.255	0.258	0.432	0.492	0.492	0.705	0.443
5	0.124	0.327	0.424	0.487	0.737	0.426	0.934	0.213

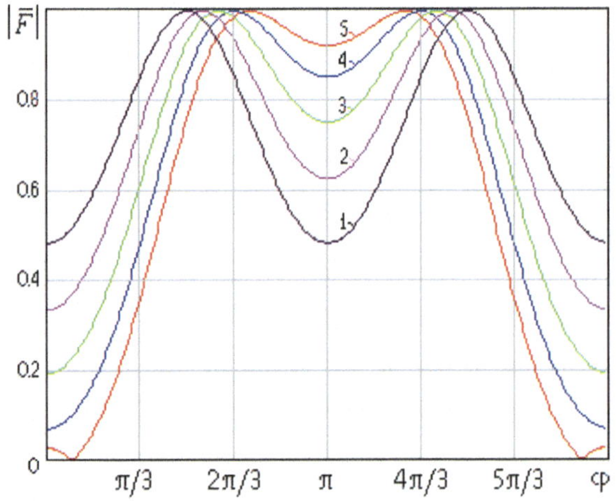

Fig. 6.3 The normalized RP of the linear impedance vibrator antenna array: 1 – $\varphi_{\max} = 90°$; 2 – $\varphi_{\max} = 100°$;3 – $\varphi_{\max} = 110°$; 4 – $\varphi_{\max} = 120°$; 5 – $\varphi_{\max} = 130°$

$\overline{Z}_{Sn}$ of the vibrator with the number $n = 1$ must be equal to zero for all angles $\varphi_{\max}$. The normalized RP, $|\overline{F}|$, obtained using (6.24) as function of the angle $\varphi \in [0, 2\pi]$ are plotted for the angle $\varphi_{\max}$ given in the Table 6.1. As can be seen from Fig. 6.3 the results of calculations confirm the possibility, to vary the shape of the RP by changing the intrinsic impedances of the vibrators. The radiation pattern can be varied over a wide range: from a quasi-homogeneous to table-like form, directed along the longitudinal axis of the array.

6.2 Impedance Synthesis for Two-Dimensional Radiating Vibrator Array

In Sect. 6.1 analytically solved the problem of impedance synthesis for a linear radiating vibrator array. In this Section the generalize methodology of the analytical solution of the of impedance synthesis problem widespread on the two-dimensional planar vibrator antenna arrays.

6.2.1 Formulation of the Impedance Synthesis Problem

Let us consider an antenna array consisting of several rows of linear symmetrical impedance vibrators located in a free space with material parameters (ε_1; μ_1). The array is placed in the plane ($x0z$) of the Cartesian coordinate system (x, y, z), which origin coincides with the array center. The geometry of the array is shown in Fig. 6.4. Let d_z be a distance between neighboring rows of vibrators, d_x be a distance between the centers of the vibrators in one row, N_z be a number of rows, and N_x be a number of vibrators in one row. We will also introduce a spherical coordinate system, whose polar axis coincides with the axis {0z}, and the angle φ is measured from the axis {0x}.

We assume that all vibrators $N = N_z \times N_x$ are of equal length $2L$ and are characterized by different variable impedances $\overline{Z}_{nm}(s)$, normalized to the wave impedance of the medium, $Z_0 = \sqrt{\mu_1 / \varepsilon_1}$ [Ohm], and integer indices $n \in [1, N_z]$ and $m \in [1, N_x]$. We also assume that the vibrator radii satisfy thin-wire approximations $\frac{\rho_{nm}}{2L} \ll 1$ and

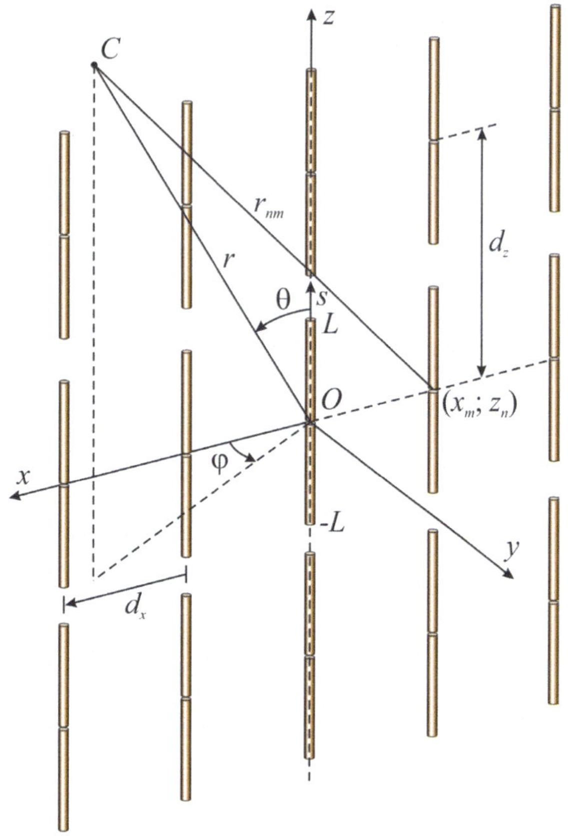

Fig. 6.4 The geometry of antenna array and accepted notations

$\frac{\rho_{nm}\sqrt{\varepsilon_1\mu_1}}{\lambda} \ll 1$, λ is the wavelength in free space. Let the vibrators be excited by monochromatic voltage-delta generators with predetermined amplitudes V_{nm} at the angular frequency ω. Expression for longitudinal electric currents on vibrators can be written as (Sect. 6.1)

$$J_{nm}(s) = -\frac{i\omega\varepsilon_1\alpha V_{nm}}{2\tilde{k}_{nm}\cos\tilde{k}_{nm}L}\sin\tilde{k}_{nm}(L-|s|)$$
$$= J_{nm}\sin\tilde{k}_{nm}(L-|s|) \tag{6.29}$$

where $\tilde{k}_{nm}$ are the wave vectors averaged over the vibrator length, defined by the relation $\tilde{k}^2_{nm}(s) = k_1^2\Big(1+\frac{i2\alpha_{nm}}{k_1\rho_{nm}}\cdot\frac{1}{2L}\int_{-L}^{L}\overline{Z}_{nm}(s)ds\Big)$, J_{nm} are the complex amplitudes of the vibrator currents, $\alpha_{nm} = \frac{1}{2\ln[\rho_{nm}/(2L)]}$ are natural small parameters, $k_1 = k\sqrt{\varepsilon_1\mu_1}$, $k = \omega/c$ are the wavenumbers, and $c \approx 2.998\cdot 10^{10}$ cm/s is the light velocity in vacuum. Taking into account the formula (6.29), we can write the main electric field component of the vibrator radiation with indices (n, m) in the wave zone as

$$E^{\theta}_{nm} = \frac{60\pi i}{\lambda}\cdot\frac{e^{-ik_1r_{nm}}}{r_{nm}}\sin\theta\int\limits_{-L}^{L}J_{nm}(s')e^{-ik_1s'\cos\theta}ds'$$
$$= \frac{120\pi i}{\lambda}\cdot\frac{e^{-ik_1r_{nm}}}{r_{nm}}J_{nm}\frac{\cos(k_1L\cos\theta)-\cos\big(\tilde{k}_{nm}L\big)}{\tilde{k}^2_{nm}-(k_1\cos\theta)^2}\tilde{k}_{nm}\sin\theta. \tag{6.30}$$

Since the far field components E_φ and E_r of the array are equal to zeros, the total radiation field of the array under condition that all vibrators are resonantly tuned by selecting intrinsic resistance of the δ-generators to compensate mutual influence of the vibrators, can be presented as the sum of the radiation fields induced by each vibrator in the observation point $C(r, \theta, \varphi)$ as follows

$$E_\theta(r,\theta,\varphi) = \frac{120\pi i}{\lambda}\sum_{n=1}^{N_z}\sum_{m=1}^{N_x}\frac{e^{-ik_1r_{nm}}}{r_{nm}}J_{nm}$$
$$\frac{\cos(k_1L\cos\theta)-\cos\big(\tilde{k}_{nm}L\big)}{\tilde{k}^2_{nm}-(k_1\cos\theta)^2}\tilde{k}_{nm}\sin\theta. \tag{6.31}$$

Thus, the impedance synthesis problem is reduced to finding impedances $\overline{Z}_{nm}(s)$ of each vibrator using the formula (6.31) that can steer the radiation maximum $(\theta_{\max}, \varphi_{\max})$ to the predefined direction in space.

6.2.2 Problem Solution

First, let us represent expression (6.30) for the radiation field of an isolated vibrator as

$$E_{nm}^{\theta} = \frac{e^{-ik_1 r_{nm}}}{r_{nm}} \cdot \frac{60J_{nm}}{\sqrt{\varepsilon_1\mu_1}\sin\theta}\Big[iFc(\theta) + \alpha_{nm}\overline{\beta}_{nm} \left(Fc(\theta)\frac{1+\cos^2\theta}{\sin^2\theta} - k_1 L\sin(k_1 L)\right)\Big], \tag{6.32}$$

where $Fc(\theta) = \cos(k_1 L\cos\theta) - \cos(k_1 L)$ and $\overline{\beta}_{nm} = \frac{1}{2Lk_1\rho_{nm}}\int_{-L}^{L}\overline{Z}_{nm}(s)ds$. Expression (6.32) is more convenient for further analysis, since the first term in square brackets determines the RP of an isolated perfectly conducting vibrator, while the second term defines variation of the vibrator field due to its impedance coating.

The expression for the field $E_{\theta}(r, \theta, \varphi)$ in the wave zone can be presented in the following form

$$E_{\theta}(\theta, \varphi) = e^{-i(N_z+1)u/2}e^{-i(N_x+1)v/2}\frac{60iFc(\theta)}{\sqrt{\varepsilon_1\mu_1}\sin\theta} \times \sum_{n=1}^{N_z}\sum_{m=1}^{N_x} J_{nm}\left[1 - i\alpha_{nm}\overline{\beta}_{nm}\left(\frac{1+\cos^2\theta}{\sin^2\theta} - \frac{k_1 L\sin(k_1 L)}{Fc(\theta)}\right)\right]e^{i(nu+mv)} \tag{6.33}$$

where $u = k_1 d_z \cos\theta$ and $v = k_1 d_x \sin\theta\cos\varphi$. This formula was derived taking into account the expression (6.32) and a difference of geometric paths from neighboring vibrators to the observation point $C(r, \theta, \varphi)$. If the vibrators are perfectly conducting and all vibrator currents are equal, $J_{nm} = J_0$, expression (6.33) is simplified to

$$E_{\theta}(\theta, \varphi) = e^{-i(N_z+1)u/2}e^{-i(N_x+1)v/2}\frac{60iFc(\theta)J_0}{\sqrt{\varepsilon_1\mu_1}\sin\theta}\sum_{n=1}^{N_z} e^{inu}\sum_{m=1}^{N_x} e^{imv} \tag{6.34}$$

This expression includes two multipliers in form of independent power series, which represent the RP with maxima directed along the axes $\{0z\}$ and $\{0x\}$, respectively. One can see that the maximum of the RP is achieved if $u = v = 0$, and it is directed along the axis $\{0y\}$, i.e., when $\theta = \pi/2$ and $\varphi = \pi/2$.

As is known from the general theory of antenna arrays (see, i.e., [16]), a linear phase shift between currents of the array radiators results in RP steering. If the phase shift between neighboring vibrators in the row and between adjacent rows are Δu and Δv, the direction of maximum array radiation ($\theta_{\max}$; $\varphi_{\max}$) can be determined by the following relations

$$\cos\theta_{\max} = \Delta u/k_1 d_z \quad \text{and} \quad \sin\theta_{\max}\cos\varphi_{\max} = \Delta v/k_1 d_x \tag{6.35}$$

Then, the expression (6.34) can be represented as

$$E_\theta(\theta,\varphi)=e^{-i(N_z+1)u/2}e^{-i(N_x+1)v/2}\frac{60iFc(\theta)J_0}{\sqrt{\varepsilon_1\mu_1}\sin\theta}\sum_{n=1}^{N_z}\sum_{m=1}^{N_x}e^{i(nu+mv)}e^{-i[(n-1)\Delta u+(m-1)\Delta v]} \tag{6.36}$$

The expressions (6.33) and (6.36) become identical if the relations

$$\left.\begin{aligned}&1-e^{-ik_1[(n-1)d_z\cos\theta+(m-1)d_x\sin\theta\cos\varphi]}\\&=i\alpha_{nm}\overline{\beta}_{nm}\left(\frac{1+\cos^2\theta}{\sin^2\theta}-\frac{k_1L\sin(k_1L)}{Fc(\theta)}\right)\end{aligned}\right|_{\theta=\theta_{\max};\,\varphi=\varphi_{\max}} \tag{6.37}$$

are valid for any n and m.

In essence, the relations (6.37) for given angles ($\theta_{\max};\ \varphi_{\max}$) can be interpreted as equations, which solution defines the matrix of unknowns $\{\overline{\beta}_{nm}\}$. It is easy to see that the formula (6.37) for the linear array (at $k_1=k$, $d_z=0$, $d_x=d$, $m=n$, $\alpha_{nm}=\alpha$, $\overline{\beta}_{nm}=\overline{\beta}_n$) can be reduced to the expression obtained in Sect. 6.1.3:

$$e^{-i(n-1)kd\sin\theta\cos\varphi}=\left.\left[1-i\alpha\overline{\beta}_n\left(\frac{1+\cos^2\theta}{\sin^2\theta}-\frac{kL\sin(kL)}{Fc(\theta)}\right)\right]\right|_{\theta=\theta_{\max};\,\varphi=\varphi_{\max}}$$

Without loss of generality, let us assume that all vibrators have different complex impedances $\overline{Z}_{nm}=\overline{R}_{nm}+i\overline{X}_{nm}$, constant along the longitudinal axes of the vibrators. Then Eqs. (6.37) can be represented in the following form

$$1-e^{-ik_1\gamma_{nm}}=i\alpha_{nm}\frac{\overline{R}_{nm}+i\overline{X}_{nm}}{k_1\rho_{nm}}\left(\frac{1+\cos^2\theta_{\max}}{\sin^2\theta_{\max}}-\frac{k_1L\sin(k_1L)}{Fc(\theta_{\max})}\right) \tag{6.38}$$

where $\gamma_{nm}=[(n-1)d_z\cos\theta_{\max}+(m-1)d_x\sin\theta_{\max}\cos\varphi_{\max}]$.

Since the parameters α_{nm} for lossless media are real, the final formula for the real and imaginary parts of the vibrator surface impedances can be written using (6.38) as

$$\left.\begin{aligned}\overline{R}_{nm}&=\frac{k_1\rho_{nm}\sin(k_1\gamma_{nm})}{\alpha_{nm}\left(\dfrac{1+\cos^2\theta_{\max}}{\sin^2\theta_{\max}}-\dfrac{k_1L\sin(k_1L)}{Fc(\theta_{\max})}\right)},\\ \overline{X}_{nm}&=\frac{-k_1\rho_{nm}[1-\cos(k_1\gamma_{nm})]}{\alpha_{nm}\left(\dfrac{1+\cos^2\theta_{\max}}{\sin^2\theta_{\max}}-\dfrac{k_1L\sin(k_1L)}{Fc(\theta_{\max})}\right)}.\end{aligned}\right\}\quad n=1,2\ldots N_x,\ m=1,2\ldots N_y. \tag{6.39}$$

The formulas (6.37) and (6.39), remain valid for arbitrary number of vibrators in the array and arbitrary distances between the vibrators.

6.2.3 Numerical Results

As an example, consider a two-dimensional array of symmetrical half-wave impedance vibrators radiating in the front half-space $0 \leq \theta \leq \pi$, $0 \leq \varphi \leq \pi$ (Fig. 6.4). The number of the array elements is 25 ($N_x = 5, N_z = 5$), the vibrator length and radius are $L = \lambda/4$ and $\rho_{nm} = L/75$. The problem consist in finding the impedances (6.39) of each vibrator such that the main lobe of the RP was oriented in a predefined direction ($\theta_{\max}$; $\varphi_{\max}$).

The results of numerical simulations are presented for two array with different geometric structures ($d_x = 0.5\lambda, d_z = 0.5\lambda$ and $d_x = 0.25\lambda, d_z = 0.5\lambda$) are shown in Figs. 6.5 and 6.6, where variations of the angle ($\theta_{\max}$; $\varphi_{\max}$) are shown symmetrically with respect to direction ($\theta = 90°$; $\varphi = 90°$) corresponding to main lobe orientation for the array with the perfectly conducting vibrators.

As can be seen from the figures, the simulation results confirm the possibility to vary the direction of the array RP maximum within the entire half-space, at least theoretically, by varying the intrinsic impedances of the vibrators. As expected, the main lobe width depends on the array aperture dimensions as seen from the observation angles, and therefore slightly varies during spatial scanning. As an example, Fig. 6.7 shows the surface and contour plots of RP of the antenna array.

It should be noted that formulas (6.39) and relations (6.37) are valid for any number of vibrators in the array and arbitrary distance between the vibrators. However, in general case, the impedances $\overline{Z}_{nm} = \overline{R}_{nm} + i\overline{X}_{nm}$ (6.39), thus calculated, cannot guarantee that the inequality $\overline{R}_{nm} \geq 0$, arising from energy considerations, holds. Such situation can be observed in Table 6.1, where the real and imaginary parts of the vibrator impedances calculated for the RP in Fig. 6.7 are shown.

The normalized RP of the array ($d_x = 0.25\lambda, d_z = 0.5\lambda$) are represented in the main planes in Fig. 6.8. The simulation results show that the real parts of the impedances remain positive if the angle $\theta_{\max} = 90°$ and the angle $\varphi_{\max}$ varies at the interval $\varphi_{\max} \in [60°, 120°]$. As example, calculated impedances are given in Table 6.2 for the angles $\theta_{\max} = 90°$ and $\varphi_{\max} = 60°$. As can be seen from the table, all positive values are observed in the case when the resulting difference in the phase currents of the extreme radiators in the lattice does not exceed 180°. If $\overline{R}_{nm} < 0$, the real part of impedances $\overline{R}_{nm}$ should be interpreted as effective physical quantities [17], which can be realized only as a result of electrodynamic interactions between the vibrators and additional scatterers. The impedances $\overline{Z}_{nm}$ with the real parts $\overline{R}_{nm} \geq 0$ which can be interpreted as traditional intrinsic impedances under conditions that the individual vibrators can be manufactured in the form of cylindrical structures with complex intrinsic structures [8]. Therefore, for practical applications, limiting scanning angles of the array RP should be determine for which the condition $\overline{R}_{nm} \geq 0$ is satisfied. From the point of view of physical interpretation, such studies can most obviously be carried out if the RP is scanned only in one plane.

The normalized RP of the array ($d_x = 0.25\lambda$, $d_z = 0.5\lambda$) are represented in the main planes in Fig. 6.8. The simulation results show that the real parts of the impedances remain positive if the angle $\theta_{\max} = 90°$ and the angle $\varphi_{\max}$ varies at

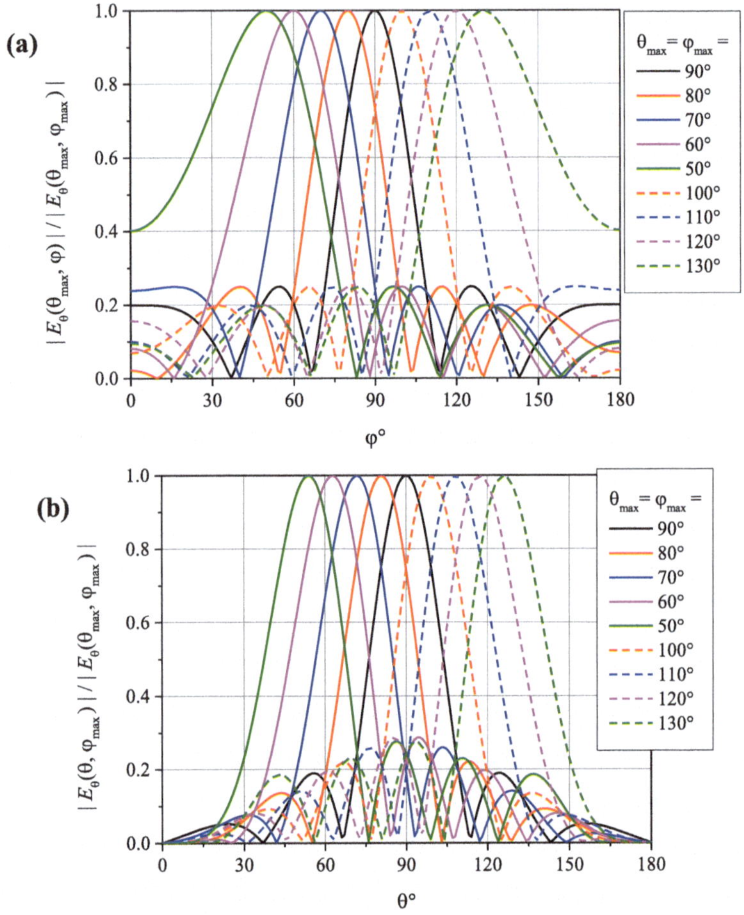

Fig. 6.5 The normalized RP of the array with the array periods $d_x = 0.5\lambda$, $d_z = 0.5\lambda$: **a** the angle $\theta_{\max}$ is fixed, **b** the angle $\varphi_{\max}$ is fixed

the interval $\varphi_{\max} \in [60°, 120°]$. As example, calculated impedances are given in Table 6.3 for the angles $\theta_{\max} = 90°$ and $\varphi_{\max} = 60°$. As can be seen from the table, all positive values are observed in the case when the resulting difference in the phase currents of the extreme radiators in the lattice does not exceed 180°. As can be seen from the Table 6.3, the positive $\overline{R}_{nm}$ are observed if in the phase shifts of the currents on the edge vibrators $\arg(J_{nm})$ do not exceed 180°. The simulation results have also shown that this condition is valid for other array geometries, particularly for the two-dimensional scanning.

The maximum angle of the RP deviation relative to the normal $\Delta\varphi_{\max} = 90° - \varphi_{\max}$ for $\overline{R}_{nm}$ to remain positive can be estimated by $\Delta\varphi_{\max} = \pm\arcsin\left(\frac{\lambda}{2d_x(N_x-1)}\right)$ or by $\Delta\varphi_{\max} \approx \pm\frac{\lambda}{2d_x(N_x-1)} \cdot \frac{180°}{\pi}$ valid for small deviations. This estimate relative

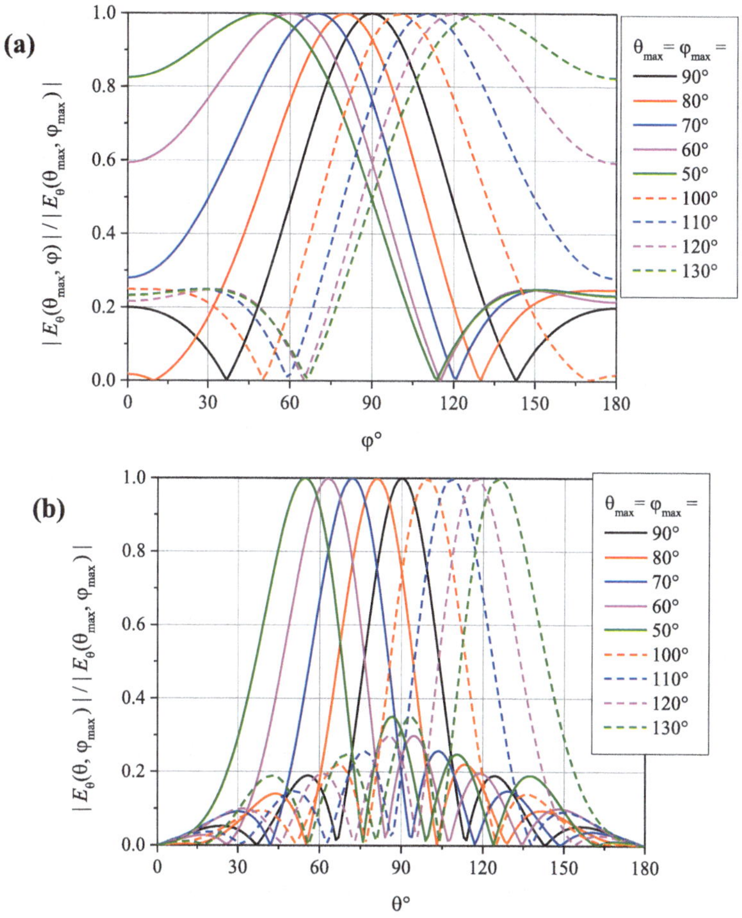

Fig. 6.6 The normalized RP of the array with the array periods $d_x = 0.25\lambda$, $d_z = 0.5\lambda$: **a** the angle θ_{max} is fixed, **b** the angle φ_{max} is fixed

to the main lobe width at half-power level $\Delta\varphi_{0.5} \approx \frac{\lambda}{d_x(N_x-1)} \cdot 51^\circ$ can be express as $\Delta\varphi_{max} \approx \pm 0.56 \cdot \Delta\varphi_{0.5}$.

6.3 Impedance Synthesis for Two-Dimensional Diffraction Vibrators Array

In this Section the problem of impedance synthesis of two-dimensional diffraction arrays of thin linear vibrators, whose geometric centers are located at the nodes of a flat rectangular grid with double periodicity is solved analytically. The problem

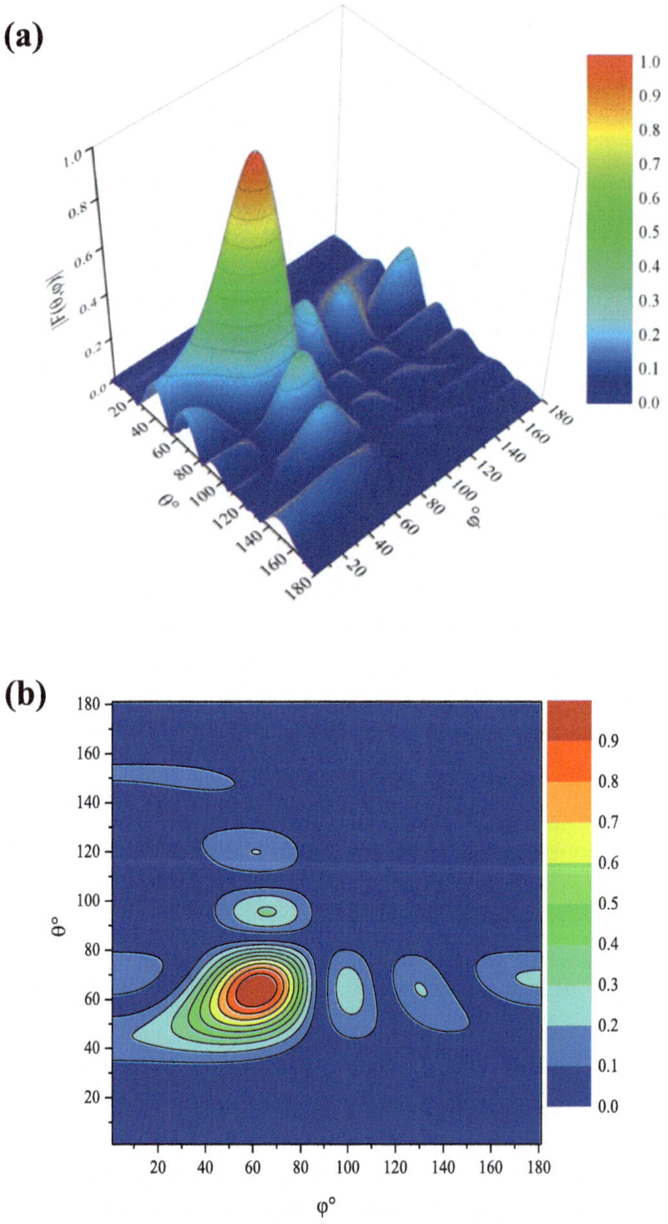

Fig. 6.7 The 3D RP of the antenna array with parameters: $d_z = d_x = 0.5\lambda$, $\theta_{\max} = \varphi_{\max} = 60°$: **a** surface plot, **b** contour plot

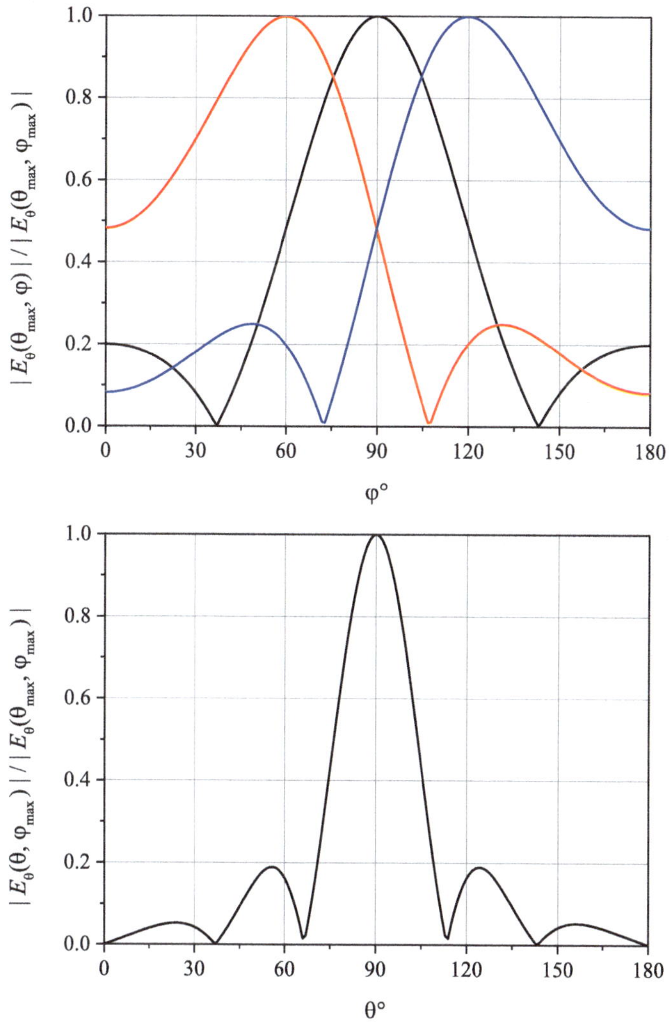

Fig. 6.8 RP in the main planes: $d_x = 0.25\lambda$, $d_z = 0.5\lambda$, $\theta_{\max} = 90°$ and $\varphi_{\max} = 60°$

Table 6.2 Calculated impedances for the RP shown in Fig. 6.7

$\overline{R}_{nm}$					$\overline{X}_{nm}$				
0	0.37	0.155	– 0.305	– 0.282	0	– 0.299	– 0.724	– 0.602	– 0.126
0.378	0.079	– 0.345	– 0.223	0.252	– 0.378	– 0.748	– 0.533	– 0.073	– 0.096
0	– 0.37	– 0.155	0.305	0.282	– 0.757	– 0.457	– 0.033	– 0.155	– 0.63
– 0.378	– 0.079	0.345	0.223	– 0.252	– 0.378	– 0.008	– 0.224	– 0.684	– 0.66
0	0.37	0.155	– 0.305	– 0.282	0	– 0.299	– 0.724	– 0.602	– 0.126

Table 6.3 Calculated values of the impedances

$\overline{R}_{nm}$					$\overline{X}_{nm}$					$\arg(J_{nm})$, [deg]				
0	0.26	0.368	0.26	0	0	− 0.108	− 0.368	− 0.628	− 0.735	0	− 45	− 90	− 135	− 180
0	0.26	0.368	0.26	0	0	− 0.108	− 0.368	− 0.628	− 0.735	0	− 45	− 90	− 135	− 180
0	0.26	0.368	0.26	0	0	− 0.108	− 0.368	− 0.628	− 0.735	0	− 45	− 90	− 135	− 180
0	0.26	0.368	0.26	0	0	− 0.108	− 0.368	− 0.628	− 0.735	0	− 45	− 90	− 135	− 180
0	0.26	0.368	0.26	0	0	− 0.108	− 0.368	− 0.628	− 0.735	0	− 45	− 90	− 135	− 180

is formulated as follows: the complex surface impedances of the vibrators should be determined which allows to steer the diffraction radiation maximum of the array to any predefined direction. The problem was solved under following assumptions: array is excited by a polarized plane wave, the RP of each vibrator element in the array coincides with that of an isolated radiator.

6.3.1 *Formulation of the Impedance Synthesis Problem*

The main advantage of antenna arrays is related to electric scanning of its RP [1, 18, 19]. The scanning can be continuous or discrete, depending on a control signal mode. Antenna arrays working in a discrete-switching mode are called step scanning antennas. The scanning antennas array can be also classified by an excitation mode of their elements. The analysis of antenna arrays with spatial excitation of an optical type known as diffraction array is still important for modern practice.

Of course, the diffraction arrays with discrete-switching can be implemented by using different types of elements, including impedance vibrators. In this case, there exists a possibility to use a surface impedance of vibrator elements as additional factor to control the array RP by varying electrical lengths of the vibrators and, therefore, the amplitude-phase distribution over the array aperture [7, 8]. It is from these positions, the authors studied the one-dimensional antenna vibrators array in [20], where the application of impedance synthesis for controlling scanning angles of the array RP maximum was justified. Moreover, the problem of an equidistant linear antenna array with symmetric vibrator excitation by voltage δ-generators was solved analytically.

Let us consider a 2D diffraction array of impedance vibrators shown in Fig. 6.9. The array is located in infinite medium with material parameters $(\varepsilon_1; \mu_1)$ in the plane $\{x0z\}$ of Cartesian coordinate system (x, y, z). The parameters of the diffraction array are: N_z is number of rows, N_x is number of vibrators in a row, d_z is the distance between adjacent rows, and d_x is the distance vibrators in the row. Let us also introduce spherical coordinates (r, θ, φ), which polar axis coincides with the axis $\{0z\}$, and the angle φ is measured from the axis $\{0x\}$.

Without losing generality, we assume that the length of all $N = N_z \times N_x$ vibrators is $2L$ (than $dz > 2L$), the radius ρ_{nm} of the vibrator with indices nm ($n \in [1, N_z]$, $m \in [1, N_x]$) satisfy the thin-wire approximation $\frac{\rho_{nm}}{2L} \ll 1$, $\frac{\rho_{nm}\sqrt{\varepsilon_1\mu_1}}{\lambda} \ll 1$ (λ is the free space wavelength). Each vibrator is characterized by the constant impedance $\overline{Z}_{nm}$ normalized to the medium wave impedance $Z_0 = \sqrt{\mu_1/\varepsilon_1}$ [Ohm]. Let us consider an array excitation by a monochromatic plane polarized wave $E_z = E_0 e^{ik_1 y}$ ($k_1 = k\sqrt{\varepsilon_1\mu_1}$),incident normally on the array from a half-space $y \leq 0$. The longitudinal electric currents $J_{nm}(s)$ on the vibrator elements can be represented by the zero approximation relative to the first degree of a natural small parameter α [8], Chap. 2

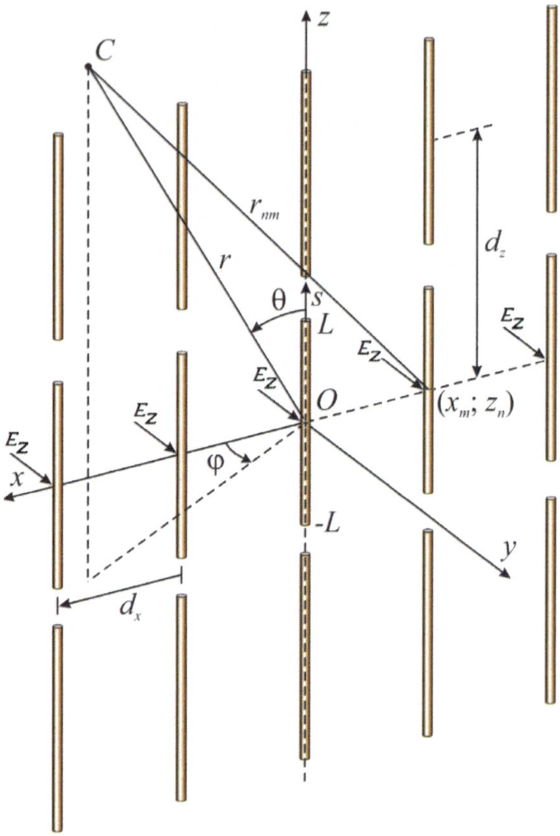

Fig. 6.9 The geometry of the diffraction antenna array

$$J_{nm}(s) = -\alpha_{nm}\frac{i\omega}{k_1\tilde{k}_{nm}}E_0\frac{\cos\tilde{k}_{nm}s - \cos\tilde{k}_{nm}L}{\cos\tilde{k}_{nm}L + \alpha_{nm}P^s[k_1\rho_{nm}, k_1L]} \tag{6.40}$$

where

$$\begin{aligned} P^s(k_1\rho_{nm}, k_1L) &= \cos k_1L \\ &\left[2\ln 2 - \gamma_{nm}(L) - (1/2)Cin(2k_1L) - (i/2)Si(4k_1L)\right] \\ &+ \sin k_1L[(1/2)Si(4k_1L) - (i/2)Cin(4k_1L)], \end{aligned}$$

$\tilde{k}^2_{nm}(s) = k_1^2\left(1 + \frac{i2\alpha_{nm}}{k_1\rho_{nm}}\overline{Z}_{nm}\right) = k_1^2\left(1 + i2\alpha_{nm}\overline{\beta}_{nm}\right)$ are the coefficients, $\alpha_{nm} = \frac{1}{2\ln[\rho_{nm}/(2L)]}$ are the natural small parameters, $\overline{\beta}_{nm} = \frac{\overline{Z}_{nm}}{k_1\rho_{nm}}$, $\gamma_{nm}(L) = \ln\frac{\rho_{nm}[2L+\sqrt{4L^2+\rho^2_{nm}}]}{4L^2}$, $Si(x)$ and $Cin(x)$ are sine and cosine integrals of the complex argument, and

$$\alpha_{nm}\int\limits_{-L}^{L}\frac{e^{-ik_1\sqrt{(L-s)^2+\rho_{nm}^2}}}{\sqrt{(L-s)^2+\rho_{nm}^2}}\cos\tilde{k}_{nm}s\,ds$$

$$\approx\alpha_{nm}\int\limits_{-L}^{L}\frac{e^{-ik_1\sqrt{(L-s)^2+\rho_{nm}^2}}}{\sqrt{(L-s)^2+\rho_{nm}^2}}\begin{bmatrix}\cos(k_1 s)\\ -ik_1Ls\alpha_{nm}\overline{\beta}_{nm}\sin(k_1 s)\end{bmatrix}ds$$

$$\approx\alpha_{nm}\int\limits_{-L}^{L}\frac{e^{-ik_1\sqrt{(L-s)^2+\rho_{nm}^2}}}{\sqrt{(L-s)^2+\rho_{nm}^2}}\cos(k_1 s)ds.$$

Then the main component of the wave zone electric field reradiated by the vibrator can be written as

$$\begin{aligned}E_{nm}^{\theta}&=\frac{60\pi i}{\lambda}\cdot\frac{e^{-ik_1 r_{nm}}}{r_{nm}}\sin\theta\int\limits_{-L}^{L}J_{nm}(s)e^{isk_1\cos\theta}ds\\&=120\pi\frac{e^{-ik_1 r_{nm}}}{\lambda k_1^2 r_{nm}}\cdot\frac{1}{\cos\theta}\cdot\frac{\alpha_{nm}\omega E_0\sin\theta}{\tilde{k}_{nm}^2-k_1^2\cos^2\theta}\\&\times\frac{k_1\cos\theta\sin\left(\tilde{k}_{nm}L\right)\cos(k_1L\cos\theta)-\tilde{k}_{nm}\cos\left(\tilde{k}_{nm}L\right)\sin(k_1L\cos\theta)}{\cos\tilde{k}_{nm}L+\alpha_{nm}P^s[k_1\rho_{nm},k_1L]}.\end{aligned}\tag{6.41}$$

The components E_φ and E_r of the electric field diffracted at single vibrator in the coordinate system (r, θ, φ) are equal to zero. If phases of fields arriving to the observation point $C(r, \theta, \varphi)$ are taken into account, the total wave zone reradiating field of the diffraction array $E_\theta(r, \theta, \varphi)$ is equal to the sum of the secondary radiation fields of each vibrator. Therefore, we can write

$$\begin{aligned}E_\theta(r,\theta,\varphi)=&\frac{120\pi\omega E_0\sin\theta}{\lambda k_1^2\cos\theta}\sum_{n=1}^{N_z}\sum_{m=1}^{N_x}\frac{e^{-ik_1 r_{nm}}}{r_{nm}}\cdot\frac{\alpha_{nm}}{\tilde{k}_{nm}^2-k_1^2\cos^2\theta}\\&\times\frac{k_1\cos\theta\sin\left(\tilde{k}_{nm}L\right)\cos(k_1L\cos\theta)-\tilde{k}_{nm}\cos\left(\tilde{k}_{nm}L\right)\sin(k_1L\cos\theta)}{\cos\tilde{k}_{nm}L+\alpha_{nm}P^s[k_1\rho_{nm},k_1L]}.\end{aligned}\tag{6.42}$$

The impedance synthesis problem can be formulated as follows: for a given direction $(\theta_{max}, \varphi_{max})$ of the maximum of the secondary radiation field in the wave zone, the impedances $\overline{Z}_{nm}$ of each vibrator element can be determined based on the expression (6.42).

6.3.2 Solution of the Impedance Synthesis Problem

First, let us convert the expression (6.41) defining the field of secondary field reradiated by an isolated vibrator [20] taking into account the relations $\tilde{k}_{nm} \approx k_1\left(1 + i\alpha_{nm}\overline{\beta}_{nm}\right)$ and $\tilde{k}_{nm}^2 - k_1^2\cos^2\theta = k_1^2\left(\sin^2\theta + 2i\alpha_{nm}\overline{\beta}_{nm}\right)$ valid due to the smallness of the parameters α_{nm}, and representations of trigonometric functions $\sin\left(\tilde{k}_{nm}L\right) \approx \sin(k_1L) + ik_1L\alpha_{nm}\overline{\beta}_{nm}\cos(k_1L)$ and $\cos\left(\tilde{k}_{nm}L\right) \approx \cos(k_1L) - ik_1L\alpha_{nm}\overline{\beta}_{nm}\sin(k_1L)$ valid up to terms of α_{nm}^2. Then we can write

$$\begin{aligned}E_{nm}^{\theta} = {} & 120\pi\frac{e^{-ik_1r_{nm}}}{\lambda k_1^3 r_{nm}}\cdot\frac{\alpha_{nm}\omega E_0}{\cos\theta\sin^3\theta}\\ & \times\left(f_1(\theta)\sin^2\theta + i\alpha_{nm}\overline{\beta}_{nm}\left[f_2(\theta)\sin^2\theta - 2f_1(\theta)\right]\right)\end{aligned}\tag{6.43a}$$

where

$$f_1(\theta) = \frac{\cos\theta\sin(k_1L)\cos(k_1L\cos\theta) - \cos(k_1L)\sin(k_1L\cos\theta)}{\cos\tilde{k}_{nm}L + \alpha_{nm}P^s[k_1\rho_{nm}, k_1L]},$$
$$f_2(\theta) = \frac{k_1L\cos\theta\cos(k_1L)\cos(k_1L\cos\theta) + \sin(k_1L\cos\theta)[k_1L\sin(k_1L) - \cos(k_1L)]}{\cos\tilde{k}_{nm}L + \alpha_{nm}P^s[k_1\rho_{nm}, k_1L]}.$$

After defining

$$\begin{aligned}C_{nm} &= \cos(k_1L) + \alpha_{nm}\mathrm{Re}\left(P^s[k_1\rho_{nm}, k_1L]\right),\\ \tilde{f}_1(\theta) &= \cos\theta\sin(k_1L)\cos(k_1L\cos\theta)\\ &\quad - \cos(k_1L)\sin(k_1L\cos\theta),\\ \tilde{f}_2(\theta) &= k_1L\cos\theta\cos(k_1L)\cos(k_1L\cos\theta)\\ &\quad + \sin(k_1L\cos\theta)[k_1L\sin(k_1L) - \cos(k_1L)]\end{aligned}$$

the expression for E_{nm}^{θ} can be written as

$$\begin{aligned}E_{nm}^{\theta} = {} & 120\pi\frac{e^{-ik_1r_{nm}}}{\lambda k_1^3 r_{nm}}\frac{\alpha_{nm}\omega E_0}{C_{nm}\cos\theta\sin\theta}\\ & \times\left\{\begin{aligned}&\tilde{f}_1(\theta) - i\alpha_{nm}\frac{\tilde{f}_1(\theta)}{C_{nm}}\mathrm{Im}\left(P^s[k_1\rho_{nm}, k_1L]\right)\\ &+ i\alpha_{nm}\overline{\beta}_{nm}\left[\frac{\tilde{f}_2(\theta)\sin^2\theta - 2\tilde{f}_1(\theta)}{\sin^2\theta} + \frac{\tilde{f}_1(\theta)}{C_{nm}}k_1L\sin(k_1L)\right]\end{aligned}\right\}\end{aligned}\tag{6.43b}$$

The expression (6.43b) is more convenient for the analysis, since the first two terms in curly bracket determine the RP of the isolated perfectly conducting vibrator, while the third term defines the field variation due to the impedance coating of the vibrator.

If the difference of geometric path for the waves propagating from neighboring vibrators and observation point $C(r, \theta, \varphi)$ is taken into account, the expression for the total secondary radiation field in the wave zone can be presented based on (6.43b) as follows [7, 16, 21], Sect. 6.2

$$E_\theta(\theta, \varphi) = \frac{120\pi\omega E_0}{k_1^3\lambda\cos\theta\sin\theta} e^{-i(N_z+1)u/2} e^{-i(N_x+1)v/2} \sum_{n=1}^{N_z}\sum_{m=1}^{N_x} \frac{\alpha_{nm}}{C_{nm}} e^{i(nu+mv)} \times \left\{ \begin{array}{l} \tilde{f}_1(\theta) - i\alpha_{nm}\frac{\tilde{f}_1(\theta)}{C_{nm}}\mathrm{Im}\left(P^s[k_1\rho_{nm}, k_1L]\right) \\ +i\alpha_{nm}\overline{\beta}_{nm}\left[\frac{\tilde{f}_2(\theta)\sin^2\theta - 2\tilde{f}_1(\theta)}{\sin^2\theta} + \frac{\tilde{f}_1(\theta)}{C_{nm}}k_1L\sin(k_1L)\right] \end{array} \right\}, \tag{6.44}$$

where $u = k_1 d_z\cos\theta$ and $v = k_1 d_x\sin\theta\cos\varphi$. If the array vibrators are perfectly conducting, $\overline{\beta}_{nm} = \overline{Z}_{nm} = 0$, and amplitudes of all vibrator currents are equal to $J_{nm} = J_0 = \frac{120\pi\alpha\omega E_0}{Ck_1^3\lambda}\left[1 - i\frac{\alpha}{C}\mathrm{Im}(P^s[k_1\rho, k_1L])\right]$, then $\alpha_{nm} = \alpha$, $\rho_{nm} = \rho$, $C = C_{nm}|_{\alpha_{nm}=\alpha, \rho_{nm}=\rho}$ and the expression (6.44) is reduced to

$$E_\theta(\theta, \varphi) = J_0 e^{-i(N_z+1)u/2} e^{-i(N_x+1)v/2} \frac{\tilde{f}_1(\theta)}{\cos\theta\sin\theta} \sum_{n=1}^{N_z}\sum_{m=1}^{N_x} e^{i(nu+mv)} \tag{6.45}$$

As can be seen, the RP maximum is directed along the axis $\{0y\}(\theta = \pi/2, \varphi = \pi/2)$ under condition $u = v = 0$. One can also see that the common array factor in expression (6.45) is a product of two independent series describing the RP of the two linear vibrator arrays which axes are directed along the axes $\{0x\}$ and $\{0z\}$.

As is known from the antenna array theory (for example, [16, 21]), the rotational displacement of the RP in space can be realized by linear phase shifts between the currents of the vibrators. If the phase shift between waves arriving to observation point from adjacent vibrator rows is Δu, and that from adjacent vibrator in a row is Δv, the direction of the antenna array RP maximum $(\theta_{\max}; \varphi_{\max})$ is determined by the following relations.

$$\cos\theta_{\max} = \Delta u / k_1 d_z \text{ and } \sin\theta_{\max}\cos\varphi_{\max} = \Delta v / k_1 d_x \tag{6.46}$$

Then the expression (6.45) can be rewritten as

$$E_\theta(\theta, \varphi) = J_0 e^{-i(N_z+1)u/2} e^{-i(N_x+1)v/2} \frac{\tilde{f}_1(\theta)}{\cos\theta\sin\theta} \sum_{n=1}^{N_z}\sum_{m=1}^{N_x} e^{i(nu+mv)} e^{-i[(n-1)\Delta u+(m-1)\Delta v]} \tag{6.47}$$

As can be seen from the above formulas, expressions (6.44) and (6.47) are identical if the following relations

$$e^{-ik_1[(n-1)d_z\cos\theta+(m-1)d_x\sin\theta\cos\varphi]}-1$$
$$=i\alpha\overline{\beta}_{nm}\left.\frac{C\tilde{f}_2(\theta)\sin^2\theta-\tilde{f}_1(\theta)\left[2C-k_1L\sin(k_1L)\sin^2\theta\right]}{\tilde{f}_1(\theta)\sin^2\theta[C-i\alpha\mathrm{Im}(P^s[k_1\rho,k_1L])]}\right|_{\theta=\theta_{\max};\,\varphi=\varphi_{\max}} \tag{6.48}$$

are valid for all indices n and m. The relations (6.48) can be used as equations allowing to find the unknowns $\{\overline{\beta}_{nm}\}$ for any angle $(\theta_{\max};\ \varphi_{\max})$.

Let us denote $\chi_{nm}=[(n-1)d_z\cos\theta_{\max}+(m-1)d_x\sin\theta_{\max}\cos\varphi_{\max}]$, as the phase parameter, $D=\left.\frac{C\tilde{f}_2(\theta)\sin^2\theta-\tilde{f}_1(\theta)\left[2C-k_1L\sin(k_1L)\sin^2\theta\right]}{\tilde{f}_1(\theta)\sin^2\theta}\right|_{\theta=\theta_{\max}}$ as the functional multiplier, and $\overline{Z}_{nm}=\overline{R}_{nm}+i\overline{X}_{nm}$ as the complex impedances, where $\overline{R}_{nm}$ and $\overline{X}_{nm}$ are the real and imaginary parts of the impedances. Then Eq. (6.48) can be written in the following form

$$1-e^{-ik_1\chi_{nm}}=-i\alpha\frac{\overline{R}_{nm}+i\overline{X}_{nm}}{k_1\rho}\frac{D}{C-i\alpha\mathrm{Im}(P^s[k_1\rho,k_1L])} \tag{6.49}$$

For lossless media, considering that the parameter α is real, and the relations (6.49) can be rewritten as

$$\begin{aligned}\frac{i\alpha D\overline{R}_{nm}-\alpha D\overline{X}_{nm}}{k_1\rho}&=-[1-\cos(k_1\chi_{nm})+i\sin(k_1\chi_{nm})]\\&\quad\times\left[C-i\alpha\mathrm{Im}\left(P^s[k_1\rho,k_1L]\right)\right]\end{aligned} \tag{6.50}$$

Substituting C defined in the relations (6.43a) into (6.50), we obtain the identity

$$\begin{aligned}&\frac{i\alpha D\overline{R}_{nm}-\alpha D\overline{X}_{nm}}{k_1\rho}\\&-[1-\cos(k_1\chi_{nm})]\left[\cos(k_1L)+\alpha\mathrm{Re}\left(P^s[k_1\rho,k_1L]\right)\right]\\&-\alpha\sin(k_1\chi_{nm})\mathrm{Im}\left(P^s[k_1\rho,k_1L]\right)\\&-i\sin(k_1\chi_{nm})\left[\cos(k_1L)+\alpha\mathrm{Re}\left(P^s[k_1\rho,k_1L]\right)\right]\\&+i\alpha\mathrm{Im}\left(P^s[k_1\rho,k_1L]\right)[1-\cos(k_1\chi_{nm})].\end{aligned} \tag{6.51}$$

Denoting the real and imaginary parts of the functional $P^s[k_1\rho_{nm},k_1L]$ defined in (6.40) as

$$\begin{aligned}\mathrm{Re}\left(P^s\right)&=\cos k_1L\left[2\ln 2-\gamma_{nm}(L)-(1/2)\mathrm{Cin}(2k_1L)\right]\\&\quad+(1/2)\mathrm{Si}(4k_1L)\sin k_1L,\\\mathrm{Im}\left(P^s\right)&=-(1/2)[\mathrm{Si}(4k_1L)\cos k_1L+Cin(4k_1L)\sin k_1L],\end{aligned} \tag{6.52}$$

we can represent the relation (6.51) in the form

$$
\begin{aligned}
\frac{i\alpha D\overline{R}_{nm} - \alpha D\overline{X}_{nm}}{k_1\rho} &= -[1-\cos(k_1\chi_{nm})] \\
&\times\left[\cos(k_1L) + \alpha\mathrm{Re}\left(P^s\right)\right] - \alpha\sin(k_1\chi_{nm})\mathrm{Im}\left(P^s\right) \\
&-i\sin(k_1\chi_{nm})\left[\cos(k_1L) + \alpha\mathrm{Re}\left(P^s\right)\right] \\
&+i\alpha\mathrm{Im}\left(P^s\right)[1-\cos(k_1\chi_{nm})].
\end{aligned} \tag{6.53}
$$

The final formulas for the real and imaginary parts of the surface impedance of the array vibrators can be obtained based on (6.53), as

$$
\begin{aligned}
\overline{R}_{nm} &= \frac{k_1\rho}{\alpha D}\left\{\begin{array}{l}\alpha\mathrm{Im}\left(P^s\right)[1-\cos(k_1\chi_{nm})] \\ -\sin(k_1\chi_{nm})\left[\cos(k_1L)+\alpha\mathrm{Re}\left(P^s\right)\right]\end{array}\right\}, \\
\overline{X}_{nm} &= \frac{k_1\rho}{\alpha D}\left\{\begin{array}{l}\alpha\sin(k_1\chi_{nm})\mathrm{Im}\left(P^s\right) \\ +[1-\cos(k_1\chi_{nm})]\times\left[\cos(k_1L)+\alpha\mathrm{Re}\left(P^s\right)\right]\end{array}\right\}, \\
n &= 1,2\ldots N_x,\ m = 1,2\ldots N_z.
\end{aligned} \tag{6.54}
$$

The formulas (6.54) are valid for any number of vibrators in the array and arbitrary distances between the vibrators. However, in the general case, the impedance $\overline{Z}_{nm} = \overline{R}_{nm} + i\overline{X}_{nm}$ defined as effective physical quantities does not guarantee that the condition $\overline{R}_{nm} \geq 0$ can be always fulfilled. Since this condition, put forward based on energy laws, determines the possibility of practical realization in the form of vibrator intrinsic impedance, the fulfillment of the condition $\overline{R}_{Sn} \geq 0$ should always be checked in the course of numerical simulations.

The total field $E_{\Sigma}^{\theta}(r,\theta,\varphi)$ in the upper half-space above the array can be represented as the sum of the excitation and secondary radiation field, that is, the incident plane wave field and the field diffracted by the antenna array. Therefore, we can write

$$
E_{\Sigma}^{\theta}(r,\theta,\varphi) = E_0 e^{-ik_1 r\cos\theta} + \sum_{n=1}^{N_z}\sum_{m=1}^{N_x} E_{nm}^{\theta}, \tag{6.55}
$$

which states that the formation of the antenna RP in the wave zone is determined only by the diffraction component.

6.3.3 Numerical Results

The electromagnetic wave incident at an obstacle whose dimensions are commensurate with the radiation field wavelength in the medium, is scattered in all direction. If the obstacle is characterized by periodic variation of any parameter influencing the wave propagation, its energy is scattered in discrete directions, known as diffraction orders. The main diffraction order in such structures can be analyzed by using scattering matrices which include the energy reflection and transmission coefficients [22]. However, the key stage is still the study of the diffraction characteristics of the structures that are changed due to variation of the vibrator impedance. Therefore, the simulation of the antenna arrays can be carried out based on the diffraction characteristics.

As an example, consider the diffraction array consisting of N symmetric resonant impedance vibrators $\left(N = 25, N_x = 5, N_z = 5\right)$, which lengths and radii are $(2L = 0.4908\lambda)$ and $\left(\rho_{nm} = L/75\right)$, for which in the front half-space $(0 \leq \theta \leq \pi,\ 0 \leq \varphi \leq \pi$, Fig. 6.9). To solve the problem, the impedance of each vibrator should be found based on the expressions (6.54) so that the RP maximum was scanned in the direction defined by the angle $(\theta_{\max};\ \varphi_{\max})$. When the procedure of impedances synthesis procedure is completed, the array RP should be computed by busing the expression (6.55).

The simulation results are presented for the two impedance arrays with the geometric parameters: $d_z = 0.5\lambda$, $d_x = 0.5\lambda$ and $d_z = 0.5\lambda$, $d_x = 0.25\lambda$. The simulation results are shown as deviations of the angle $(\theta_{\max};\ \varphi_{\max})$ symmetrical relative to the array normal $(\theta = 90°;\ \varphi = 90°)$, which corresponds to the direction of the RP maxima of the array with perfectly conducting vibrators.

The cross sections of the normalized array RP plotted, base on the expression (6.55), as function of the RP maxima deviation in the plane $\theta_{\max} = 90°$ are shown in Fig. 6.10. The plots of the RP as functions of the angular coordinates $\varphi°$ and $\theta°$ are shown in Figs. 6.10a and 6.10b.

Three curves plotted using the parameter $\varphi_{\max}$ equal to $90° - \Delta\varphi_{\max}$, $90°$, and $90° + \Delta\varphi_{\max}$ $(\Delta\varphi_{\max} = 14.4775°)$ are shown in Fig. 6.10a. The simulation results, that is, the impedance defined by the formula (6.54) and current arguments (6.40) on the vibrators are presented in Tables 6.4 and 6.5 for the angles $\varphi_{\max} = 104.4775°$ and $\varphi_{\max} = 75.5225°$. The row and column indices in the tables, n and m, correspond to the positions of the vibrator scatterer in the array structure (Fig. 6.9).

The simulation data presented in the Tables confirm that the absolute value of the $\overline{R}_{nm}$ and $\overline{X}_{nm}$ are far less than that for active arrays when the vibrators are excited by concentrated electromotive forces [20]. As can be seen, if $\varphi_{\max} = 75.5225°$, the negative values $\overline{R}_{nm}$ and $\arg(J_{nm})$ are observed simultaneously. Thus, the physically correct values of the impedance $\overline{R}_{nm}$ satisfying the conditions $\overline{R}_{nm} \geq 0$ can be obtained by changing in (6.45) the direction of the vibrator's currents on the opposite ones. Since the proposed diffraction antenna array has no elements to control the vibrator currents, this requirement cannot be fulfilled. In other words, if the vibrator array is interpreted from a functional point of view as a thin dielectric layer with effective material parameters, the interval of refraction angles of the incident field satisfying the condition $\overline{R}_{nm} \geq 0$ determines the refraction angle zone for the material layer, while the interval of angles corresponding to $\overline{R}_{nm} < 0$ define an anomalous beam deflection observed, for example, for metamaterials [23, 24].

We have found that the positive values of $\overline{R}_{nm}$ are observed if the realizable phase difference between the currents on the edge vibrator elements does not exceed 180°. The maximal angle by which the RP can deviate from the normal is defined by the

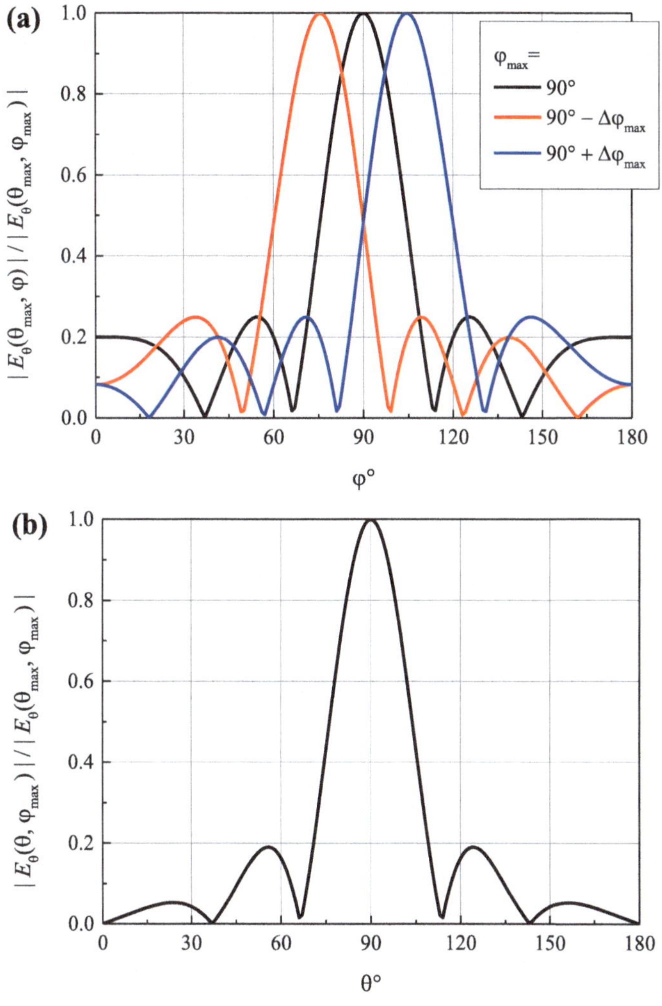

Fig. 6.10 The array RP in the plane $\theta_{\max} = 90°$ with $d_z = 0.5\lambda$, $d_x = 0.5\lambda$, and $\Delta\varphi_{\max} = 14.4775°$ as functions of: **a** the angle $\varphi°$, **b** the angle $\theta°$

formula

$$\Delta\varphi_{\max} = \arcsin\left(\frac{\pi}{kd_x(N_x - 1)}\right) = \arcsin\left(\frac{\lambda}{2d_x(N_x - 1)}\right) \tag{6.56}$$

or $\Delta\varphi_{\max} \approx \frac{\lambda}{2d_x(N_x-1)} \cdot \frac{180°}{\pi}$ if the deviations are small. If we compare the value of $\Delta\varphi_{\max}$ defined by the formula (6.56) with the width of the array RP at the half power level $\Delta\varphi_{0.5} \approx \frac{\lambda}{d_x(N_x-1)} \cdot 51°$, the theoretical level of the maximum deviations are $\Delta\varphi_{\max} \approx \pm 0.56 \cdot \Delta\varphi_{0.5}$.

Table 6.4 Simulation results ($\theta_{max} = 90°$; $\varphi_{max} = 104.4775°$)

$\overline{R}_{nm}$					$\overline{X}_{nm}$					$\arg(J_{nm})$, [deg]				
0	0.0056	0.00792	0.0056	0	0	0.00232	0.00792	0.01352	0.01584	0	45	90	135	180
0	0.0056	0.00792	0.0056	0	0	0.00232	0.00792	0.01352	0.01584	0	45	90	135	180
0	0.0056	0.00792	0.0056	0	0	0.00232	0.00792	0.01352	0.01584	0	45	90	135	180
0	0.0056	0.00792	0.0056	0	0	0.00232	0.00792	0.01352	0.01584	0	45	90	135	180
0	0.0056	0.00792	0.0056	0	0	0.00232	0.00792	0.01352	0.01584	0	45	90	135	180

Table 6.5 Simulation results ($\theta_{\max} = 90°$; $\varphi_{\max} = 75.5225°$)

$\overline{R}_{nm}$					$\overline{X}_{nm}$					$\arg(J_{nm})$, [deg]				
0	− 0.0056	− 0.00792	− 0.0056	0	0	0.00232	0.00792	0.01352	0.01584	0	− 45	− 90	− 135	− 180
0	− 0.0056	− 0.00792	− 0.0056	0	0	0.00232	0.00792	0.01352	0.01584	0	− 45	− 90	− 135	− 180
0	− 0.0056	− 0.00792	− 0.0056	0	0	0.00232	0.00792	0.01352	0.01584	0	− 45	− 90	− 135	− 180
0	− 0.0056	− 0.00792	− 0.0056	0	0	0.00232	0.00792	0.01352	0.01584	0	− 45	− 90	− 135	− 180
0	− 0.0056	− 0.00792	− 0.0056	0	0	0.00232	0.00792	0.01352	0.01584	0	− 45	− 90	− 135	− 180

If the angle $\varphi_{\max} = 90°$, the condition (6.56) is transformed to

$$\Delta\theta_{\max} = \arcsin\left(\frac{\lambda}{2d_z(N_z - 1)}\right) \tag{6.57}$$

In this case, the theoretical estimate $\Delta\theta_{\max} \approx \pm 0.56 \cdot \Delta\theta_{0.5}$ is also valid.

Similar conclusions can be made by analyzing plots of the normalized array RP defined by the formula (6.55) if the maxima are deviated in the plane $\varphi_{\max} = 90°$. The plots are shown in Fig. 6.11.

The cross sections of the normalized array RP as functions of the angular coordinate $\varphi°$ and $\theta°$ are plotted in Fig. 6.11a, b. Three curves corresponding to the parameter $\theta_{\max}$ equal to $\theta_{\max} = 75.5225° = 90° - \Delta\theta_{\max}$, $90°$, and $\theta_{\max} = 104.4775° = 90° + \Delta\theta_{\max}$, are shown in Fig. 6.11b. The simulation results are presented in Tables 6.4 and 6.5 for the angles $\theta_{\max} = 104.4775°$ and $\theta_{\max} = 75.5225°$.

It is quite clear, that the RP maximum of the proposed diffraction antenna array can be scanned in any predefined directions including those lying in diagonal planes by means of varying the impedance $\overline{R}_{nm}$ of all vibrators. To confirm this conclusion, the main sections of the normalized RP obtained under the condition $\varphi_{\max} = \theta_{\max}$ are shown in Fig. 6.12. We will not present here the corresponding table of the impedances, but we simply indicate that the angular zone of the natural array refraction is within the intervals $90° \leq \varphi_{\max}, \theta_{\max} \leq 97°$.

The relations (6.56) allows us to state that the smaller is the ratio d_x/λ, the wider is the angular zone of natural refraction of the diffraction array in the plane $\theta_{\max} = 90°$. As the parameter d_x/λ is decreased while d_z/λ is constant (Fig. 6.9), the array RP gets more pronounced rectangular shape asymmetrically extended in the plane of the smaller side of the diffraction apertures. The plots of the normalized cross sections of the array RP with parameters $d_z = 0.5\lambda$ and $d_x = 0.25\lambda$ as functions of the angular coordinate $\varphi°$ and $\theta°$ are shown in Fig. 6.13a, b. Three curves plotted for the parameter $\varphi_{\max}$ equal to $60°$, $90°$, and $120°$ are shown in Fig. 6.13a. The simulation results for the angles $\varphi_{\max} = 120°$ and $\varphi_{\max} = 60°$ are also presented in Tables 6.6 and 6.7.

In this case, the simulation results have shown that the real part of the impedance $\overline{Z}_{nm}$ is always positive if the angle $\theta_{\max} = 90°$, while the angle $\varphi_{\max}$ varies at the interval $\varphi_{\max} \in [90°, 120°]$. When the direction of the RP maximum varies in the plane $\varphi_{\max} = 90°$, the effects similar to those shown in Fig. 6.11b are observed, that is, the angular zone of natural refraction of the diffraction array is $90° \leq \theta_{\max} \leq 97°$. The plots of the main sections of the normalized RP obtained under the condition $\varphi_{\max} = \theta_{\max}$ are shown in Fig. 6.13. The simulation results confirm that the sector of the angular zone of the natural refraction for the array is in the range $(90° \leq \varphi_{\max} \leq 120°;\ 90° \leq \theta_{\max} \leq 97°)$.

The materials of this Chapter are contained in the authors' publications [20, 25, 26].

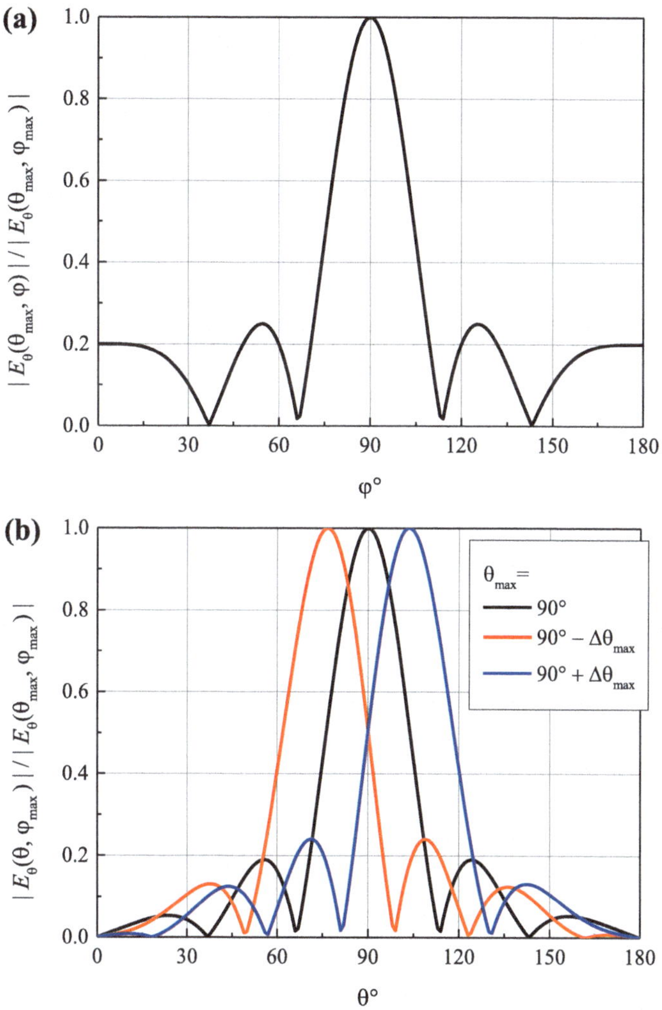

Fig. 6.11 The array RP with $d_z = 0.5\lambda$, $d_x = 0.5\lambda$, $\varphi_{\max} = 90°$, $\Delta\theta_{\max} = 14.4775°$ as functions of: **a** the angle $\varphi°$, **b** the angle $\theta°$

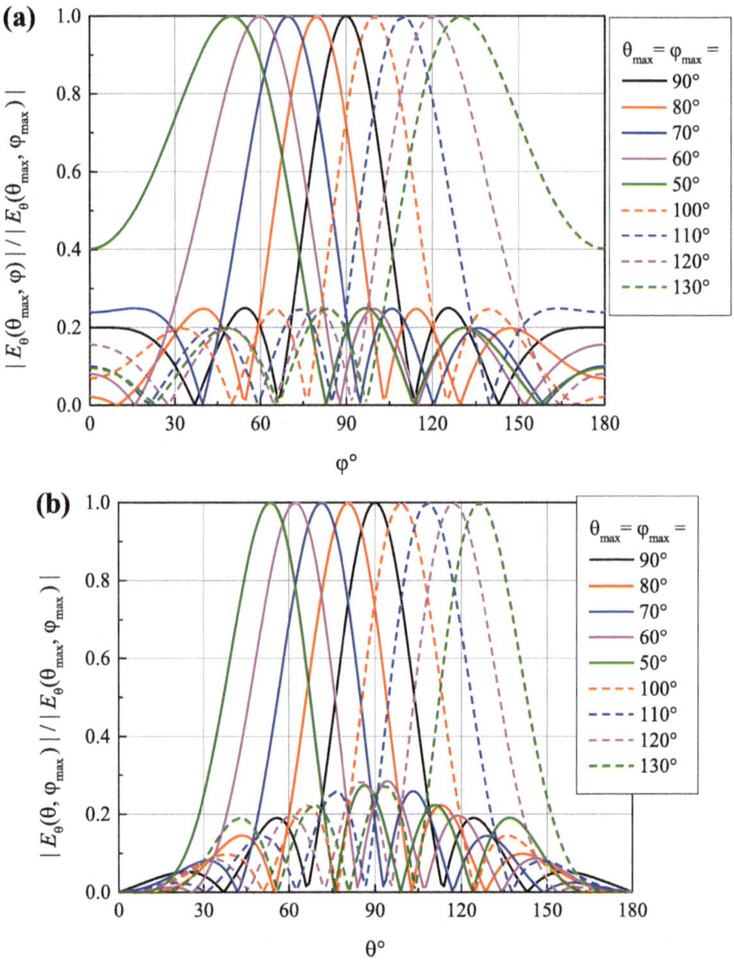

Fig. 6.12 The array RP with $d_z = 0.5\lambda$, $d_x = 0.5\lambda$, $\varphi_{max} = \theta_{max}$ as functions of: **a** the angle $\varphi°$, **b** the angle $\theta°$

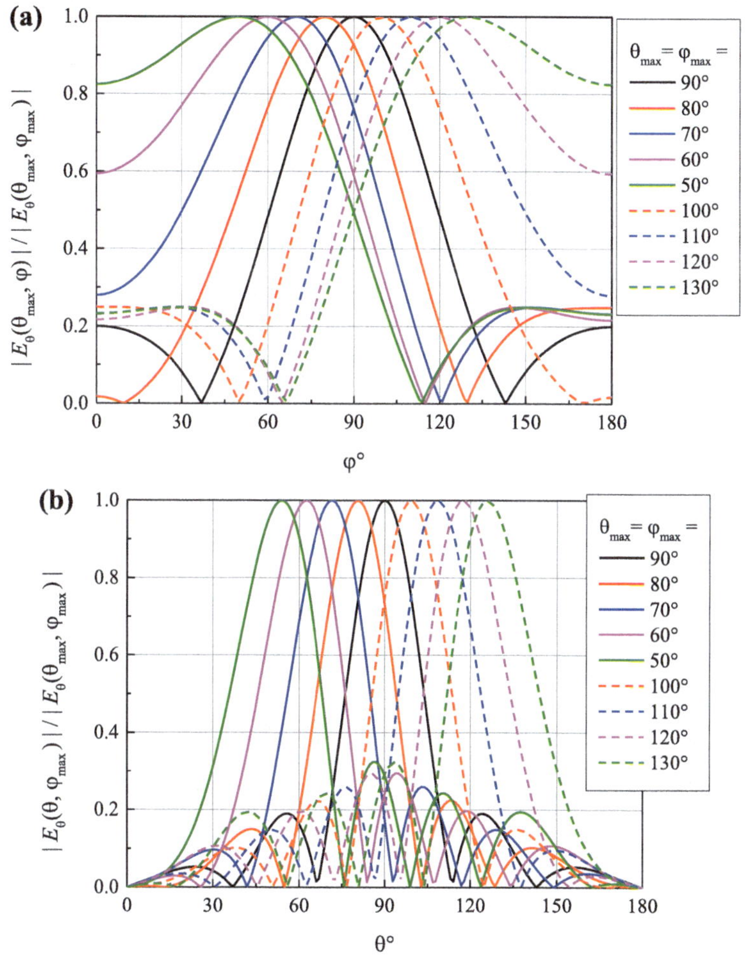

Fig. 6.13 The array RP with $d_z = 0.5\lambda$, $d_x = 0.25\lambda$, and $\varphi_{max} = \theta_{max}$ as functions of: **a** the angle $\varphi°$, **b** the angle $\theta°$

Table 6.6 Simulation results obtained for $\left(\varphi_{\max} = 120°; \theta_{\max} = 90°\right)$

$\overline{R}_{nm}$					$\overline{X}_{nm}$					$\arg(J_{nm}), [\text{deg}]$				
0	0.0056	0.00792	0.0056	0	0	0.00232	0.00792	0.01352	0.01584	0	45	90	135	180
0	0.0056	0.00792	0.0056	0	0	0.00232	0.00792	0.01352	0.01584	0	45	90	135	180
0	0.0056	0.00792	0.0056	0	0	0.00232	0.00792	0.01352	0.01584	0	45	90	135	180
0	0.0056	0.00792	0.0056	0	0	0.00232	0.00792	0.01352	0.01584	0	45	90	135	180
0	0.0056	0.00792	0.0056	0	0	0.00232	0.00792	0.01352	0.01584	0	45	90	135	180

Table 6.7 Simulation results obtained for $(\varphi_{\max} = 60°; \theta_{\max} = 90°)$

$\overline{R}_{nm}$					$\overline{X}_{nm}$					$\arg(J_{nm})$, [deg]				
0	– 0.0056	– 0.00792	– 0.0056	0	0	0.00232	0.00792	0.01352	0.01584	0	– 45	– 90	– 135	– 180
0	– 0.0056	– 0.00792	– 0.0056	0	0	0.00232	0.00792	0.01352	0.01584	0	– 45	– 90	– 135	– 180
0	– 0.0056	– 0.00792	– 0.0056	0	0	0.00232	0.00792	0.01352	0.01584	0	– 45	– 90	– 135	– 180
0	– 0.0056	– 0.00792	– 0.0056	0	0	0.00232	0.00792	0.01352	0.01584	0	– 45	– 90	– 135	– 180
0	– 0.0056	– 0.00792	– 0.0056	0	0	0.00232	0.00792	0.01352	0.01584	0	– 45	– 90	– 135	– 180

References

1. Bakhrakh LD, Kremenetsky SD (1974) Synthesis of radiation systems (Theory and methods of calculation), Sov. Radio, Moscow (in Russian)
2. Chaplin AF (1969) Synthesis of passive vibrator arrays. Izv Vuzov Radioelectron 559–562 (in Russian)
3. Panduro MA, Reynoso-Hernandez JA (2012) A method for determining the characteristic impedance of transmission lines embedded in transitions. Int J Electron Commun AEUE Elsevier 66(3):185–188
4. Ravelo B (2015) Radiated near-field emission extraction on 3D curvilinear surfaces from 2D data. Prog Electromagn Res M 44:191–201
5. Ravelo B, Liu Y, Jastrzebski AK (2015) PCB near-field transient emission time-domain model. IEEE Trans Electromagn Compatib 57(6):1320–1328
6. Liu Y, Ravelo B, Jastrzebski AK (2016) Time-domain magnetic dipole model of PCB near-field emission. IEEE Trans Electromagn Compatib 58(5):1561–1569
7. Berdnik SL, Katrich VA, Nesterenko MV, Penkin YuM (2016) Electromagnetic waves radiation by a vibrators system with variable surface impedance. Prog Electromagn ResM 51:157–163
8. Nesterenko MV, Katrich VA, Penkin YuM, Dakhov VM, Berdnik SL (2011) Thin impedance vibrators. Theory and applications. Springer Science+Business Media, New York
9. Chaplin AF, Makkaveeva VF (1977) Synthesis of the impedance thread arrays. Izv Vuzov Radiophysics 274–279 (in Russian)
10. Chaplin AF (1987) Analysis and synthesis of antenna arrays. Vysshaya Shkola, L'vov (in Russian)
11. Nesterenko MV (2010) Analytical methods in the theory of thin impedance vibrators. Prog Electromagn Res B 21:299–328
12. Bogolyubov NN, Mitropolsky UA (1974) Asymptotic methods in the theory of nonlinear fluctuations. Nauka, Moskow (in Russian)
13. Philatov AN (1974) Asymptotic methods in the theory of differential and integral-differential equations. PHAN, Tashkent (in Russian)
14. Wu TT, King RWP (1965) The cylindrical antenna with nonreflecting resistive loading. IEEE Trans Antennas Propag 13:369–373
15. Glushkovskiy EA, Israylit AB, Levin BM, Rabinovich EYa (1967) Linear antennas with variable surface impedance. Antennas 2:154–165 (in Russian)
16. Amitay N, Galindo V, Wu CP (1972) Theory and analysis of phased array antennas. Wiley, New York
17. Penkin YuM, Katrich VA, Nesterenko MV (2016) Development of fundamental theory of thin impedance vibrators. Prog Electromagn Res M 45:185–193
18. Voskresensky DI (2003) Microwave devices and antennas. Designing phased antenna arrays. Radiotekhnika, Moscow (in Russian)
19. Vendik OG, Parnes MD, Bakhrakh LD (2001) Antennas with electric scanning. Science Press, New York
20. Penkin YuM, Katrich VA, Nesterenko MV (2017) Formation of radiation fields of linear vibrator arrays by using impedance synthesis. Prog Electromagn Res M 57:1–10
21. Eisenberg GZ (1985) Shortwave antennas. Radio i Svyaz', Moscow (in Russian)
22. Lee R, Mittra SW (1972) Analytical techniques in the theory of guided waves. Macmillan Series in Electrical Science, New York
23. Penkin YuM, Katrich VA, Nesterenko MV (2019) Impedance synthesis of 2D antenna arrays of slotted spherical radiators. Prog Electromagn Res L 81:93–100
24. Lagarkov AN, Semenenko VN, Basharin AA, Balabukha NP (2008) Abnormal radiation pattern of metamaterial waveguide. PIERS Online 4(6):641–644
25. Penkin YuM, Katrich VA, Nesterenko MV, Berdnik SL (2018) Analytical solution of impedance synthesis problem for a 2D array of thin vibrators. Prog Electromagn Research M 65:43–49
26. Penkin YuM, Katrich VA, Nesterenko MV, Berdnik SL, Pshenichnaya SV (2020) Impedance synthesis of plane diffraction vibrator arrays. Prog Electromagn Res M 89:31–41

Chapter 7
Impedance Synthesis for Slot Antenna Arrays

7.1 Impedance Synthesis for Two-Dimensional Slot Array with Magneto-Dielectric Filling

In this Section a problem of impedance synthesis for a 2D antenna arrays consisting of narrow linear slots with magneto-dielectric filling is solved. The slots are cut in a perfectly conducting screen, which geometric centers coincide with nodes of a rectangular grid with double periodicity. The slots are excited by voltage δ-generators in the center of the slot apertures. The slot cavities are represented by rectangular resonators with magneto-dielectric filling which control the induced magnetic impedance of the radiators by varying the material parameters of the slots filling. The problem solution is based on the approach proposed by the authors for the impedance synthesis of vibrator arrays (Chap. 6) and based on the electrodynamic principle of permutation duality (Sect. 1.4.2).

As is known, the slot AA are widely used in practice (Chap. 4). A number of publications (see, for example [1–3]) considered the characteristics of single slot impedance loads which have been developed based on a single mathematical model. This model is based on solving a problem related to coupling of two volumes through a hole in an infinite perfectly conducting plane. In this case, the concept of slot impedance was used in the classical interpretation [4]. However, any references concerning the problem solution of the impedance synthesis of slot arrays have not been found.

M. V. Nesterenko et al., *Vibrator and Slot Antenna Arrays for Modern Applications*,
Lecture Notes in Electrical Engineering 1244,
https://doi.org/10.1007/978-3-031-68346-6_7

7.1.1 Electromagnetic Principle of Permutation Duality

7.1.1.1 Extended Theoretical Background

Maxwell's equations for time-harmonic oscillations depending on time t as $e^{i\omega t}$ can be presented in the following form

$$\begin{cases} \operatorname{rot} \vec{E} = -ik\mu\vec{H} - \dfrac{4\pi}{c}\vec{j}^m, \\ \operatorname{rot} \vec{H} = ik\varepsilon\vec{E} + \dfrac{4\pi}{c}\vec{j}^e, \end{cases} \tag{7.1}$$

where ω is circular frequency of the monochromatic process, $\vec{E}$ and $\vec{H}$ are complex vector amplitudes of electric and magnetic field intensities, ε and μ are permittivity and permeability of homogeneous medium, $\vec{j}^{e(m)}$ are volume densities of electrical and magnetic extraneous currents marked by superscripts e and m, $k = \omega/c_0$ is wave number, and $c_0 = 2.998 \cdot 10^{10}$ (cm/s) is the velocity of light in vacuum. Equation (7.1) is written in the dually symmetric form proposed by Heaviside in 1893. The possible permutations are shown in Table 7.1.

These permutations, along with the permutations $\vec{B} \Leftrightarrow \vec{D}$ and $\rho^e \Leftrightarrow \rho^m$ are considered to be basic. Here $\vec{B}$ and $\vec{D}$ are complex amplitude of the magnetic and electric induction vectors, $\rho^{e(m)}$ are volume densities of electrical and magnetic charges. Other dual permutations between notions arising when various approaches are applied to the solution of the Maxwell's Equation (7.1) are classified as non-basic ones. A complete list of non-basic dual permutation cannot be presented now, since new theoretical concepts are constantly arising when electrodynamic methods are developed.

The dual symmetry of Eq. (7.1) is based on the invariance of the Maxwell's equations with respect to linear transformations of fields according to the following rules

$$\begin{aligned} &\vec{E}^* = \vec{E}\cos\theta + \vec{H}\sin\theta, \quad \vec{H}^* = \vec{H}\cos\theta + \vec{E}\sin\theta, \\ &\vec{D}^* = \vec{D}\cos\theta + \vec{B}\sin\theta, \quad \vec{B}^* = \vec{B}\cos\theta + \vec{D}\sin\theta, \end{aligned} \tag{7.2}$$

where θ is an arbitrary angular parameter. In particular, identical transformations are obtained if $\theta = 0$, and the standard transformations of permutation duality are obtained if $\theta = \pi/2$. Hence, the permutations presented in Table 7.1 should be classified as basic standard transformations. For convenience, we use the Gaussian absolute system of units, where vectors of electromagnetic fields $\vec{E}, \vec{D}, \vec{H}$, and $\vec{B}$

Table 7.1 The possible permutations

Original notions	$\vec{E}$	$\vec{H}$	ε	μ	$\vec{j}^e$	$\vec{j}^m$
Dual notions	$\vec{H}$	$\vec{E}$	$-\mu$	$-\varepsilon$	$-\vec{j}^m$	$-\vec{j}^e$

have the same dimensions, ($g^{1/2}$ $cm^{-1/2}s^{-1}$). Therefore, in vacuum, we can avoid using additional constants by virtue of relations $\vec{E} = \vec{D}$ and $\vec{H} = \vec{B}$, so that the dimensionless permittivity and permeability become $\varepsilon_0 = 1, \mu_0 = 1$.

The electrodynamic problems for the electrically thin antennas are formulated as boundary value problems based on the boundary conditions for the electric and magnetic fields

$$\begin{aligned} \left[\vec{n}_1, \vec{E}_1\right] + \left[\vec{n}_2, \vec{E}_2\right] &= \frac{4\pi}{c_0}\vec{j}^m, \\ \left[\vec{n}_1, \vec{H}_1\right] + \left[\vec{n}_2, \vec{H}_2\right] &= -\frac{4\pi}{c_0}\vec{j}^e, \end{aligned} \tag{7.3}$$

where $\vec{n}_{1(2)}$ are the unit vectors of the normal to the boundary surface between adjacent electrodynamic regions 1 and 2, facing the interior of the region where the field is determined. For these regions, the material parameters are assumed to be $\varepsilon_{1(2)}$ and $\mu_{1(2)}$. As can be seen, Eq. (7.1) and boundary conditions (7.3) have the dual symmetric form with respect to the basic permutations.

Functional relationship between extraneous currents and excited fields can be established by introduction of auxiliary vector Hertz potentials, $\vec{\prod}^e$ and $\vec{\prod}^m$, related to the vectors $\vec{E}$ and $\vec{H}$ as

$$\begin{aligned} \vec{E} &= \text{grad div} \vec{\prod}^e + k^2\varepsilon\mu\vec{\prod} - ik\mu \text{ rot} \vec{\prod}^m, \\ \vec{H} &= \text{grad div} \vec{\prod}^m + k^2\varepsilon\mu\vec{\prod} - ik\mu \text{ rot} \vec{\prod}^e \end{aligned} \tag{7.4}$$

These vector potentials can be found as solutions of inhomogeneous Helmholtz equations

$$\begin{aligned} \Delta\vec{\prod}^e + k^2\varepsilon\mu\vec{\prod}^e &= -\frac{4\pi}{i\omega\varepsilon}\vec{j}^e, \\ \Delta\vec{\prod}^m + k^2\varepsilon\mu\vec{\prod}^m &= -\frac{4\pi}{i\omega\varepsilon}\vec{j}^m, \end{aligned} \tag{7.5}$$

where $\Delta \equiv$ graddiv – rotrot is the Laplace operator. The solution of the equations can be obtained using the Green's functions

$$\vec{\prod}^{e(m)}(\vec{r}) = -\frac{1}{ik\varepsilon(\mu)} \int\limits_V \vec{j}^{e(m)}(\vec{r}\,')\hat{G}^{e(m)}(\vec{r}; \vec{r}\,')dv \tag{7.6}$$

where V is the volume where the extraneous currents are enclosed, $\vec{r}, \vec{r}\,'$ are the radius-vectors of observation point and source, $\hat{G}^e(\vec{r}; \vec{r}\,')$ and $\hat{G}^m(\vec{r}; \vec{r}\,')$ are the tensor Green's functions of electric and magnetic type for the corresponding spatial region. As can be seen, the dual symmetry principle is satisfied for the non-basic correspondences

$\vec{\prod}^e \Leftrightarrow \vec{\prod}^m$ and $\hat{G}^e \Leftrightarrow \hat{G}^m$ in expressions (7.4)–(7.6) under the condition of correct permutations of the eigenfunctions in the Green's functions.

In general, the solution of boundary value problems requires that the electromagnetic fields were determined in both adjacent regions. However, this problem can be reduced to the problem of the field determination only in one volume using single-sided boundary condition at the interface between the volumes. That is, approximate Schukin-Leontovich boundary conditions (Sects. 1.3 and 1.4), also known as impedance boundary conditions, can be used instead of relations (7.3). These conditions can be presented as

$$\left[\vec{n}, \vec{E}\right] = -z_i\left[\vec{n}\left[\vec{n}, \vec{H}\right]\right]\Big|_S \tag{7.7}$$

where $\vec{n}$ is the unit normal vector to the surface directed the impedance body interior, and z_i is the distributed complex impedance of the surface S. Substituting the relations (7.3) into (7.7), we can obtain expressions for the equivalent surface currents

$$\left[\vec{E}, \vec{n}\right] = -\frac{4\pi}{c_0}\vec{j}_S^m, \left[\vec{n}, \vec{H}\right] = -\frac{4\pi}{c_0}\vec{j}_S^e, \vec{j}_S^m = -z_i\left[\vec{n}, \vec{j}_S^e\right]\Big|_S, \tag{7.8}$$

where $\vec{j}_S^{e(m)}$ are the surface current densities. Since the conditions (7.7) and (7.8) are not symmetric with respect to the basic dual permutations, application of the permutation duality principle to the theory of electrically thin antennas is somewhat limited.

7.1.1.2 Formulation of Impedance Boundary Conditions

Up to the present, the integral equations for radiators of electric and magnetic types in a dually symmetric form were not applied to the theory of thin antennas. Therefore, the boundary-value problems for vibrator and slot radiators were solved by using different techniques (e.g., 4, 5), but it is often stated that absolute majority of the solution methods are mathematically equivalent or similar. This situation can be caused by two main reasons: the absence of the permutation duality in the formulation of the impedance condition (7.7), and by the difference of electrodynamic volumes for the vibrator and slot radiators (slot is cut into some kind of screen surface). Now we will analyze the influence of these factors on the representation of the integral equations with the aim of their symmetrization with respect to basic permutations.

Let us first consider the classical example of the plane electromagnetic wave incidence at a plane interface between two media [6], taking into account the specifics of the goals in this paper. Let the plane $(x, y, 0)$ in a rectangular coordinate system be the interface between two media with parameters $(\varepsilon_1, \mu_1, \sigma_1)$ and $(\varepsilon_2, \mu_2, \sigma_2)$. The conductivity of the second medium is $\sigma_2 >> 1$. Suppose that the plane electromagnetic wave with the electric vector directed along the axis $\{0x\}$ incidents on the interface from the first region. Let us also define reduced electromagnetic fields as follows

$$\tilde{\vec{E}}_1 = \sqrt{\varepsilon_1'}\vec{E}_1, \tilde{\vec{H}}_1 = \sqrt{\mu_1}\vec{H}_1, \tilde{\vec{E}}_2 = \sqrt{\varepsilon_2'}\vec{E}_2, \tilde{\vec{H}}_2 = \sqrt{\mu_2}\vec{H}_2, \tag{7.9}$$

where $\{\vec{E}_1, \vec{H}_1\}$ and $\{\vec{E}_2, \vec{H}_2\}$ are the true field, $\varepsilon_1 = \varepsilon_1' - i\sigma_1/\omega$ and $\varepsilon_2 = \varepsilon_2' - i\sigma_2/\omega$ are the complex permittivity's of the media. If incident angle θ_1 is measured from the normal to the interface, the plane wave can be represented in the form $\tilde{\vec{E}}_1 = (\tilde{E}_1, 0, 0)$ with $\tilde{E}_1 = E_0 e^{ik_1(y\sin\theta_1 - z\cos\theta_1)}$. For the plane wave we can write

$$k_1\tilde{\vec{H}}_1 = -i\mathrm{rot}\tilde{\vec{E}}_1 = [\vec{k}_1, \tilde{\vec{E}}_1], \tag{7.10}$$

therefore, $\vec{k}_1 = (0, \sin\theta_1, -\cos\theta_1)$, $k_1 = \omega\sqrt{\varepsilon_1'\mu_1}$, and $\left|\tilde{\vec{E}}_1\right| = \left|\tilde{\vec{H}}_1\right|$ The plane wave which polarization corresponds to that of the incident wave is excited in the second medium, while the density of the surface current is zero. Then

$$\tilde{\vec{E}}_2 = (\tilde{E}_2, 0, 0), k_2\tilde{\vec{H}}_2 = [\vec{k}_2, \tilde{\vec{E}}_2], \tag{7.11}$$

where $\vec{k}_2 = (0, \sin\theta_2, -\cos\theta_2)$, $k_2 = \omega\sqrt{\varepsilon_2'\mu_2}$, and $\left|\tilde{\vec{E}}_2\right| = \left|\tilde{\vec{H}}_2\right|$.

The tangential components of the electromagnetic field in the second medium are $\tilde{E}_{2\tau} = \tilde{E}_{2x} = \tilde{E}_2$ and $\tilde{H}_{2\tau} = \tilde{H}_{2y} = \tilde{H}_2\cos\theta_2$, where θ_2 is the propagation angle in this medium. Therefore, we can immediately determine the ratio of the tangential components of the electromagnetic field in the second medium as

$$\tilde{E}_{2x}/\tilde{H}_{2y} = \tilde{E}_2/(\tilde{H}_2\cos\theta_2) = 1/\cos\theta_2 \tag{7.12}$$

The value of $\cos\theta_2$ can be determined by Snell's law, $\frac{\sin\theta_1}{\sin\theta_2} = n = \sqrt{\frac{\varepsilon_2'\mu_2}{\varepsilon_1'\mu_1}}$, where n is the refractive index. Since $\sigma_2 >> 1$ and, hence, $|\varepsilon_2'| >> 1$ and $n >> 1$, we can obtain that

$$\cos\theta_2 = \sqrt{1 - \sin^2\theta_2} = \sqrt{1 - \sin^2(\theta_1)/n^2}. \tag{7.13}$$

When the angle $\theta_1 \approx 0$ or the inequality $\left|\frac{\sin\theta_1}{n}\right| << 1$ holds, we obtained that $\cos\theta_2 \approx 1$ and the relation (7.12) transfers to

$$\tilde{E}_{2x}/\tilde{H}_{2y} \approx 1. \tag{7.14}$$

Returning to the original fields, we find that

$$\tilde{E}_{2x}/\tilde{H}_{2y} = \sqrt{\mu_2/\varepsilon_2'}. \tag{7.15}$$

The boundary condition for the fields in the first medium obtained by using the continuity of the tangential components of the electric and magnetic fields under condition that the surface currents are absent can be written as

$$E_{1x} = z_i^e H_{1y}, \tag{7.16}$$

where $z_i^e = \sqrt{\mu_2 / \varepsilon_2'}$ is the conventional surface impedance of electrical type for the second medium. Since, $\varepsilon_2' = \varepsilon_2 + i\sigma_2/\omega$ is the complex value, the square root branch in the expression (7.16) should be selected so that the imaginary part $\mathrm{Im}\sqrt{\varepsilon_2'} > 0$ ensuring attenuation of the waves propagating in the second medium. For the orthogonal polarization we obtain

$$E_{1y} = -z_i^e H_{1x}. \tag{7.17}$$

Generalizing (7.16) and (7.17) to the arbitrary polarization, we can easily obtain the boundary conditions (7.7) in which $z_i = z_i^e$.

The relations (7.14) and (7.15) can be represented as $\tilde{H}_{2y} / \tilde{E}_{2x} \approx 1$ and $\tilde{H}_{2y} / \tilde{E}_{2x} = \sqrt{\varepsilon_2' / \mu_2}$. Then we can write

$$H_{1y} = z_i^m E_{1x}, \tag{7.18}$$

where $z_i^m = \sqrt{\varepsilon_2' / \mu_2}$ is the surface impedance for the second medium. For the orthogonal polarization of the incident wave we obtain

$$H_{1x} = -z_i^m E_{1y}. \tag{7.19}$$

Then, for the arbitrary polarization we can write

$$\left[\vec{n}, \vec{H}\right] = -z_i^m \left[\vec{n}\left[\vec{n}, \vec{E}\right]\right]\Big|_S. \tag{7.20}$$

Taking into account the relations $\left[\vec{E}, \vec{n}\right] = -\frac{4\pi}{c_0}\vec{j}_S^m$ and $\left[\vec{n}, \vec{H}\right] = -\frac{4\pi}{c_0}\vec{j}_S^e$, we arrive at the following identity

$$\vec{j}_S^e = z_i^m \left[\vec{n}, \vec{j}_S^m\right]\Big|_S. \tag{7.21}$$

Direct comparison of the expressions (7.7) and (7.20), (7.8) and (7.21) shows that the application of the surface impedance (7.18) allowed us to obtain a dually symmetric form of the impedance boundary conditions. In this case, the permutation $z_i^e \Leftrightarrow z_i^m$ is based on the set of basic dual correspondences listed in Table 7.1.

7.1.1.3 Analysis of Representations of Integral Equations for Currents

Let us first consider the electrically thin vibrator consisting of a circular cylindrical wire of finite length. The vibrator radius and length, r and $2l$, are such that inequalities $\frac{r}{2l} << 1$ and $\frac{r}{\lambda} << 1$ hold (λ is wavelength in the external medium). Then we require that the boundary conditions (7.3) should be satisfied on the vibrator surface under condition that the extraneous magnetic current is equal to zero. The influence of the internal vibrator field $\vec{E}_1$ can be taken into account by the equivalent surface magnetic current (7.8), which can be replaced by the electric current. Taking into account the impedance condition, we obtain

$$\left[\vec{n},\vec{E}_2\right] - \bar{z}_i^e\left[\vec{n},\vec{j}_S^e\right]\Big|_S = 0, \tag{7.22}$$

where $\bar{z}_i^e = \frac{4\pi}{c_0} z_i^e$, S is the vibrator surface, $\vec{n}$ is the normal to surface S directed to the region 2. Then we isolate the excitation field $\vec{E}_0$, take into account the relations (7.4)–(7.5), and write, on the basis of (7.22), the equation with respect to the current on the impedance vibrator in the following form

$$-\frac{1}{ik\varepsilon_2}\left[\vec{n},\left(\mathrm{grad}\,\mathrm{div} + k^2\varepsilon_2\mu_2\right)\int\limits_V \vec{j}^e(\vec{r}\,')\hat{G}_2^e(\vec{r},\vec{r}\,')dv\right] + \bar{z}_i^e\left[\vec{n},\vec{j}_S^e\right]\Big|_S = \left[\vec{n},\vec{E}_0\right]. \tag{7.23}$$

In the general case, the outer region 2 must be considered as an area bounded in space by an additional surface S_2, hence, the Green's function for this region can be represented as the sum of two terms [4, 7]

$$\hat{G}_2^e(\vec{r},\vec{r}\,') = \hat{I}G(\vec{r},\vec{r}\,') + \hat{G}_2^{ev}(\vec{r},\vec{r}\,'), \tag{7.24}$$

where $\hat{I}$ is the unit tensor, $G(\vec{r},\vec{r}\,') = \frac{e^{-i|\vec{r}-\vec{r}\,'|k\sqrt{\varepsilon_2\mu_2}}}{|\vec{r}-\vec{r}\,'|}$ is the Green's function of an unbounded space, and $\hat{G}_2^{ev}(\vec{r},\vec{r}\,')$ is the function regular everywhere, which satisfies the homogeneous equation

$$\Delta\hat{G}_2^{ev}(\vec{r},\vec{r}\,') + k^2\varepsilon_2\mu_2\hat{G}_2^{ev}(\vec{r},\vec{r}\,') = 0 \tag{7.25}$$

and ensures, together with $\hat{I}G(\vec{r},\vec{r}\,')$, fulfillment of the boundary conditions on the surface S_2 for the source field located at the point $\vec{r}\,'$. Then (7.23) can be rewritten in the following form

$$-\frac{1}{ik\varepsilon_2}\left[\vec{n},\,(\text{grad div} + k^2\varepsilon_2\mu_2)\int\limits_V \vec{j}^e(\vec{r}')(\hat{I}G(\vec{r},\vec{r}') + \hat{G}_2^{ev}(\vec{r},\vec{r}'))dv\right]$$
$$+\tilde{z}_i^e\left[\vec{n},\vec{j}_S^e\right]\Big|_S = \left[\vec{n},\vec{E}_0\right] \tag{7.26}$$

If we introduce the external effective induced impedance of an electric type z_{ext}^e based on the relation

$$z_{ext}^e\vec{j}_S^e = -(\text{grad div} + k^2\varepsilon_2\mu_2)\frac{1}{ik\varepsilon_2}\int\limits_V \vec{j}^e(\vec{r}')\hat{G}_2^{ev}(\vec{r},\vec{r}')dv, \tag{7.27}$$

Equation (7.26) can be written in the form

$$\frac{1}{ik\varepsilon_2}\left[\vec{n},\,(\text{grad div} + k^2\varepsilon_2\mu_2)\int\limits_V \vec{j}^e(\vec{r}')\hat{I}G(\vec{r},\vec{r}')dv\right]$$
$$+Z^e\left[\vec{n},\vec{j}_S^e\right]\Big|_S = \left[\vec{n},\vec{E}_0\right], \tag{7.28}$$

where $Z^e = \left(z_{ext}^e + \tilde{z}_i^e\right)$ is the total effective impedance of electrical type distributed on the surface of the vibrator.

Let us now consider a narrow slot cut in the boundary surface S_2 of the screen separating regions 1 and 2. The slot length and width, $2l$ and d, are such that inequalities $\frac{d}{2l} << 1$ and $\frac{d}{\lambda} << 1$ hold. Without loss of generality, we assume that the surface S_2 is infinitely thin. Let the boundary conditions (7.3) be satisfied on the slot surface under condition that the extraneous electric current is equal to zero, i.e., $\left[\vec{n}_1,\vec{H}_1\right] = -\left[\vec{n}_2,\vec{H}_2\right]$. Then we represent the slot field $\vec{H}_1$ by using the equivalent electric surface current $\left[\vec{n},\vec{H}\right] = -\frac{4\pi}{c_0}\vec{j}_S^e$, and replace it by the magnetic current (7.21), taking into account the impedance condition. As a result, we obtain:

$$\left[\vec{n},\vec{H}_2\right] + \tilde{z}_i^m\left[\vec{n},\vec{j}_S^m\right]\Big|_S = 0, \tag{7.29}$$

where $\tilde{z}_i^m = \frac{4\pi}{c_0}z_i^m$, S is the slot surface, $\vec{n}$ is the normal to this surface directed to the region 2. Here, we assume that the equivalent currents flow over the slot aperture S is covered by the material that coincides with that of the screen surface. Then we isolate the exciting field $\vec{H}_0$, take into account the relations (7.4)–(7.5), and write, on the basis of (7.29), the equation for the magnetic current in the slot

$$\frac{1}{ik\mu_2}\left[\vec{n},\,(\text{grad div} + k^2\varepsilon_2\mu_2)\int\limits_V \vec{j}^m(\vec{r}')\hat{G}_2^m(\vec{r},\vec{r}')dv\right] + \tilde{z}_i^m\left[\vec{n},\vec{j}_S^m\right]\Big|_S = -\left[\vec{n},\vec{H}_0\right]. \tag{7.30}$$

The magnetic Green's function for region 2 can be represented as the sum of two terms (similarly with (7.24), [5])

$$\hat{G}_2^m(\vec{r},\vec{r}') = \hat{I}G(\vec{r},\vec{r}') + \hat{G}_2^{mv}(\vec{r},\vec{r}'), \tag{7.31}$$

where $\hat{G}_2^{mv}(\vec{r},\vec{r}')$ is the function regular everywhere, which satisfies the homogeneous equation

$$\Delta\hat{G}_2^{mv}(\vec{r},\vec{r}') + k^2\varepsilon_2\mu_2\hat{G}_2^{mv}(\vec{r},\vec{r}') = 0 \tag{7.32}$$

and ensures in conjunction with $\hat{I}G(\vec{r},\vec{r}')$ fulfillment of the boundary conditions on the surface S_2 for the field of a point source located at a point $\vec{r}'$. Then (7.30) can be rewritten in the following form

$$\frac{1}{ik\mu_2}\left[\vec{n},\left(\text{grad div} + k^2\varepsilon_2\mu_2\right)\int\limits_V \vec{j}^m(\vec{r}')\left(\hat{I}G(\vec{r},\vec{r}') + \hat{G}_2^{mv}(\vec{r},\vec{r}')\right)dv\right] + \bar{z}_i^m\left[\vec{n},\vec{j}_S^m\right]\Big|_S = -\left[\vec{n},\vec{H}_0\right]. \tag{7.33}$$

Then we introduce into consideration the external effective impedance of magnetic type, induced on the slot aperture based on the relation

$$z_{ext}^m\vec{j}_S^m = \left(\text{grad div} + k^2\varepsilon_2\mu_2\right)\frac{1}{ik\mu_2}\int\limits_V \vec{j}^m(\vec{r}')\hat{G}_2^{mv}(\vec{r},\vec{r}')dv, \tag{7.34}$$

and rewrite Eq. (7.33) in the following form

$$\frac{1}{ik\mu_2}\left[\vec{n},\left(\text{grad div} + k^2\varepsilon_2\mu_2\right)\int\limits_V \vec{j}^m(\vec{r}')\hat{I}G(\vec{r},\vec{r}')dv\right] + Z^m\left[\vec{n},\vec{j}_S^m\right]\Big|_S = -\left[\vec{n},\vec{H}_0\right], \tag{7.35}$$

where $Z^m = \left(z_{ext}^m + \bar{z}_i^m\right)$ is the total effective impedance of magnetic type, distributed at the slot aperture.

Thus, new notions: the external effective induced impedances of electric and magnetic types, $z_{ext(i)}^{e(m)}$, were introduced in the electrodynamic analysis of electrically thin antennas. The notions can be effectively used to formulate integral equations for radiators of electric and magnetic types in a dually symmetric form. Up to the present, the integral equations for the currents in this form are never used in classical theory of antennas. The proposed approach made it possible to eliminate two existing methodological problems: the nonsymmetrical in the impedance boundary conditions with respect to the dual basic permutations and using of heterogeneous electrodynamic regions in boundary-value problems for vibrator and slot antennas. As a result, the uniform form of integral equations for currents in electrically thin antennas was derived in the form

$$\frac{1}{ik\begin{pmatrix}\varepsilon_2\\ \mu_2\end{pmatrix}}\left[\vec{n},\left(\text{grad div}+k^2\varepsilon_2\mu_2\right)\int\limits_V\begin{pmatrix}\vec{j}^e(\vec{r}')\\ -\vec{j}^m(\vec{r}')\end{pmatrix}\hat{I}G(\vec{r},\vec{r}')dv\right]$$
$$+\begin{pmatrix}Z^e\\ Z^m\end{pmatrix}\left[\vec{n},\begin{pmatrix}\vec{j}_S^e\\ -\vec{j}_S^m\end{pmatrix}\right]\Bigg|_S=\left[\vec{n},\begin{pmatrix}\vec{E}_0\\ \vec{H}_0\end{pmatrix}\right] \qquad (7.36)$$

which is symmetric with respect to the basic and additional dual permutations. Equation (7.36) is the methodological basis for the unification of analysis methods applicable for the radiators of electric and magnetic types.

7.1.2 Impedance Model of the Slot Radiator

Before solving the impedance synthesis problem for the slot AA, a model of the single slot radiator will be considered, and the features of the constructive implementation of surface impedance will be analyzed. Consider a narrow linear rectangular slot cut in an infinite perfectly conduction screen. The slot length and width are $2L$ and d. Let us introduce a Cartesian coordinate system related to the screen and slot as shown in Fig. 7.1

Let the slot be loaded by a cavity of a rectangular resonator which transverse dimension are define by the slot aperture S. The resonator with perfectly conducting wall is uniformly filled by a magneto-dielectric with material parameters (ε, μ). The internal volume of the resonator is v, and its wall thickness is c. The half-space above

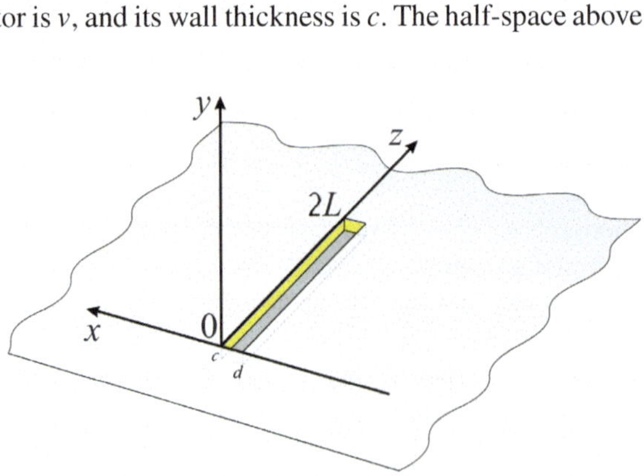

Fig. 7.1 The slot element geometry

the screen mark by index V is assumed to be filled by a homogeneous medium with material parameters (ε_1, μ_1). Such structures excited by an incident wave propagating from the external half-space are termed as passive slot impedance loads [8–10]. If the voltage δ-generator is located between the slot edges in the geometric slot center, then such structure represent the slot radiator with concentrated excitation.

The excitation problem of the slot radiator can be solved as the coupling problem between two electrodynamic volumes through the slot hole, taking into account the primary field $\{\vec{E}_0(\vec{r}),\ \vec{H}_0(\vec{r})\}$ of the δ-generator ($\vec{r}$ is the radius-vector of the observation point). The electromagnetic fields $\{\vec{E}^V(\vec{r}),\ \vec{H}^V(\vec{r})\}$ and $\{\vec{E}^v(\vec{r}),\ \vec{H}^v(\vec{r})\}$ satisfying the boundary conditions on the screen and resonator wall should be determined in the coupling volumes.

The continuity condition for the magnetic fields at the slot aperture can be written as the functional equation for the slot electric field $\vec{e}_S$

$$\vec{H}_\tau^V(\vec{e}_S) + \vec{H}_{0\tau}^V = \vec{H}_\tau^v(\vec{e}_S), \tag{7.37}$$

where $\vec{H}_\tau^V(\vec{e}_S)$ and $\vec{H}_\tau^v(\vec{e}_S)$ are the components of the magnetic field vector tangential to the slot plane, $\vec{H}_{0\tau}^V$ is the tangential component of the magnetic field vector which is not perturbed by the primary field of the δ-generator. The solution will be obtained under conditions that the inequalities $\frac{d}{2L} \ll 1$ and $\frac{d}{\lambda} \ll 1$ hold (λ is the wavelength in free space). The slot length $2L$ is commensurate with $\lambda/2$, the field in the slot can be approximated by a basis function $f(z)$ with an unknown complex amplitude V_0 [5]

$$\vec{e}_S = \vec{x}^0 V_0 f(z), \tag{7.38}$$

where $\vec{x}^0$ is the unit vector of the coordinate axis $\{0, x\}$. Then the expression (7.37) can be represented as follows

$$V_0\vec{H}_\tau^V\left(\vec{x}^0 f(z)\right) + \vec{H}_{0\tau}^V = V_0\vec{H}_\tau^v\left(\vec{x}^0 f(z)\right). \tag{7.39}$$

Since the longitudinal magnetic current in the slot $\vec{j}_S^m = \left[\vec{e}_S, \vec{n}\right]$, where the unit vector $\vec{n}$ is the inward normal to the slot surface of the corresponding volume, Eq. (7.39) relative to the magnetic current can be represented as

$$V_0\vec{H}_\tau^V\left(\vec{z}^0 f(z)\right) + V_0\vec{H}_\tau^v\left(\vec{z}^0 f(z)\right) = -\vec{H}_{0\tau}^V, \tag{7.40}$$

where $\vec{z}^0$ is the unit vector of the coordinate axis $\{0, z\}$.

If the slot surface is conditionally metallized, the magnetic fields in adjacent volumes defined by (7.40) can be determined using magnetic tensor Green's functions for the vector potential $\hat{G}_V^m(\vec{r}, \vec{r}')$ and $\hat{G}_v^m(\vec{r}, \vec{r}')$ for the half-space above the screen and rectangular resonator [5]. Then the expression (7.40) can be written in the following form

$$\frac{V_0}{i\omega\mu_1}\left[\vec{n},\left\{\text{grad div}+k_V^2\right\}\int\limits_S \vec{z}^0 f\left(z'\right)\hat{G}_V^m(\vec{r},\vec{r}')d\vec{r}'\right]$$
$$+\frac{V_0}{i\omega\mu}\left[\vec{n},\left\{\text{grad div}+k_v^2\right\}\int\limits_S \vec{z}^0 f\left(z'\right)\hat{G}_v^m(\vec{r},\vec{r}')d\vec{r}'\right]$$
$$=-\left[\vec{n},\vec{H}_0^V(\vec{r})\right]. \tag{7.41}$$

where $k_V = k\sqrt{\varepsilon_1\mu_1}$, $k_v = k\sqrt{\varepsilon\mu}$,$k = 2\pi/\lambda$ are the wave numbers in the corresponding volumes, $\vec{r}$ and $\vec{r}'$ are the radius-vectors of the observation points and sources, respectively.

On the other hand, Eq. (7.41) relative the magnetic current in the slot can be written based on the duality principle in the form

$$\frac{V_0}{i\omega\mu_1}\left[\vec{n},\left\{\text{grad div}+k_V^2\right\}\int\limits_S \vec{z}^0 f\left(z'\right)\hat{G}_V^m(\vec{r},\vec{r}')d\vec{r}'\right]$$
$$+\frac{4\pi}{c}z_i^m(z)V_0 f(z)\vec{z}^0=-\left[\vec{n},\vec{H}_0^V(\vec{r})\right], \tag{7.42}$$

where $z_i^m(z)$ is the internal effective linear magnetic impedance induced on the slot aperture.

The linear impedance in (7.42) which, in the general case, can be variable will be presented as a constant averaged integral parameter $\overline{z}_i^m = \frac{1}{2L}\int_0^{2L} z_i^m(z)dz$ [4]. Then, based on a comparative analysis of solution Eqs. (7.41) and (7.42) obtained by the Galerkin method, it is not difficult to write

$$\overline{z}_i^m = \frac{c_0\int\limits_S\left[\vec{z}^0 f(z),\left\{\text{grad div}+k_v^2\right\}\int\limits_S \vec{z}^0 f\left(z'\right)\hat{G}_v^m(\vec{r},\vec{r}')d\vec{r}'\right]d\vec{r}}{4\pi i\omega\mu\int\limits_S f^2(z)dxdz}. \tag{7.43}$$

If the narrow slot is resonantly exited, the approximation (7.38) can be presented in the form of a trigonometric function [5]

$$\vec{e}_S = \frac{V_0}{d}\sin\left(\frac{\pi z}{2L}\right)\vec{x}^0. \tag{7.44}$$

Then, the final expression for the impedance $\overline{z}_i^m$ can be obtained based on (7.43) in the form

$$\overline{z}_i^m = \frac{\sqrt{k^2(\varepsilon\mu) - (\pi/(2L))^2}}{ikd\tan\left(c\sqrt{k^2(\varepsilon\mu) - (\pi/(2L))^2}\right)}\sqrt{\frac{\varepsilon}{\mu}}. \tag{7.45}$$

This formula can be interpreted as a justification of constructive implementation of the magnetic type impedance on the aperture of the slot radiator. As can be seen from (7.45), the required impedance can be obtained by fitting the electrical parameters of the slot load in the form of the resonator filled with the magneto-dielectric.

7.1.3 Solution of the Impedance Synthesis Problem

Consider a flat AA consisting of several rows of narrow linear slots which are symmetrically excited by voltage δ-generators radiating into the half-space with material parameters (ε_1; μ_1) and placed above the perfectly conducting screen and located in the plane {x0z} of the Cartesian coordinate system (x, y, z) as shown in Fig. 7.2.

Let the origin of the coordinate system be related to the center of the array. Define the following array parameters: d_z, the distance between the adjacent slot rows, d_x, the distance between the geometric centers of slots located in a row, N_z, the number of rows, and N_x, the number of slots in a row. Let us also introduce the spherical coordinate system which polar axis coincides with the axis {0*z*}, and the angle φ is measured from the axis {0*x*}. Without loss of generality, assume that the length of all $N = N_z \times N_x$ slots is $2L$ and the slots are characterized by variable induced impedances $\overline{Z}_{nm}^m = \overline{z}_i^m / Z_0$ normalized to the wave resistance of the medium $Z_0 = \sqrt{\mu_1/\varepsilon_1}$ [Ohm], where $n \in [1, N_z]$ and $m \in [1, N_x]$ are integers. We also assume that the slot widths d_{nm} satisfy the inequalities $\frac{d_{nm}}{2L} << 1$. The coordinates of geometric centers of slot apertures are (x_m; z_n).

The magnetic currents $J_{nm}^m(s)$ in the slots can be approximated in the form

$$J_{nm}^m(s) = J_{nm} \sin \tilde{k}_{nm}(L - |s|), \tag{7.46}$$

where $\tilde{k}_{nm}^2(s) = k_1^2\left(1 + \frac{i8\alpha_{nm}\overline{Z}_{nm}^m}{k_1 d_{nm}}\right)$, $\alpha_{nm} = \frac{1}{2\ln[d_{nm}/(8L)]}$ and local coordinates s are defined on Fig. 7.2.

Suppose that slot radiators are resonantly tuned varying the internal resistances of the δ-generators, then the mutual influence of the slot radiators can be compensated. Since the components of the magnetic field in the wave zone $H_\varphi = 0$ and $H_r = 0$, the total field $H_\theta(r, \theta, \varphi)$ radiated by the array is equal to the sum of the fields radiated by each slot under conditions that the phases of these fields in the observation point $C(r, \theta, \varphi)$. This field can be written as

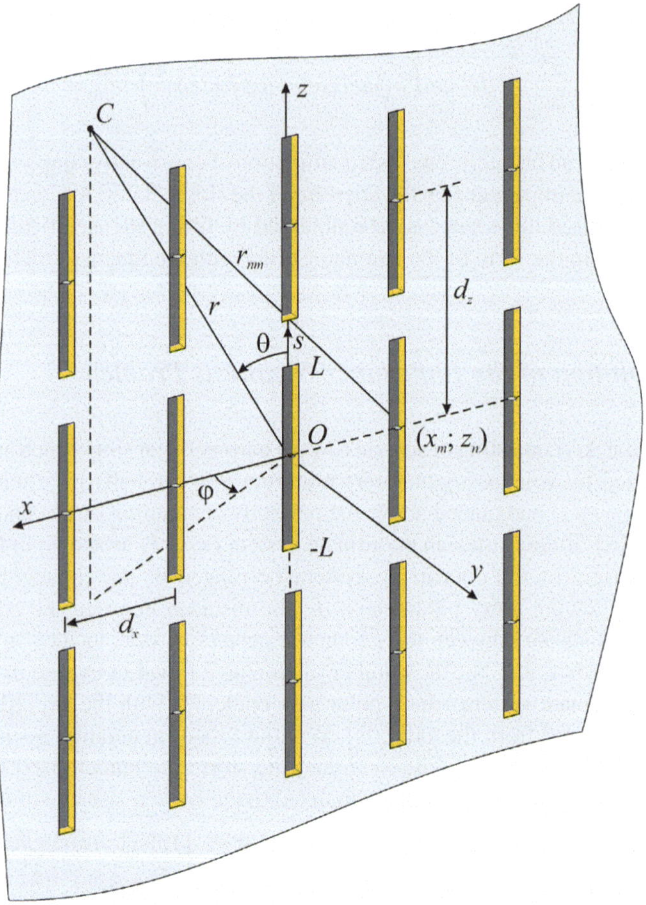

Fig. 7.2 The geometry of the slot AA

$$H_\theta(r, \theta, \varphi) = \frac{120\pi i}{\lambda} \sum_{n=1}^{N_z} \sum_{m=1}^{N_x} \frac{e^{-ik_1 r_{nm}}}{r_{nm}} J_{nm}$$

$$\frac{\cos(k_1 L \cos\theta) - \cos\left(\tilde{k}_{nm} L\right)}{\tilde{k}_{nm}^2 - (k_1 \cos\theta)^2} \tilde{k}_{nm} \sin\theta. \tag{7.47}$$

The impedance synthesis problem solution consists in determining the impedances $\overline{Z}_{nm}^{m} = \overline{R}_{nm}^{m} + i\overline{X}_{nm}^{m}$ of each radiating element required to direct the RP maximum in a predefine direction $(\theta_{\max}, \varphi_{\max})$. This can be done based on the expression (7.47). The final analytical expressions defining the slot impedances, based on the results obtained in (Chap. 6), can be written as:

$$\overline{R}_{nm}^{m} = \frac{k_1 d_{nm} \sin(k_1 \gamma_{nm})}{4\alpha_{nm}\left(\frac{1+\cos^2\theta_{\max}}{\sin^2\theta_{\max}} - \frac{k_1 L \sin(k_1 L)}{Fc(\theta_{\max})}\right)}, \quad (7.48a)$$

$$\overline{X}_{nm}^{m} = \frac{-k_1 d_{nm}\left[1 - \cos(k_1 \gamma_{nm})\right]}{4\alpha_{nm}\left(\frac{1+\cos^2\theta_{\max}}{\sin^2\theta_{\max}} - \frac{k_1 L \sin(k_1 L)}{Fc(\theta_{\max})}\right)}, \quad (7.48b)$$

Where $Fc(\theta_{\max}) = \cos(k_1 L \cos\theta_{\max}) - \cos(k_1 L)$, $\gamma_{nm} = [(n-1)d_z \cos\theta_{\max} + (m-1)d_x \sin\theta_{\max} \cos\varphi_{\max}]$.

7.1.4 Numerical Results

The possibility of steering the RP maximum of the AA within the entire front half-space $0 \le \theta \le \pi$, $0 \le \varphi \le \pi$ by varying the radiator internal impedances was validated in (Chap. 6). This was done based on the example of the square AA consisting of twenty-five $(N = 25,\ N_x = 5,\ N_z = 5)$ half-wave radiators $(2L = \lambda/2)$ with equal transverse dimensions $(d_{nm} = 4L/75)$. As expected, the RP main lobe width turned out to be dependent upon the aperture dimensions and the observation angles, therefore, the width can vary during scanning.

Formula (7.48) are valid for any number of slots in the array and an arbitrary distance between the radiators. However, it turned out that, in an arbitrary case, the calculated impedance values cannot guarantee the positivity of their real parts, a requirement that follows from energy considerations (Chap. 6). It was found that positive values of the real parts of the impedances can be guaranteed if the resulting phase difference of the vibrator currents at array ends does not exceed 180°.

Let us analyze the application of this approach for the slot radiators within the framework of their constructive implementation. The results of such studies are presented for the AA characterized by positive values of the real parts of the slot impedances which is scanned only in the E-plane when angle $\theta_{\max} = 90°$, while the angle $\varphi_{\max}$ varies in the range $\varphi_{\max} \in [60°, 120°]$.

In Fig. 7.3 the normalized H-plane RP are plotted by the dashed curve; the E-plane RP are shown by the curves plotted for the angle: 1 – $\varphi_{\max} = 90°$, 2 – $\varphi_{\max} = 60°$, and 3 – $\varphi_{\max} = 120°$. These curves were plotted under condition that the impedances $\overline{Z}_{1m}^{m} = \overline{Z}_{2m}^{m} = \overline{Z}_{3m}^{m} = \overline{Z}_{4m}^{m} = \overline{Z}_{5m}^{m}$ of the radiators located in vertical rows and oriented along the axis $\{0z\}$ are equal to each other. The calculated impedances in the horizontal rows are different, for example, when $\varphi_{\max} = 60°$, the slot impedances are $\overline{Z}_{11}^{m} = 0$, $\overline{Z}_{12}^{m} = 0.26 - i0.108$, $\overline{Z}_{13}^{m} = 0.368 - i0.368$, $\overline{Z}_{14}^{m} = 0.26 - i0.628$, $\overline{Z}_{15}^{m} = -i0.735$.

The simple formula for estimating the slot permeability can be obtained based on (7.9) assuming that condition $c << \lambda$ fulfils and the slot is filled by the magnetics with $\varepsilon = 1$. This formula can be written as

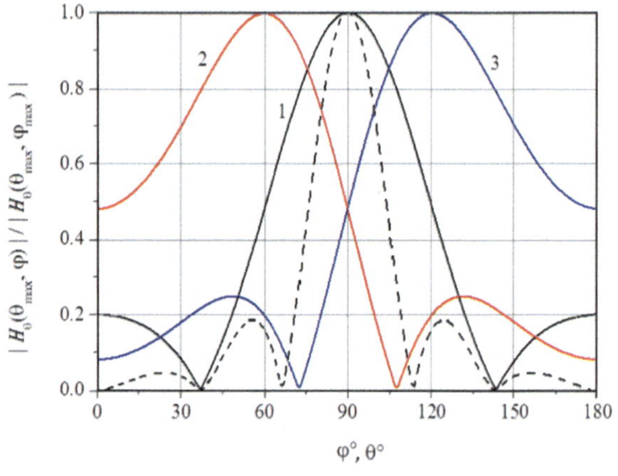

Fig. 7.3 The RP of the AA in the main polarization planes: E-plane: 1—$\varphi_{\max} = 90°$, 2—$\varphi_{\max} = 60°$, 3—$\varphi_{\max} = 120°$; H-plane—dashed curve

$$\mu = -1 \Big/ \left(kdcZ_0\overline{Z}_{nm}^{m}\right)^2, \tag{7.49}$$

If $\lambda = 0.03$ m and $c = \lambda/100$, this formula allows to obtain $\mu_{11} = 626$, $\mu_{12} = (9.917 - i9.957) \times 10^4$, $\mu_{13} = -i4.113 \times 10^4$, $\mu_{14} = (1.706 - i1.704) \times 10^4$, $\mu_{15} = 5.887 \times 10^4$. It should be noted that the impedance $\overline{Z}_{11}^{m} = 0$ can be realized if $c = \lambda/(4\sqrt{\varepsilon\mu - 1})$, that is, if the tangent in the denominator (7.45) tend to infinity. In this case, the slot impedance required for RP scanning can be obtained by filling slot cavities with a ferromagnetic.

In conclusion, we note that this technique is valid for the problems of impedance synthesis in the case of combined vibrator-slot structures [11].

7.2 Impedance Synthesis for Two-Dimensional Array of Slotted Spherical Radiators

In this Section an impedance synthesis problem of 2D antenna array consisting of slotted spherical radiators, whose geometric centers are located at nodes of a flat rectangular grid with double periodicity, has been presented. The problem is formulated as follows: to determine complex impedances distributed over surfaces of the spherical radiators which allows us to steer the RP of the antenna array to given directions. Analytical solution of the impedance synthesis problem (as an alternative to numerical solution) was obtained under the assumption that spherical radiators are excited by axially symmetric magnetic currents with equal amplitudes.

It should be noted that the electrodynamic characteristics of a slotted spherical antenna consisting of spherical scatterer with a narrow equatorial slot can be analyzed using the theory of conformal antennas [12]. However, such an analysis by direct numerical methods is difficult to accomplish even for an isolated ideally conducting sphere and especially for an array of spherical radiators with impedance coatings.

Arrays of small impedance spherical resonators can also be considered as elements of tunable metamaterials for monitoring terahertz (THz) radiation [13]. For example, such materials based on dielectric resonators which do not use conductive materials and, hence, eliminate some loss); they can be important for THz applications [14, 15]. Such structures with innovative impedance coatings can be used for the detection of THz waves, and they can be used in integrated elements of photodetectors [16]. In any case, such applications confirm the practical significance of the results presented in this Section.

7.2.1 Formulation of the Impedance Synthesis Problem

Let a circular slot be cut in a sphere parallel to its equatorial section as shown in Fig. 7.4. The sphere radius is R, the slot width measured along the slot ark is d. The coordinate of the slot center is $\theta' = \theta_0$. In the spherical coordinate system (r', θ', φ'), the slot occupies the area defined by coordinates $\varphi' \in [0\ ;\ 2\pi]$ and $\theta' \in [\theta_0 - d/(2R)\ ;\ \theta_0 + d/(2R)]$. The sphere surface is characterized by distributed isotropic impedance $\overline{Z}_S$ normalized to free space resistance $Z_0 = 120\pi\ \Omega$.

Let us consider the most interesting for practical application axially symmetric excitation of narrow ($d << R$) circular slot. We may assume that the slot electric field has only a meridional component with a constant distribution along the slot axis. Then, within the framework of the above approximations, the density of equivalent magnetic $\vec{J}^m$ and electric $\vec{J}^e$ currents on the slot aperture can be represented as [17]

$$\vec{J}^m = -\frac{V_0}{d}\delta(r' - R)\vec{\varphi}^0, \vec{J}^e = \frac{V_0}{Z_S d}\delta(r' - R)\vec{\theta}^0, \tag{7.50}$$

where $\vec{\varphi}^0$ and $\vec{\theta}^0$ are unit vectors of the spherical coordinate system, $\delta(r' - R)$ is Dirac delta function, V_0 is the complex voltage amplitude, and $Z_S = \overline{Z}_S Z_0$.

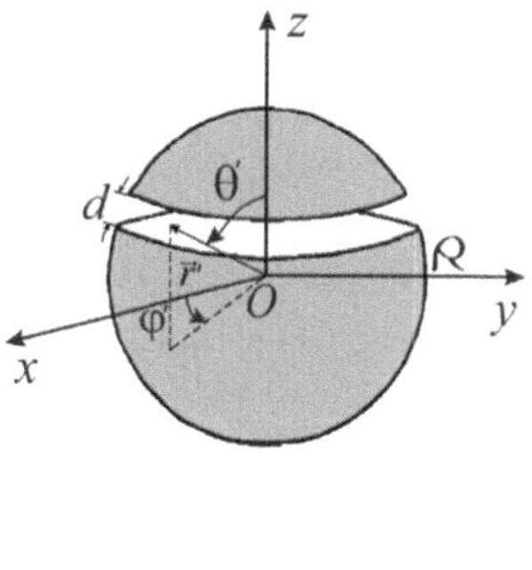

Fig. 7.4 The slotted spherical antenna

The electromagnetic fields excited by impedance spherical antennas in free space can be obtained based on components of the Green's functions by using pair representations for currents (7.50) on the spherical scatterer surface [17]. In the case of monochromatic excitation, when the time t dependence is $e^{i\omega t}$ (ω is the circular frequency), electric field components in the spherical coordinate system (r, θ, φ) can be written as

$$\begin{aligned} E_r(r,\theta) &= \frac{V_0 R \sin\theta_0}{2rd}\sum_{j=1}^{\infty} A_j\left[\cot\theta\, P_j^1(\cos\theta) + \frac{dP_j^1(\cos\theta)}{d\theta}\right] \\ &\quad\times\left\{h_j^{(2)}(kr) + \frac{\overline{Z}_S}{ikr}\left[(j-1)h_j^{(2)}(kr) - kr\, h_{j+1}^{(2)}(kr)\right]\right\}, \\ E_\theta(r,\theta) &= \frac{V_0 kR\sin\theta_0}{2d}\sum_{j=1}^{\infty} A_j\left\{P_j^1(\cos\theta)\left[h_{j+1}^{(2)}(kr) - h_j^{(2)}(kr)\left(i\overline{Z}_S + \frac{j+1}{kr}\right)\right]\right. \\ &\quad\left. -\frac{\overline{Z}_S}{i(kr)^2}h_j^{(2)}(kr)\left[\begin{array}{l} j(j+1)\left(P_j^1(\cos\theta) + \dfrac{dP_j^1(\cos\theta)}{d\theta}\right) \\ -\dfrac{1}{\sin^2\theta}\left(\dfrac{dP_j^1(\cos\theta)}{d\theta} - 2\cot\theta\, P_j^1(\cos\theta)\right)\end{array}\right]\right\}, \\ E_\varphi(r,\theta) &= 0, \end{aligned} \tag{7.51}$$

where $k = 2\pi/\lambda$ is the wave number, λ is the free space wavelength, $h_j^{(2)}(kr) = \sqrt{\frac{\pi}{2kr}}H_{j+1/2}^{(2)}(kr)$ are the Hankel's spherical functions of the second kind, $P_j^l(\cos\theta)\Big|_{l=0,1}$ are the Legendre associated functions of the first kind, and the coefficients A_j are defined as follows

$$A_j = \frac{(2j+1)}{j\,(j+1)}\times\frac{P_j(\cos(\theta_0 + d/(2R))) - P_j(\cos(\theta_0 - d/(2R)))}{Q_j\left(h_j^{(2)}(kR)\right)}, \tag{7.52}$$

where $Q_j(h_j^{(2)}(kR)) = \frac{(j+1)\left[j\overline{Z}_S - ikR\,(1+\overline{Z}_S^2)\right] - 2\overline{Z}_S k^2R^2}{ikR\,(1+\overline{Z}_S^2)}\, h_j^{(2)}(kR) + kRh_{j+1}^{(2)}(kR)$. The formulas (7.51) will be used for further analysis.

Consider the 2D antenna array consisting of the slot spherical antennas (Fig. 7.4) located in free space at the plane $\{x0y\}$ of the Cartesian coordinate system (x, y, z) as shown in Fig. 7.5. We will also define a spherical coordinate system, which polar axis coincides with the axis $\{0z\}$, and the angle φ is measured from the axis $\{0x\}$. Let d_y be the distance between adjacent rows of the array, d_x be the distance between the centers of spherical elements in a row, N_y be the number of rows, and N_x be the number of elements in a row. Without loss of generality, we will assume that the parameters of the spherical radiators: sphere radii R, d, and θ_0 are equal for all

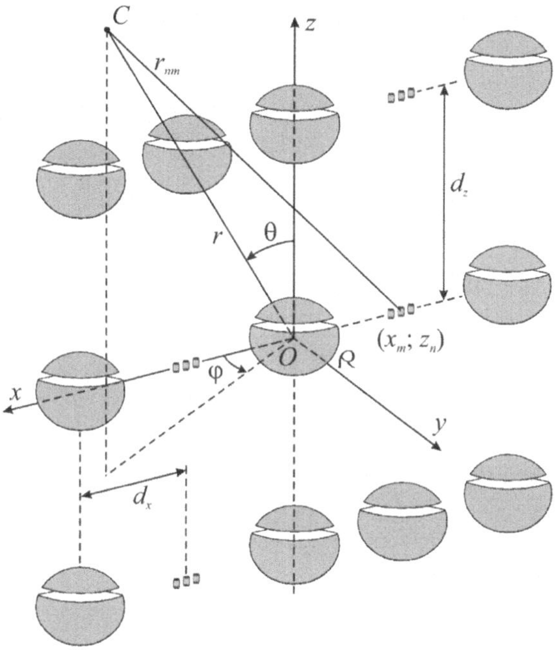

Fig. 7.5 The geometry of the 2D antenna array

$N = N_z \times N_x$. The sphere impedances, $\overline{Z}_{S,nm}$ ($n \in [1, N_z]$ and $m \in [1, N_x]$), are different.

Taking into account the formulas (7.51) and the asymptotics of the spherical functions $h_j^{(2)}(kr) \approx \frac{1}{kr} i^{j+1} e^{-ikr}$ under conditions $|kr| \gg j$, $(kr) \to \infty$, we can write the main component of the radiated electric field in the wave zone by the spherical antenna with indices (n, m) in the following form

$$E_{nm}^{\theta} = -\frac{V_{nm} kR \sin\theta_0}{2d} \times \frac{e^{-ikr_{nm}}}{kr_{nm}} \sum_{j=1}^{\infty} A_{j,nm} P_j^1(\cos\theta) i^j \left[1 - \overline{Z}_{S,nm}\right], \tag{7.53}$$

where $A_{j,nm} = A_j\big|_{\overline{Z}_S \to \overline{Z}_{S,nm}}$.

If terms proportional to $(1/r)^2$ and $(1/r)^3$ are eliminated from the formulas (7.51), the components of the electric fields E_φ and E_r of a single spherical radiator in the wave zone are equal to zero. Let all spherical radiators in the array be resonantly tuned by selecting their internal impedances and parameters d_x/λ, R/d_x, d_y/λ, R/d_y. In this case, the mutual influence of elements can be compensated, and the total radiation field of the array $E_\theta(r, \theta, \varphi)$, can be presented as the sum of the radiation fields of all radiator. If the phases of these fields arriving to the observation point $C(r, \theta, \varphi)$ are taken into account, the resulting field can be written as

$$E_\theta(r,\theta,\varphi) = -\frac{kR\sin\theta_0}{2d}\sum_{n=1}^{N_z}\sum_{m=1}^{N_x}\frac{e^{-ikr_{nm}}}{r_{nm}}V_{nm}$$

$$\sum_{j=1}^{\infty} A_{j,nm}P_j^1(\cos\theta)i^j\left[1-\overline{Z}_{S,nm}\right]. \tag{7.54}$$

Thus, the problem of impedance synthesis is reduced to defining the impedances of each radiators, $\overline{Z}_{S,nm}$, that for a given direction of the maximum radiation in the wave zone $(\theta_{\max}, \varphi_{\max})$,, is determined based on formula (7.54).

7.2.2 Solution of the Impedance Synthesis Problem

First of all, let us convert the expression (7.53) for the radiation field of the single spherical radiator in another form. For perfectly conducting spherical radiators ($\overline{Z}_S = 0$), the coefficients (7.52) can be presented as

$$A_j\big|_{\overline{Z}_S=0} = A_j^0 = \frac{(2j+1)}{j\,(j+1)} \times \frac{P_j\left(\cos\left(\theta_0 + d/(2R)\right)\right) - P_j\left(\cos\left(\theta_0 - d/(2R)\right)\right)}{-(j+1)h_j^{(2)}(kR) + kRh_{j+1}^{(2)}(kR)}. \tag{7.55}$$

Since according to the impedance concept $\left|\overline{Z}_{S,nm}\right| \ll 1$, the expression for the functional $Q_j(h_j^{(2)}(kR))$ up to terms proportional $\overline{Z}_{S,nm}^2$ can be can rewritten as

$$Q_j(h_j^{(2)}(kR)) \approx -(j+1)h_j^{(2)}(kR) + kRh_{j+1}^{(2)}(kR) - i\frac{\overline{Z}_{S,nm}}{kR}\left(j(j+1) - 2k^2R^2\right)h_j^{(2)}(kR).$$

Multiplying the numerator and denominator of the expression (7.52) by $\left[-(j+1)h_j^{(2)}(kR) + kRh_{j+1}^{(2)}(kR) + i\frac{\overline{Z}_{S,nm}}{kR}\left(j(j+1) - 2k^2R^2\right)h_j^{(2)}(kR)\right]$. we obtain with accuracy up to $\overline{Z}_{S,nm}^2$

$$A_{j,nm} \approx A_j^0\left[1 + i\frac{\overline{Z}_{S,nm}}{kR} \times \frac{j(j+1) - 2k^2R^2}{kRh_{j+1}^{(2)}(kR) - (j+1)h_j^{(2)}(kR)}h_j^{(2)}(kR)\right]. \tag{7.56}$$

Then, the formula (7.53) takes the following form

$$E_{nm}^{\theta} \approx -\frac{V_{nm}kR\sin\theta_0}{2d} \times \frac{e^{-ikr_{nm}}}{kr_{nm}} \sum_{j=1}^{\infty} A_j^0 P_j^1(\cos\theta) i^j$$
$$\times \left[1 - \overline{Z}_{S,nm}\left(1 - \frac{ih_j^{(2)}(kR)}{kR} \frac{j(j+1) - 2k^2R^2}{kRh_{j+1}^{(2)}(kR) - (j+1)h_j^{(2)}(kR)}\right)\right]. \quad (7.57)$$

The formula (7.57) can be presented as sum of two terms. The first term defines the RP of the annular slot cut in a perfectly conducting sphere [17, 18],

$$Fs(\theta) = -\frac{kR\sin\theta_0}{2kd} \sum_{j=1}^{\infty} A_j^0 P_j^1(\cos\theta) i^j, \quad (7.58)$$

while the second defines variation of the radiation field of the spherical antenna, due to influence of the sphere impedance coating. Thus, we obtain

$$E_{nm}^{\theta} = V_{nm}\frac{e^{-ikr_{nm}}}{r_{nm}}\left[Fs(\theta) - \overline{Z}_{S,nm}(Fs(\theta) + \Delta Fs(\theta))\right], \quad (7.59)$$

where the term $\Delta Fs(\theta) = \frac{\sin\theta_0}{2kd} \sum_{j=1}^{\infty} A_j^0 \frac{j(j+1) - 2k^2R^2}{kRh_{j+1}^{(2)}(kR) - (j+1)h_j^{(2)}(kR)} P_j^1(\cos\theta) h_j^{(2)}(kR) i^{j+1}$. Finally, we can write

$$E_\theta(r, \theta, \varphi) = \sum_{n=1}^{N_z}\sum_{m=1}^{N_x} \frac{e^{-ikr_{nm}}}{r_{nm}} V_{nm}$$
$$\left[Fs(\theta) - \overline{Z}_{S,nm}(Fs(\theta) + \Delta Fs(\theta))\right]. \quad (7.60)$$

Taking into account the geometric path difference for the wave propagating from neighboring array elements to the observation point $C(r, \theta, \varphi)$, we can rewrite the expression (7.60) for the field $E_\theta(r, \theta, \varphi)$ in wave zone as

$$E_\theta(\theta, \varphi) = e^{-i(N_z+1)u/2} e^{-i(N_x+1)v/2}$$
$$\sum_{n=1}^{N_z}\sum_{m=1}^{N_x} V_{nm}\left[Fs(\theta) - \overline{Z}_{S,nm}(Fs(\theta) + \Delta Fs(\theta))\right] e^{i(nu+mv)}. \quad (7.61)$$

where $u = kd_z \cos\theta$ and $v = kd_x \sin\theta\cos\varphi$. If the spherical radiators are perfectly conducting, $\overline{Z}_{S,nm} = 0$, and amplitudes of excitation voltages are equal, $V_{nm} = V_0$, the expression (7.61) can be reduced to

$$E_\theta(\theta, \varphi) = V_0 e^{-i(N_z+1)u/2} e^{-i(N_x+1)v/2} Fs(\theta) \sum_{n=1}^{N_z} e^{inu} \sum_{m=1}^{N_x} e^{imv}. \tag{7.62}$$

In this case, the maximum of the RP of the array is reached under conditions $u = v = 0$, and it is directed along the axis $\{0y\}$ ($\theta = \pi/2, \varphi = \pi/2$). As can be seen, the expression (7.62) is the product of two multipliers consisting of independent series. These multipliers present the RP of the linear array of spherical antennas which axes are directed along the axes $\{0z\}$ and $\{0x\}$.

As is known from the antenna array theory (see, e.g., [19]), scanning of the array RP in space can be achieved by linear phase shifts between currents of the adjacent array radiators. If the phase shift between adjacent rows and adjacent radiators in a row are equal to Δv and Δu, the direction of the maximum array radiation can be determined by the angles ($\theta_{\max}$; $\varphi_{\max}$) defined by the formulas.

$$\cos\theta_{\max} = \Delta u/(kd_z) \text{ and } \sin\theta_{\max}\cos\varphi_{\max} = \Delta v/(kd_x). \tag{7.63}$$

Then the expression (7.62) can be represented as

$$E_\theta(\theta, \varphi) = V_0 e^{-i(N_z+1)u/2} e^{-i(N_x+1)v/2} Fs(\theta)$$
$$\sum_{n=1}^{N_z} \sum_{m=1}^{N_x} e^{i(nu+mv)} e^{-i[(n-1)\Delta u+(m-1)\Delta v]}. \tag{7.64}$$

As can be seen, the formulas (7.61) and (7.64) become identical if the relations

$$1 - e^{-ik[(n-1)d_z\cos\theta+(m-1)d_x\sin\theta\cos\varphi]} = \overline{Z}_{S,nm}\left(1 + \frac{\Delta Fs(\theta)}{Fs(\theta)}\right) \tag{7.65}$$

are valid for all indices n and m.

If the impedances of the array radiators are complex quantities $\overline{Z}_{S,nm} = \overline{R}_{nm} + i\overline{X}_{nm}$, Eq. (7.65) can be represented as follows

$$1 - \cos(k\gamma_{nm}) + i\sin(k\gamma_{nm}) = \left(\overline{R}_{nm} + i\overline{X}_{nm}\right)\left(1 + \frac{\Delta Fs(\theta_{\max})}{Fs(\theta_{\max})}\right), \tag{7.66}$$

where $\gamma_{nm} = [(n-1)d_z\cos\theta_{\max} + (m-1)d_x\sin\theta_{\max}\cos\varphi_{\max}]$. The relations (7.65) are equations for the array beam scanning, which solution uniquely defines the matrix of unknown impedances $\{\overline{Z}_{S,nm}\}$ for the predefined angles ($\theta_{\max}$; $\varphi_{\max}$). It is easy to verify that if $d_z = 0$ or $d_x = 0$, Eq. (7.65) can be reduced to equations for the one-dimensional arrays.

Taking into account (7.58) and (7.60), we obtain based on (7.66) the final formulas for the real and imaginary parts of the surface impedances

$$\overline{R}_{nm} = \frac{C_2 \sin(k\gamma_{nm}) + C_1\left[1 - \cos(k\gamma_{nm})\right]}{C_1^2 + C_2^2},$$
$$n = 1, 2 \ldots N_z;\; m = 1, 2 \ldots N_x;$$
$$\overline{X}_{nm} = \frac{C_1 \sin(k\gamma_{nm}) - C_2\left[1 - \cos(k\gamma_{nm})\right]}{C_1^2 + C_2^2}, \tag{7.67}$$

where $C_1 = \mathrm{Re}\left\{1 + \frac{\Delta Fs(\theta_{max})}{Fs(\theta_{max})}\right\}$ and $C_2 = \mathrm{Im}\left\{1 + \frac{\Delta Fs(\theta_{max})}{Fs(\theta_{max})}\right\}$.

It should be noted that formulas (7.67) are valid for any number of radiators and for arbitrary distance between them. However, in the general case, the values of the impedances $\overline{Z}_{S,nm} = \overline{R}_{nm} + i\overline{X}_{nm}$ thus obtained as effective physical quantities cannot guarantee that the conditions $\overline{R}_{nm} \geq 0$ hold true. These conditions, put forward by energy considerations, determine the possibility of practical realization in the form of the internal impedance $\overline{Z}_{S,nm} = \overline{R}_{nm} + i\overline{X}_{nm}$ of the spherical radiator. In such cases, the conditions $\overline{R}_{nm} \geq 0$ can be fulfilled after substitutions $V_{nm} \to -V_{nm}$ and $\overline{Z}_{S,nm} \to -\overline{Z}_{S,nm}$.

7.2.3 Numerical Results

In order to verify the approach, we present simulation results for a one-dimensional equidistant array consisting of five identical spherical radiators located along the axis $\{0z\}$ at the distance $d_z = \lambda/2$ from each other (Fig. 7.5). The array parameters were the following: $N_z = 5$, $d_x = 0$, $d_z = \lambda/2$, $\theta_0 = \pi/2$, $R = \lambda/20$, and $d = \pi\lambda/500$. Such array geometry simplifies the analysis and demonstration of beam scanning, since, in this case, the RP does not depend on the angular coordinate φ. In other words, the array radiation fields should be calculated only in the E-plane, considering the geometric difference of the wave paths from the adjacent radiators to the observation point, $d_z \cos\theta$, in the wave zone.

Simulation results, the value of required complex impedances $\overline{Z}_{S,n} = \overline{R}_{S,n} + i\overline{X}_{S,n}$, are listed in Table 7.2 for the angles $\theta_{max} = 90°; 70°; 50°; 30°$. The results were obtained based on (7.67), taking into account only ten terms in series of the formulas (7.58) and (7.59) that was sufficient for the spherical antennas of small radii.

Table 7.2 Estimated values of complex impedances $\overline{Z}_{S,n} = \overline{R}_{S,n} + i\overline{X}_{S,n}$ for various θ_{max}

$n\backslash\theta_{max}$	$\theta_{max} = 30°$	$\theta_{max} = 50°$	$\theta_{max} = 70°$
1	$0 + i0$	$0 + i0$	$0 + i0$
2	$0.117 - i0.28$	$0.177 - i0.192$	$0.148 - i0.054$
3	$-0.104 - i0.072$	$-0.072 - i0.268$	$0.171 - i0.211$
4	$0.183 - i0.171$	$-0.033 - i0.01$	$0.044 - i0.305$
5	$-0.119 - i0.198$	$0.183 - i0.157$	$-0.099 - i0.239$

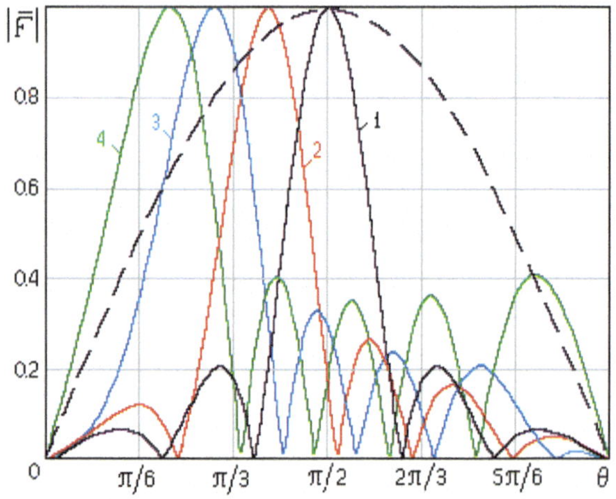

Fig. 7.6 The E-plane RP of linear AA: 1—$\theta_{\max} = 90°$; 2—$\theta_{\max} = 70°$; 3—$\theta_{\max} = 50°$; 4—$\theta_{\max} = 30°$; dotted line presents the RP of the single radiator

As expected, all spherical radiators should be perfectly conducting, i.e., $\overline{Z}_{S,n} = 0$ when $\theta_{\max} = 90°$. The normalized RP for this case is shown in Fig. 7.6 as solid curve 1. As can be seen, the basis array element, the radiator with index $n = 1$, should be perfectly conducting for any other $\theta_{\max}$. A similar effect was also observed for vibrator arrays (Chap. 6).

As can be seen from Fig. 7.6, the shape of the angular dependence of the electric field intensity $|\overline{F}|$ varies slightly when the RP maximum is varied: the greater is the deviation of the $\theta_{\max}$ angle from the vertical, the more noticeable are these variations. The simulation results have shown that impedances of the array elements calculated for the sector $\theta_{\max} \in [90° \pm 45°]$ allows accurately steer the array RP in any predefined direction in this sector. In other angular sectors, the direction defined by the angle $\theta_{\max}$ and the direction of the RP maximum may differ due to the fact that the radiation of a single spherical slot antenna is ineffective in these angular sectors (dashed curve 1 in Fig. 7.6).

7.3 Impedance Synthesis for Ring Slotted Radiators on Hemispherical Ledges Above Screen

7.3.1 Formulation and Solution of the Impedance Synthesis Problem

Consider the antenna array that consists of the row of slotted hemispherical antennas located on the perfectly conducting screen along the axis $\{0x\}$ of a Cartesian coordinate system (x, y, z) as shown in Fig. 7.7. The origin is superposed with the center of the array. Let d_x be the distance between the geometric centers of the neighboring

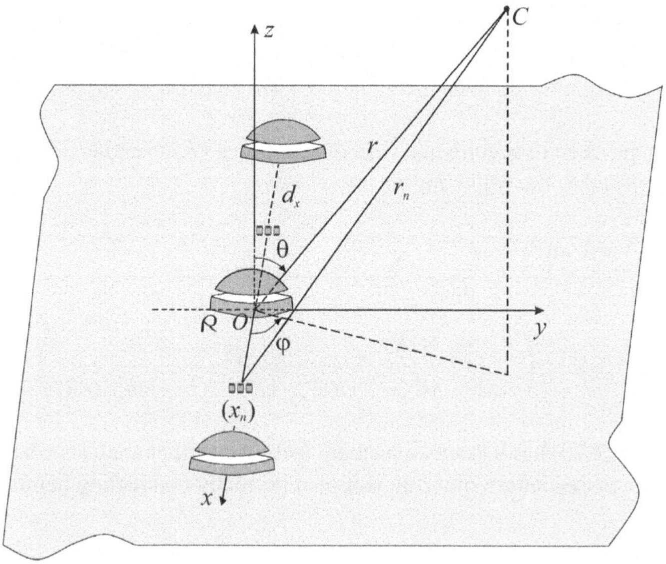

Fig. 7.7 The geometry of the linear AA on hemispherical ledges above screen

radiators of the array, and N is their quantity. We also introduce a spherical coordinate system, the polar axis of which coincides with the axis $\{0z\}$, and the angle φ is measured from the axis $\{0x\}$. Without loss of generality, we assume that all of the hemispherical ledges have the same radius R, the same parameters of the ring slots d and θ_0, but, generally, different impedances $\overline{Z}_{S,n}$, where the integer index $n \in [1, N]$.

We can write the main component of the electrical field of radiation of the array element (x_n) for the wave zone in the following form:

$$E_n^{\theta} = -\frac{V_n kR \sin\theta_0}{2d} \times \frac{e^{-ikr_n}}{kr_n} \sum_{j=1,3\ldots}^{\infty} A_{j,n} P_j^1(\cos\theta) i^j \left[1 - \overline{Z}_{S,n}\right], \tag{7.68}$$

where $A_{j,n} = A_j\big|_{\overline{Z}_S \to \overline{Z}_{S,n}}$.

Assuming that the all hemispherical radiators in the array have the same geometry of slots ($\theta_0 = Const$) and they are tuned to resonance due to selection of their internal impedances and parameter values d_x/λ and R/d_x, allowing to compensate the mutual influence of elements, the total radiation field of the array $E_\theta(r, \theta, \varphi)$ is equal to the sum of the radiated fields of each element taking into account the phases of fields that arrive to the observation point $C(r, \theta, \varphi)$:

$$E_\theta(r, \theta, \varphi) = -\frac{kR \sin\theta_0}{2d} \sum_{n=1}^{N} \frac{e^{-ikr_n}}{r_n} V_n \sum_{j=1,3\ldots}^{\infty} A_{j,n} P_j^1(\cos\theta) i^j \left[1 - \overline{Z}_{S,n}\right]. \tag{7.69}$$

Thus, the problem of impedance synthesis will include the determination of the impedances $\overline{Z}_{S,n}$ for each of hemispherical elements for a given direction of the maximum radiation of the linear array in the wave zone $(\theta_{\max}, \varphi_{\max})$ on the basis of (7.69).

Next, we proceed in accordance with the formulas (7.55) and (7.56). As a result, formula (7.68) takes the following form:

$$E_n^{\theta} \approx -\frac{V_n kR \sin\theta_0}{2d} \times \frac{e^{-ikr_n}}{kr_n} \sum_{j=1,3\ldots}^{\infty} A_j^0 P_j^1(\cos\theta) i^j$$
$$\times \left[1 - \overline{Z}_{S,n}\left(1 - \frac{ih_j^{(2)}(kR)}{kR} \cdot \frac{j(j+1) - 2k^2R^2}{kRh_{j+1}^{(2)}(kR) - (j+1)h_j^{(2)}(kR)}\right)\right] \tag{7.70}$$

Expression (7.70) has a more convenient form for further analysis where the first term determines the pattern of a ring slot on a perfectly conducting hemisphere:

$$Fs(\theta) = -\frac{kR \sin\theta_0}{2kd} \sum_{j=1,3\ldots}^{\infty} A_j^0 P_j^1(\cos\theta) i^j, \tag{7.71}$$

and the second one defines the changes in the radiation field of the hemispherical antenna due to its impedance coating. Thus, we get the formula

$$E_n^{\theta} = V_n \frac{e^{-ikr_n}}{r_n}\left[Fs(\theta) - \overline{Z}_{S,n}(Fs(\theta) + \Delta Fs(\theta))\right], \tag{7.72}$$

where the term

$$\Delta Fs(\theta) = \frac{\sin\theta_0}{2kd} \sum_{j=1,3\ldots}^{\infty} A_j^0 \frac{j(j+1) - 2k^2R^2}{kRh_{j+1}^{(2)}(kR) - (j+1)h_j^{(2)}(kR)}$$
$$P_j^1(\cos\theta) h_j^{(2)}(kR) i^{j+1}$$

and, respectively

$$E_{\theta}(r, \theta, \varphi) = \sum_{n=1}^{N} \frac{e^{-ikr_n}}{r_n} V_n\left[Fs(\theta) - \overline{Z}_{S,n}(Fs(\theta) + \Delta Fs(\theta))\right] \tag{7.73}$$

Next, we proceed in accordance with the formulas (7.61)–(7.66). As a result, we obtain the final calculation formulas for the real and imaginary parts of the surface impedances of spherical radiators:

$$\overline{R}_n = \frac{C_2 \sin(k\gamma_n) + C_1\left[1 - \cos(k\gamma_n)\right]}{C_1^2 + C_2^2},$$

$$n = 1, 2...N;$$
$$\overline{X}_n = \frac{C_1 \sin(k\gamma_n) - C_2[1 - \cos(k\gamma_n)]}{C_1^2 + C_2^2}, \quad (7.74)$$

where the constants are $C_1 = \mathrm{Re}\left\{1 + \frac{\Delta Fs(\theta_{max})}{Fs(\theta_{max})}\right\}$, $C_2 = \mathrm{Im}\left\{1 + \frac{\Delta Fs(\theta_{max})}{Fs(\theta_{max})}\right\}$ and $\gamma_n = (n-1)d_x \sin\theta_{max} \cos\varphi_{max}$.

7.3.2 Numerical Results

In order to confirm the correctness of the formulas obtained, we present the results of calculations for a linear equidistant array of five identical hemispherical radiators $(N = 5, \theta_0 = 9\pi/20, d = \pi\lambda/500)$ with radii $R = \lambda/20$ located along the axis $\{0x\}$ at a distance $d_x = \lambda/2$ (Fig. 7.7). An example of the results will be presented for the section $\theta = 9\pi/20$. By successively fixing the values of the angle $\varphi_{max} = \pi/2; \pi/3; \pi/6; 0$ at $\theta_{max} = 9\pi/20$, we obtain the calculated values $\overline{Z}_{S,n} = \overline{R}_n + i\overline{X}_n$ of the complex impedances on the basis of (7.74) (see Table 7.3).

As one would expect, all spherical radiators become the perfectly conducting ones with $\overline{Z}_{S,n} = 0$ when the condition $\varphi_{max} = \pi/2$ is fulfilled. The normalized pattern for this case is shown in Fig. 7.8 by solid curve 1. Here the patterns are represented in the angular sector $0 \leq \varphi \leq \pi$ taking into account their symmetry about the axis $\{0x\}$. Note that, as in the case of the vibrator array (Chap. 6), at any other value φ_{max} for the basis element of the array (here with the number $n = 1$) the impedance value is preserved $\overline{Z}_{S,1} = 0$.

As can be seen from Fig. 7.8, by varying the direction of the maximum radiation of the array the shape of the angular dependence of the modulus of the electrical field strength of the radiation $|\overline{F}|$ also slightly changes. Moreover, the biggest width of the lobe of the pattern at the level of 0.5 is observed in the direction of the axis $\{0y\}$. However, there is a good agreement between the given value of the angle φ_{max} and the direction of the maximum of the radiation pattern of the synthesized array in the whole sector. This result is confirmed by modeling for arrays with a different number of elements N and other conditions for specifying the direction $(\theta_{max}; \varphi_{max})$.

Table 7.3 Estimated values of complex impedances $\overline{Z}_{S,n} = \overline{R}_{S,n} + i\overline{X}_{S,n}$ for various φ_{max}

$n \backslash \varphi_{max}$	$\varphi_{max} = \pi/3$	$\varphi_{max} = \pi/6$	$\varphi_{max} = 0$
1	$0 + i0$	$0 + i0$	$0 + i0$
2	$-0.137 + i0.174$	$-0.031 + i0.306$	$0.034 + i0.314$
3	$0.034 + i0.314$	$0.131 + i0.045$	$0.012 - i0.0011$
4	$0.178 + i0.146$	$-0.129 + i0.209$	$0.022 + i0.315$
5	$0.012 - i0.0011$	$0.177 + i0.176$	$0.024 - i0.0013$

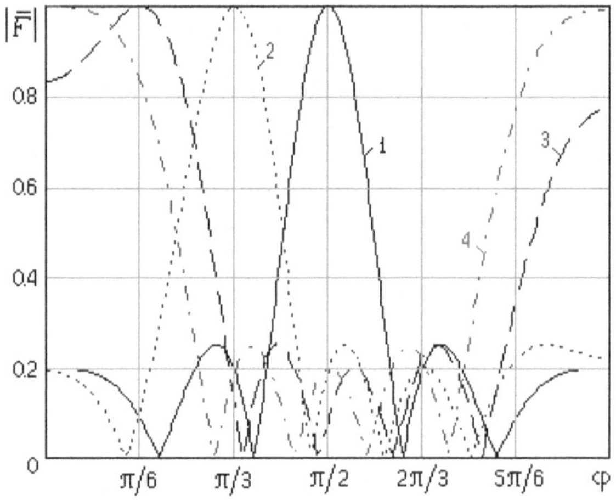

Fig. 7.8 RP of AA in cross-section $\theta = 9\pi/20$ for: 1—$\varphi_{\max} = \pi/2$, 2—$\varphi_{\max} = \pi/3$, 3—$\varphi_{\max} = \pi/6$, 4—$\varphi_{\max} = 0$

The materials of this Chapter are contained in the authors' publications [20–22].

References

1. Bakhrakh LD, Kremenetsky SD (1974) Synthesis of radiation systems (Theory and methods of calculation), Sov. Radio, Moscow (in Russian)
2. Skobelev SP (2011) Phased array antennas with optimized element patterns. Artech House, Boston
3. Vendik OG, Parnes MD, Bakhrakh LD (2001) Antennas with electric scanning. Science Press, New York
4. Nesterenko MV, Katrich VA, Penkin YuM, Dakhov VM, Berdnik SL (2011) Thin impedance vibrators. Theory and applications. Springer Science+Business Media, New York
5. Nesterenko MV, Katrich VA, Penkin YuM, Berdnik SL (2008) Analytical and hybrid methods in theory of slot-hole coupling of electrodynamic volumes. Springer Science+Business Media, New York
6. Brekhovskikh LM (1980) Waves in layered media. Academic Press, New York
7. Penkin YuM, Katrich VA, Nesterenko MV (2016) Development of fundamental theory of thin impedance vibrators. Prog Electromagn Res M 45:185–193
8. Manuilov MB, Lerer VA, Sinyavskii GP (2008) An efficient method for the electromagnetic analysis of slotted-waveguide antenna arrays. J Commun Technol Electron 53:1380–1390
9. Tsaliev TA, Cherenkov VS (1985) Single groove excitation and equivalent surface impedance of ribbed structures. Radiotekhnika i Electronika 30:1689–1694 (in Russian)
10. Koshkid'ko VG (2000) Equivalent surface impedance of slot impedance loads incorporated into infinite gratings. J Commun Technol Electron 45:773–783
11. Nesterenko MV, Katrich VA, Penkin YuM, Berdnik SL, Dumin OM (2020) Combined vibrator-slot structures: theory and applications. Springer Nature Swizerland AG, Cham, Swizerland
12. Josefson L, Persson P (2006) Conformal array antenna theory and design. IEEE Press Series on Electromagnetic Wave Theory, A Wiley—Intersience Publication
13. Vendik IB, Vendik OG, Odit MA, Kholodnyak DV, Zubko SP, Sitnikova MF, Turalchuk PA, Zemlyakov KN, Munin IV, Kozlov DS, Turgaliev VM, Ustinov AB, Park Y, Kihm C-W (2012) Tunable metamaterials for controlling THz radiation. IEEE Trans Terahertz Sci Technol 2(5):538–549

14. Odit M, Vendik I, Kozlov D, Munin I, Turgaliev V (2013) Tunable metamaterial structures for controlling THz radiation. In: Proceedings of symposium on physics and engineering of microwaves, millimeter, and submillimeter waves Kharkiv, Ukraine, pp 40–44
15. Vendik I, Odit M, Kozlov D (2009) 3D metamaterial based on a regular array of resonant dielectric inclusions. Radioeng Proc Czech Slovak Techn Univ URSI Commities 18(2):111–116
16. Viti L, Hu J, Coquillat D, Politano A, Knap W, Vitiello MS (2016) Efficient Terahertz detection in black-phosphorus nano-transistors with selective and controllable plasma-wave, bolometric and thermoelectric response. Scientific Reports 6, Article number: 20474
17. Penkin YuM, Katrich VA, Nesterenko MV, Berdnik SL, Dakhov VM (2019) Electromagnetic fields excited in volumes with spherical boundaries. Springer Nature Switzerland AG, Cham, Switzerland
18. Mushiake Y, Webster RE (1957) Radiation characteristics with power gain for slots on a sphere. IRE Trans Antennas Propag 5(1):47–55
19. Amitay N, Galindo V, Wu CP (1972) Theory and analysis of phased array antennas. Wiley, New York
20. Penkin Yu, Katrich V, Nesterenko M, Berdnik S (2020) Impedance synthesis of flat slot antenna arrays. In: Proceedings of the 2020 IEEE Ukrainian Microwave Week Kharkiv, Ukraine, pp 89–92
21. Penkin YuM, Katrich VA, Nesterenko MV (2019) Impedance synthesis of 2D antenna arrays of slotted spherical radiators. Prog Electromagn Res L 81:93–100
22. Penkin YuM, Katrich VA, Nesterenko MV, Dumin OM, Pshenichnaya SV (2019) Impedance synthesis for the ring slotted radiators on hemispherical ledges above screen. In: Proceedings of IIth Ukraine conference on electrical and computer engineering Lviv, Ukraine, pp 68–72

Chapter 8
Large-Aperture Rectenna Arrays of Wireless Power Transmission Systems

Rectennas are part of wireless power transmission (WPT) systems and are intended to receive microwave energy and convert it to DC energy. Efficiency of the WTP system is largely determined by the rectenna, which is an unphased array of receiver-rectifier elements (RREs), each of which consists of a receiving antenna and a rectifier circuit. Parameters of RRE mainly determine such indices of rectenna quality as efficiency, level of over-radiation at frequencies of higher harmonics, reliability, cost and suitability for serial production. Today, the optimization of both the parameters of individual RRE and the characteristics of the entire array as a whole remains relevant for the array. Of course, such optimization can be implemented only as a result of mathematical modeling.

This chapter describes the main results of the authors' research in field of development of theory of large-aperture arrays, suitable for the analysis of the periodic mode of excitation of a sufficiently wide class of both single retention elements and large-aperture arrays based on them, taking into account the nonlinear processes occurring in them.

8.1 Large-Aperture Rectenna Arrays at Uniform Excitation

Presence of elements with nonlinear characteristics (rectifier diodes) in the rectenna array (Fig. 8.1) leads to appearance of such nonlinear effects as nonlinear dependence of rectenna parameters on level of input influence, and formation of new spectral components in rectenna response, which are absent in input influence. Obtaining nonlinear effects depends on scheme and design of the rectenna, parameters of radiators, and linear and nonlinear elements included in its schema. Therefore, a correct consideration of a large-aperture rectenna (LAR) array with analysis of its inherent

M. V. Nesterenko et al., *Vibrator and Slot Antenna Arrays for Modern Applications*, Lecture Notes in Electrical Engineering 1244,
https://doi.org/10.1007/978-3-031-68346-6_8

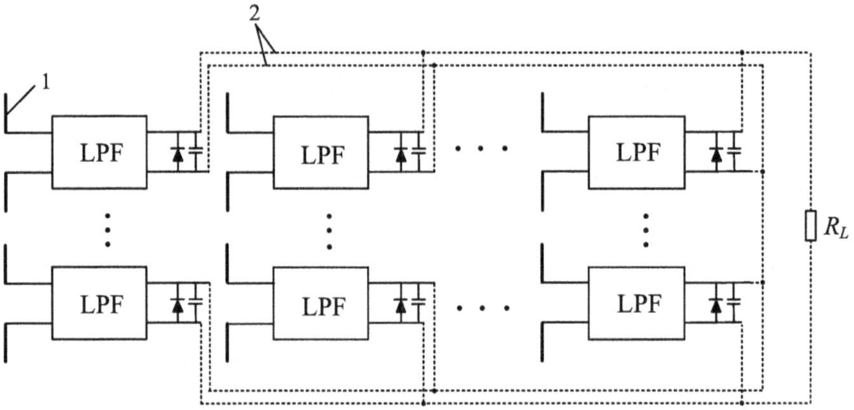

Fig. 8.1 Schematic diagram of the rectenna array fragment: 1—radiator; 2—wires of power collection circuit; LPF—low-pass filter

nonlinear effects requires a comprehensive approach, in which, along with the properties of the circuit containing nonlinear elements, properties of radiators, as well as the relationships between them and the ways of implementing DC power collection scheme are taken into account. In other words, the rectenna must be considered as a single complex device with nonlinear characteristics.

For a complete analysis of totality of electromagnetic processes occurring in LARs, we apply approach developed by the authors [1–5], which is based on a unified idea, at the electrodynamic level, of functioning WPT systems of wide class and purpose, which include antennas and their power paths containing nonlinear elements (NE). Within the frames of the given approach the nonlinear mathematical model of electrodynamic level of WPT system is developed, according to which all WPT system, including in general case transmitting subsystem, receiving subsystem (rectenna) and system of all possible scatterers is considered as the single multi-input antenna system with nonlinear characteristics.

Nonlinear mathematical model of the electrodynamic level of the WPT system

- it is universal because it allows to analyze and optimize WPT systems that use different technologies of energy transmission [6, 7] (transmitting and receiving subsystems can be located in near zone, Fresnel zone, Fraunhofer zone);
- allows to take into account totality of nonlinear effects that occur in the WPT systems through the presence in their composition of antennas and tracts of their supply with nonlinear characteristics (transmitting active phased antenna arrays, parasitic nonlinearities in transmitting antennas, rectennas);
- takes into account interconnections in the WPT system itself (internal system processes), as well as its electrodynamic interaction with other radio-electronic systems (RES) and interaction of other RES with it (intersystem interaction), what allows to solve correctly problems of electromagnetic coupling (EMC) at stage

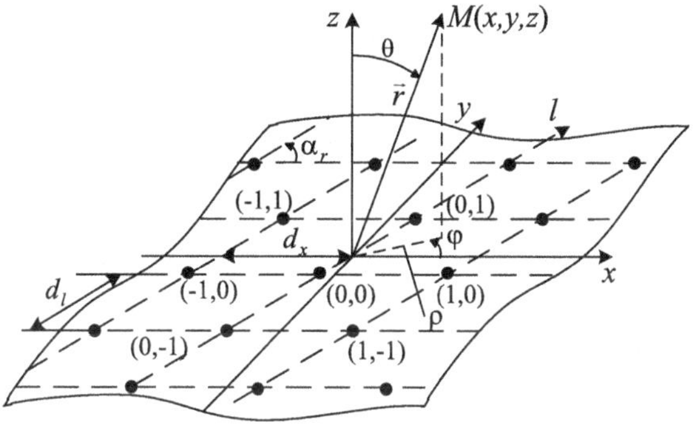

Fig. 8.2 Geometry of the rectenna array

of modeling WPT systems, as well as to solve problems of designing new classes of RES with a single transfer of information and energy;
- is flexible, because at analysis includes possibility of changing the configuration of the WPT system depending on its purpose (energy transmission technology) and the possibility of analyzing its individual subsystems, devices, units.

In this section, based on a nonlinear mathematical model of the electrodynamic level of the WPT system, considering solution of a part problem—analysis of LARs of WPT systems by a microwave beam.

Consider a flat rectenna array consisting of identical RREs of arbitrary configuration, which are equally located in the nodes of the oblique array with double periodicity along the x and l axes (Fig. 8.2). Period along the x-axis is d_x, period along the l-axis is d_l. l-axis forms an angle α_r with the x-axis. Cell with index $(0, 0)$ will be called central. Position of each RRE is defined by the indices (p, q), to which corresponds a geometric point with the radius-vector $\vec{\rho}_{pq} = pd_x\vec{x}_0 + qd_l\vec{l}_0$, where $\vec{x}_0$, $\vec{l}_0$ are unit vectors in direction of the x and l axes.

Assume that the array is excited at a frequency ω_0 (wavelength λ_0) by a plane electromagnetic wave coming from direction θ, φ (Fig. 8.2). Electric field strength vector of exciting plane wave at an arbitrary point with coordinates (x, y, z) has the form

$$\vec{E}(x, y, z) = \vec{E}_0 e^{jk_0(x\sin\theta\cos\varphi + y\sin\theta\sin\varphi + z\cos\theta)}, \tag{8.1}$$

where $k_0 = 2\pi/\lambda_0$, $\vec{E}_0$—amplitude of the wave.

8.1.1 Schema of the Large-Aperture Rectenna

According to [1], scheme shown in Fig. 8.3 is a general enough structural scheme to describe various rectenna arrays. It highlights the linear subcircuits (LS) LS-1 and LS-2 common to the whole rectenna, the first of which corresponds to the radiator

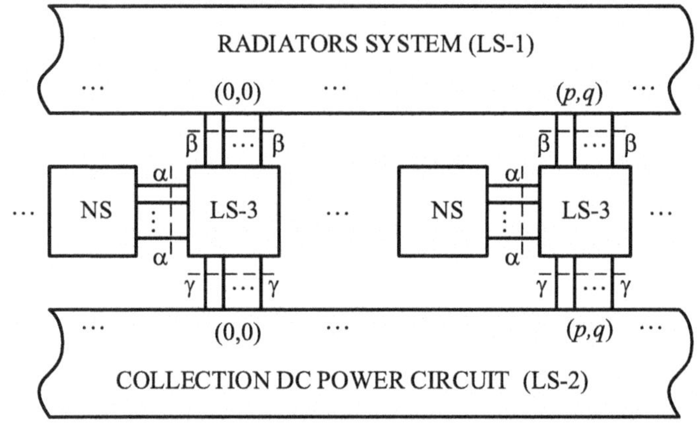

Fig. 8.3 Structural diagram of the LAR

system, and the second to power acquisition circuit and load. In addition, linear (LS-3) and nonlinear (NS) subcircuits, which integrate all linear and nonlinear elements of each SPE, respectively, are identified for each RRE.

Such splitting structural scheme into subcircuits is due to several reasons [1], of which to note the followings. Firstly, in a number of cases, depending on required depth of analysis and accuracy of obtained results, it is necessary to form rectenna models of different level of rigor and, accordingly, complexity. Secondly, characteristics of linear and nonlinear elements of rectenna have essentially different description.

The NS element will be represented as a non-linear 2-N_α multipole (NM) (Fig. 8.4a), described by the mapping $\Re$, transforming vector of voltages $\boldsymbol{u}_{nl}(t) = \left(u_{nl1}(t), u_{nl2}(t), ..., u_{nlN_\alpha}(t)\right)^T$ at its inputs into vector of currents $\boldsymbol{i}_{nl}(t) = \left(i_{nl1}(t), i_{nl2}(t), ..., i_{nlN_\alpha}(t)\right)^T$, i.e.

$$\mathbf{i}_{nl}(t) = \Re\{\mathbf{u}_{nl}(t)\} \tag{8.2a}$$

or vice versa

$$\mathbf{u}_{nl}(t) = \Re^{-1}\{\mathbf{i}_{nl}(t)\} \tag{8.2b}$$

In these relations N_α are the number of inputs connecting NM with the multi-pole corresponding to the subcircuit LS-3 (section $\alpha - \alpha$); index T stands the transpose operation. For each linear subcircuit, to assign linear multipoles LM-1, LM-2, and LM-3 (Fig. 8.4b–d). Unlike NM, linear multipoles are more convenient to characterize in the frequency domain. Before proceeding to selecting particular system of parameters describing linear rectenna multipoles, the following need to be noted. Entire procedure of rectenna analysis can be divided into several stages. This division, as well as the contents of the individual stages, is to some extent conditional,

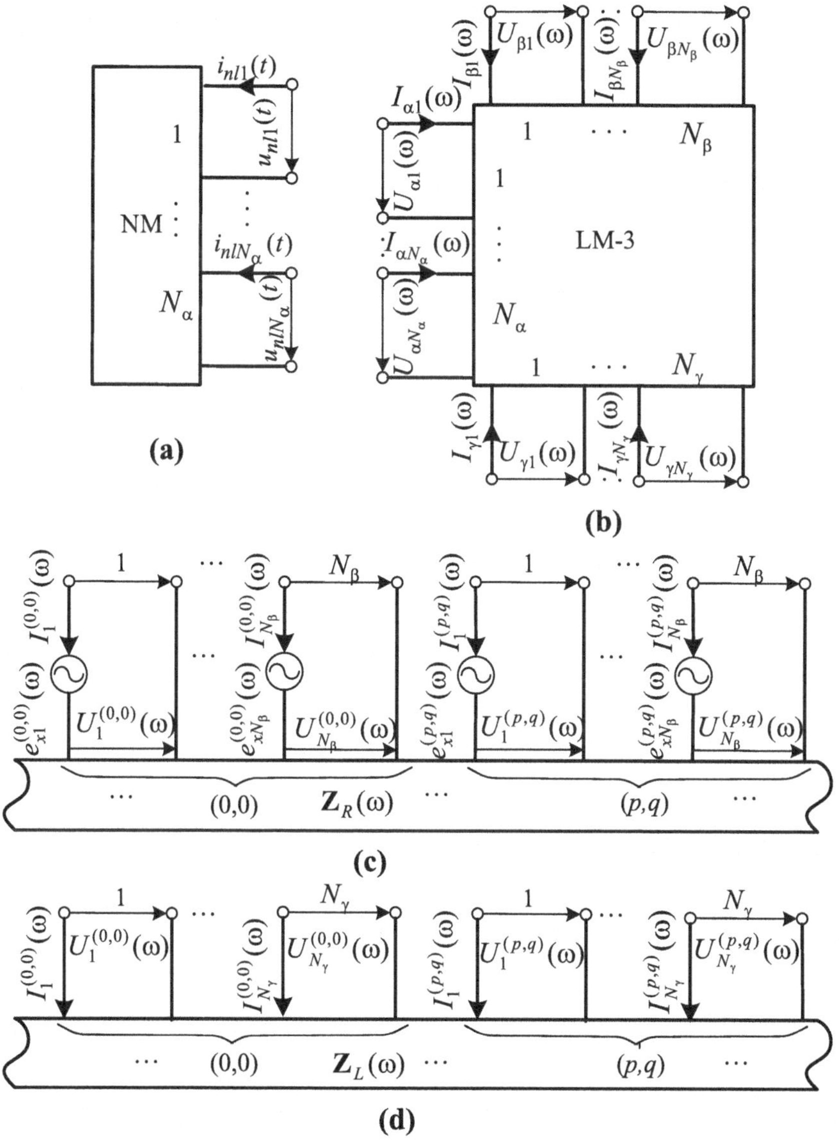

Fig. 8.4 Representation of rectenna subcircuits as multipoles at state variables definition stage

depending on approach applied to the analysis. However, in any approach, stage of calculating the operation mode of the device under analysis and stage of determining its external parameters are unavoidable. Peculiarities of the mentioned stages in analysis of LARs are as follows.

According to infinite array approximation, which we will use, it is assumed that mode of operation of each of RRE is the same as it would be if the given element was part of an infinite array. Consequently, the first of the above steps should consider an infinite ($N_x, N_l \to \infty$) array of RREs and, hence, the multipoles LM-1, LM-2 have an infinite number of inputs. In this case, the parameters characterizing the linear multipoles from the side of those inputs which are connected to each other, i.e., with respect to the cross sections $\alpha - \alpha$, $\beta - \beta$ and $\gamma - \gamma$.

At the second stage, the stage of determining the external parameters, the rectenna is considered as an array with a finite number of RREs ($N_x \times N_l$). Consequently, the multipoles LM-1 and LM-2 are described by matrices of finite dimensionality. In addition, at this stage, parameters characterizing its relation with the external space should also be known for the radiator system, i.e., a more complete system of parameters describing LM-1 is required. Therefore, it is natural that different systems of parameters describing linear multipoles can be chosen at different stages of the analysis. Main criterion determining using these or those parameters is convenience of their use in solving specific problems.

8.1.2 Analysis of the Nonlinear Mode of Large-Aperture Rectenna

Determining the mode of a nonlinear device consists in solving its equations of state [1, 8]. Aspects of forming such system and its peculiarities for case of LAR are considered below. To form a system of state equations, it is convenient to describe the multipole LM-1, which characterizes an infinite periodic system of radiators, at frequency ω by a matrix of eigen and mutual resistances $\mathbf{Z}_R(\omega)$ and vector of complex amplitudes of electromotive forces (EMF) $\mathbf{e}_\beta(\omega)$, induced at the radiator inputs by a plane wave field (Fig. 8.4c). In case under consideration (flat periodic infinite array of radiators, excited by plane wave field) matrix $\mathbf{Z}_R(\omega)$ is an infinite-dimensional block-block Teplice (BBT) matrix, and the external influence at the radiators' inputs is represented by system of EMF sources with double periodicity. This is due to the fact that if the external excitation is described by Eq. (8.1), then field strength of plane wave at point M_{00} belonging to central cell and field strength at point M_{pq} having the same coordinates relative to cell borders (p, q) as the coordinates of point M_{00} relative to the cell borders with number (0,0) are related by the equation

$$\vec{E}\left(M_{p,q}\right) = \vec{E}\left(M_{0,0}\right)e^{j(ap+bq)}, \tag{8.3}$$

where $a = k_0 d_x \sin\theta\cos\varphi$, $b = k_0 d_l(\sin\theta\cos\varphi\cos\alpha_r + \sin\theta\sin\varphi\sin\alpha_r)$.

Therefore, vectors of complex amplitudes of EMF induced by plane wave field at radiator inputs located in cells (0, 0) and (p, q) and recalculated to the cross sections (($\beta - \beta$) Fig. 8.3) are connected by equation

$$\mathbf{e}_{\beta}^{(p,q)}(\omega_0) = \mathbf{e}_{\beta}^{(0,0)}(\omega_0)\mathbf{e}^{j(ap+bq)}. \tag{8.4}$$

Here $\mathbf{e}_{\beta}^{(p,q)}(\omega_0)$ and $\mathbf{e}_{\beta}^{(0,0)}(\omega_0)$ are vectors of complex amplitudes of radiators' EMFs located at nodes of periodic array with numbers (p, q) and $(0, 0)$, respectively. Such dependence describes system of EMF sources with double periodicity.

The multipole LM-2 (Fig. 8.4d) has an infinite number of inputs and is described by the resistance matrix $\mathbf{Z}_L(\omega)$, which relates vectors of complex amplitudes of currents $\mathbf{I}(\omega)$ and voltages at $\mathbf{U}(\omega)$ its inputs (sections $\gamma - \gamma$):

$$\mathbf{U}(\omega) = \mathbf{Z}_L(\omega)\mathbf{I}(\omega). \tag{8.5}$$

Here vectors $\mathbf{I}(\omega)$ and $\mathbf{U}(\omega)$ are formed as follows:

$$\begin{aligned}\mathbf{I}(\omega) = \Big(&..., I_1^{(-1,0)}, I_2^{(-1,0)}, ..., I_{N_\gamma}^{(-1,0)}, I_1^{(0,0)}, I_2^{(0,0)},\\ &..., I_{N_\gamma}^{(0,0)}, I_1^{(1,0)}, I_2^{(1,0)}, ..., I_{N_\gamma}^{(1,0)}, ...\Big)^T,\\ \mathbf{U}(\omega) = \Big(&..., U_1^{(-1,0)}, U_2^{(-1,0)}, ..., U_{N_\gamma}^{(-1,0)}, U_1^{(0,0)},\\ &U_2^{(0,0)}, ..., U_{N_\gamma}^{(0,0)}, U_1^{(1,0)}, ..., U_{N_\gamma}^{(1,0)}, ...\Big)^T.\end{aligned}$$

Next, we will assume that the RREs are loaded with the same resistances. Multipole LM-3 (Fig. 8.4b) corresponding to LS-3 subcircuits will be described by resistance matrix that connects complex amplitudes of currents and voltages in the sections $\alpha - \alpha$, $\beta - \beta$ and $\gamma - \gamma$. Let us write this equation in block form:

$$\begin{pmatrix}\mathbf{U}_\alpha(\omega)\\ \mathbf{U}_\beta(\omega)\\ \mathbf{U}_\gamma(\omega)\end{pmatrix} = \begin{pmatrix}\mathbf{Z}'_{\alpha\alpha}(\omega) & \mathbf{Z}'_{\alpha\beta}(\omega) & \mathbf{Z}'_{\alpha\gamma}(\omega)\\ \mathbf{Z}'_{\beta\alpha}(\omega) & \mathbf{Z}'_{\beta\beta}(\omega) & \mathbf{Z}'_{\beta\gamma}(\omega)\\ \mathbf{Z}'_{\gamma\alpha}(\omega) & \mathbf{Z}'_{\gamma\beta}(\omega) & \mathbf{Z}'_{\gamma\gamma}(\omega)\end{pmatrix}\begin{pmatrix}\mathbf{I}_\alpha(\omega)\\ \mathbf{I}_\beta(\omega)\\ \mathbf{I}_\gamma(\omega)\end{pmatrix}. \tag{8.6}$$

In the above equation $\mathbf{I}_j(\omega) = \left(I_{j1}(\omega), I_{j2}(\omega), ..., I_{jN_j}(\omega)\right)^T$ and $\mathbf{U}_j(\omega) = \left(U_{j1}(\omega), U_{j2}(\omega), ..., U_{jN_j}(\omega)\right)^T$ are vectors whose elements are the complex amplitudes of currents and voltages at LM-3 terminals connected to the other multipoles (sections $\alpha - \alpha$, $\beta - \beta$ and $\gamma - \gamma$).

Thus, Eqs. (8.2)–(8.6) fully characterizing all the multipoles included in the generalized rectenna circuit and are necessary for forming the equations of state. For this purpose, it is necessary to determine properties of the matrix of eigen and mutual resistances and system of EMF vectors describing the whole linear subcircuit relative to sections $\alpha - \alpha$ where it connects with the nonlinear multipoles. These properties will further make it possible, using the infinite array approximation, to simplify solving the system of equations of state for the large-aperture array.

Let's transform the linear rectenna subcircuit including three linear subcircuits (Fig. 8.3) so that linear LS-3 subcircuit remains unchanged, and LS-1 and LS-2

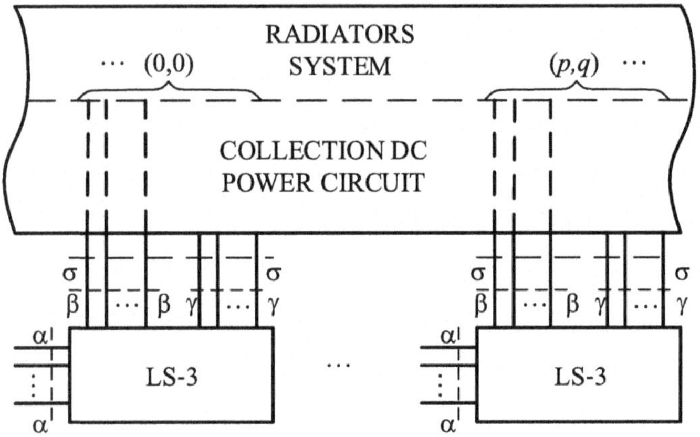

Fig. 8.5 Modified linear rectenna subcircuit

subcircuits form a new united subcircuit connected to LS-3 subcircuit in section $\sigma - \sigma$ (Fig. 8.5).

Let us denote by $\mathbf{Z}^{\Sigma}(\omega)$ the matrix of eigen resistances and mutual resistances of the combined multipole corresponding to the radiator system and the DC power collection scheme with respect to the cross sections $\sigma - \sigma$. Since the multipole matrix LM-1 and the multipole matrix LM-2 are BBT matrices, the matrix $\mathbf{Z}^{\Sigma}(\omega)$ has the same properties due to the assumption made about identical RRE loads.

Let us represent the matrices of the LM-3 multipoles in block form:

$$\mathbf{Z}(\omega) = \begin{pmatrix} \mathbf{Z}'_{\alpha\alpha}(\omega) & \mathbf{Z}'_{\alpha\sigma}(\omega) \\ \mathbf{Z}'_{\sigma\alpha}(\omega) & \mathbf{Z}'_{\sigma\sigma}(\omega) \end{pmatrix}, \tag{8.7}$$

where $\mathbf{Z}'_{ij}(\omega)$ are the blocks describing interaction of the inputs belonging to section i with the inputs belonging to sections j $(i, j = \alpha, \sigma)$. Taking into account accepted notations, matrix of eigen resistances and mutual resistances $\mathbf{Z}_{\alpha\alpha}(\omega)$ with respect to inputs to which nonlinear multipoles are connected (Fig. 8.5), and vector of complex EMF amplitudes $\mathbf{e}_x(\omega)$ at the same inputs can be determined from the equations

$$\begin{aligned} \mathbf{Z}_{\alpha\alpha}(\omega) &= \mathbf{Z}''_{\alpha\alpha}(\omega) - \mathbf{Z}''_{\alpha\sigma}(\omega)\left(\mathbf{Z}''_{\sigma\sigma}(\omega) + \mathbf{Z}^{\Sigma}(\omega)\right)^{-1}\mathbf{Z}''_{\sigma\alpha}(\omega), \\ \mathbf{e}_x(\omega) &= -\mathbf{Z}''_{\alpha\sigma}(\omega)\left(\mathbf{Z}''_{\sigma\sigma}(\omega) + \mathbf{Z}^{\Sigma}(\omega)\right)^{-1}\mathbf{e}_{x\sigma}(\omega), \end{aligned} \tag{8.8}$$

where $\mathbf{Z}''_{ij}(\omega)$ are infinite-dimensional block-diagonal matrices. Elements of these matrices are blocks $\mathbf{Z}'_{ij}(\omega)$ of matrixes $\mathbf{Z}(\omega)$. In (8.8) vector $\mathbf{e}_{x\sigma}(\omega)$ is an infinite-dimensional vector formed as follows:

$$\mathbf{e}_{x\sigma}(\omega) = \Big(\ldots, \mathbf{e}_{\beta}^{(-1,-q)}(\omega), \mathbf{0}, \mathbf{e}_{\beta}^{(0,-q)}(\omega), \mathbf{0}, \ldots, \mathbf{e}_{\beta}^{(-1,0)}(\omega), \mathbf{0}, \mathbf{e}_{\beta}^{(0,0)}(\omega), \mathbf{0}, \ldots,$$

$$\ldots, \mathbf{e}_{\beta}^{(-1,q)}(\omega), \mathbf{0}, \mathbf{e}_{\beta}^{(0,q)}(\omega), \mathbf{0}, \ldots\Big)^{T}. \tag{8.9}$$

Here $\mathbf{e}_{\beta}^{(p,q)}(\omega)$ is vector defined by Eq. (8.4); $\mathbf{0}$ is zero vector with a dimension equal to number of inputs of one RRE in section $\gamma - \gamma$. Presence of the zero vector in (8.9) reflects fact that there are no independent sources in the LM-2 multipole. Since the diagonal blocks in each of the matrices $\mathbf{Z}''_{ij}(\omega)$ are the same, these matrices are a special case of BBT matrices that, in turn, have properties of circular matrices. In particular, circular matrices with complex elements have the following properties:

- if the circular matrix is nondegenerate, then the matrix inverse to it is circular;
- set of circular matrices is a commutative ring;
- set of nondegenerate circular matrices is a commutative group on multiplication.

These properties allow us to argue that matrices $\mathbf{Z}_{\alpha\alpha}(\omega)$ and $\mathbf{Z}''_{\alpha\sigma}(\omega)\left(\mathbf{Z}''_{\sigma\sigma}(\omega) + \mathbf{Z}^{\Sigma}(\omega)\right)^{-1}\mathbf{Z}''_{\sigma\alpha}(\omega)$ are BBT matrices, and vector $\mathbf{e}_{x}(\omega)$ is an infinite-dimensional vector whose elements, like the elements of vector $\mathbf{e}_{\beta}(\omega)$, obey Eq. (8.4), i.e.

$$\mathbf{e}_{x}^{(p,q)}(\omega) = \mathbf{e}_{x}^{(0,0)}(\omega)e^{j(ap+bq)}. \tag{8.10}$$

In this equation, $\mathbf{e}_{x}^{(p,q)}(\omega)$ and $\mathbf{e}_{x}^{(0,0)}(\omega)$ are vectors of complex amplitudes of EMF induced by the incident wave field at terminals of radiating system and recalculated to the RRE sections $\alpha - \alpha$, which are located in cells with numbers (p, q) and $(0, 0)$, respectively. Hence, for EMF vector in time domain we obtain

$$\mathbf{e}_{x}^{(p,q)}(t) = \mathbf{e}_{x}^{(0,0)}\left[t + (a/\omega)p + (b/\omega)q\right]. \tag{8.11}$$

Let us show how these properties of linear subcircuit and EMF vector can be used to solve equations of state for a LAR in infinite-array approximation of RRE.

In deriving equations of state, perform following transformations:

(a) denote by $\mathbf{Z}_{p,q}^{k,l}(\omega)$ block of matrix $\mathbf{Z}_{\alpha\alpha}(\omega)$, which describes interaction between groups of inputs to which the nonlinear RRE multipoles are connected (Fig. 8.6), relating to cells with numbers (p, q) and (k, l). We assume that upper indices characterize row and lower ones characterize column at the intersection of which block is located $\mathbf{Z}_{p,q}^{k,l}(\omega)$;
(b) from the elements of the zero row ($k = 0,\ l = 0$) of matrix $\mathbf{Z}_{\alpha\alpha}(\omega)$ matrix sequence $\{\mathbf{Z}_{p,q}\}$ as follows:

$$\begin{aligned}&\{\ldots, \mathbf{Z}_{-1,-q}(\omega), \mathbf{Z}_{0,-q}(\omega), \mathbf{Z}_{1,-q}(\omega), \ldots, \mathbf{Z}_{-1,0}(\omega), \mathbf{Z}_{0,0}(\omega), \mathbf{Z}_{1,0}(\omega), \ldots,\\ &\ldots, \mathbf{Z}_{-1,q}(\omega), \mathbf{Z}_{0,q}(\omega), \ldots, \mathbf{Z}_{1,q}(\omega), \ldots\},\end{aligned} \tag{8.12}$$

where $\mathbf{Z}_{p,q}(\omega) = \mathbf{Z}_{p,q}^{0,0}(\omega)$;
(c) assume that nonlinear multipoles of RRE contain $m = N_{\alpha}$ inputs (Fig. 8.6) and are described by characteristic of form (8.2a).

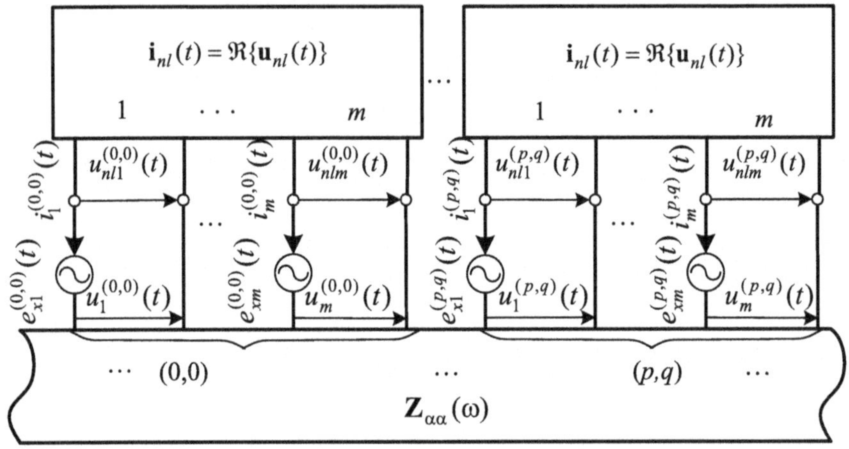

Fig. 8.6 Representation of the diagram of the infinite periodic array of the RRE

Let us consider the representation of vectors of currents $\mathbf{i}^{(p,q)}(t)$ and voltages $\mathbf{u}^{(p,q)}(t)$ related to the cell (p, q), which describe mode of group of LS inputs connected to nonlinear multipole. At periodic input action variable states (currents or voltages at nonlinear elements) are represented in form of Fourier series

$$\mathbf{i}^{(p,q)}(t) = \sum_{n=-\infty}^{\infty} \delta_n \mathbf{I}_n^{(p,q)} e^{jn\omega_0 t}, \tag{8.13}$$

where $\mathbf{I}_n^{(p,q)}$ is vector of dimension m, elements of which are complex amplitudes of currents of n-th harmonic for groups of LS inputs with numbers (p, q) (sections $\alpha - \alpha$); $\delta_n = 1$ at $n = 0$ and $\delta_n = 1/2$ at $n \neq 0$. Since condition (8.11) is satisfied for EMF vector, and matrix of linear subcircuit is a BBT matrix, then

$$\mathbf{i}^{(p,q)}(t) = \mathbf{i}^{(0,0)}\left[t + (a/\omega_0)p + (b/\omega_0)q\right]. \tag{8.14}$$

From comparison of formulas (8.13) and (8.14) we obtain connection of vector of complex amplitude currents of n-th harmonic of LS inputs with numbers (p, q) with vector of complex amplitude currents of n-th harmonic for input with zero number $(0, 0)$:

$$\mathbf{I}_n^{(p,q)} = \mathbf{I}_n^{(0,0)} e^{jn(ap+bq)}. \tag{8.15}$$

For time dependence of voltage vector at the same LS inputs we have

$$\mathbf{u}^{(p,q)}(t) = \sum_{n=-\infty}^{\infty} \delta_n \mathbf{U}_n^{(p,q)} e^{jn\omega_0 t}, \tag{8.16}$$

where $\mathbf{U}_n^{(p,q)}$ is the vector of complex amplitudes of voltages of the n-th harmonic at LS inputs with numbers (p, q). The vectors $\mathbf{U}_n^{(p,q)}$ and are $\mathbf{I}_n^{(p,q)}$ relating by equation

$$\mathbf{U}_n^{(p,q)} = \sum_{k=-\infty}^{\infty} \sum_{l=-\infty}^{\infty} \mathbf{Z}_{k,l}^{p,q}(n\omega_0)\mathbf{I}_n^{(k,l)} = \tilde{\mathbf{Z}}(n\omega_0)\mathbf{I}_n^{(p,q)}, \tag{8.17a}$$

where

$$\begin{aligned}\tilde{\mathbf{Z}}(n\omega_0) &= \sum_{k=-\infty}^{\infty} \sum_{l=-\infty}^{\infty} \mathbf{Z}_{k,l}^{p,q}(n\omega_0)\mathrm{e}^{jn[a(k-p)+b(l-q)]} \\ &= \sum_{k=-\infty}^{\infty} \sum_{l=-\infty}^{\infty} \mathbf{Z}_{k,l}(n\omega_0)\mathrm{e}^{jn[ak+bl]}.\end{aligned} \tag{8.17b}$$

As a consequence of this

$$\mathbf{u}^{(p,q)}(t) = \sum_{n=-\infty}^{\infty} \delta_n \tilde{\mathbf{Z}}(n\omega_0)\mathbf{I}_n^{(p,q)}\mathrm{e}^{jn\omega_0 t}. \tag{8.18}$$

Resulting representation of vectors $\mathbf{i}^{(p,q)}(t)$ and $\mathbf{u}^{(p,q)}(t)$ allows to write system of equations of state of rectenna in the following form:

$$\begin{aligned}&\sum_{n=-\infty}^{\infty} \delta_n \mathbf{I}_n^{(p,q)}\mathrm{e}^{jn\omega_0 t} \\ &\quad + \Re\left\{\sum_{n=-\infty}^{\infty} \delta_n \tilde{\mathbf{Z}}(n\omega_0)\mathbf{I}_n^{(p,q)}\mathrm{e}^{jn\omega_0 t} + \mathbf{e}^{(p,q)}(t)\right\} = 0, \\ &\quad \forall p, q \in -\infty, \infty; \quad t \in [0, T].\end{aligned} \tag{8.19}$$

Solutions of this system of equations are vectors of harmonic current amplitudes $\mathbf{I}_n^{(p,q)}$ for all groups of LS inputs. All equations of the system are defined for one time interval $t \in [0, T]$. Taking into account Eqs. (8.14), (8.15) and (8.16), transform the equations of the resulting system for vector of currents of inputs with zero number (central cell)

$$\begin{aligned}&\sum_{n=-\infty}^{\infty} \delta_n \mathbf{I}_n^{(0,0)}\mathrm{e}^{jn\omega_0[t+(a/\omega_0)p+(b/\omega_0)q]} \\ &\quad + \Re\left\{\sum_{n=-\infty}^{\infty} \delta_n \tilde{\mathbf{Z}}(n\omega_0)\mathbf{I}_n^{(0,0)}\mathrm{e}^{jn\omega_0[t+(a/\omega_0)p+(b/\omega_0)q]}\right. \\ &\quad \left. + \mathbf{e}^{(0,0)}[t + (a/\omega_0)p + (b/\omega_0)q]\right\} = 0,\end{aligned}$$

$$\forall p, q \in -\infty, \infty; \quad t \in \left[\Delta t^{(p,q)}, T + \Delta t^{(p,q)}\right] \tag{8.20}$$

where $\Delta t^{(p,q)} = (a/\omega_0)p + (b/\omega_0)q$.

It can be seen that determination of currents from system $\mathbf{I}_n^{(p,q)}$ ($\forall\, p, q \in -\infty, \infty$) (8.19), each of equations of which was determined on interval $t \in [0, T]$, reduced to solution of system of equations with respect to only one vector $\mathbf{I}_n^{(0,0)}$, determined at different time intervals $(a/\omega_0)p + (b/\omega_0)q \leq t \leq T + (a/\omega_0)p + (b/\omega_0)q$. Since these intervals are identical and equal to period of oscillation of fundamental frequency, it is sufficient to solve Eq. (8.20) only for one of them, for example, for $p = q = 0$:

$$\sum_{n=-\infty}^{\infty} \delta_n \mathbf{I}_n^{(0,0)} e^{jn\omega_0 t} + \Re\left\{ \sum_{n=-\infty}^{\infty} \delta_n \tilde{\mathbf{Z}}(n\omega_0) \mathbf{I}_n^{(0,0)} e^{jn\omega_0 t} + \mathbf{e}^{(0,0)}(t) \right\}$$
$$= 0, \ \forall\, t \in [0, T], \tag{8.21}$$

Thus, with the indicated properties of the LS matrix and excitation source system, analysis of the circuit in Fig. 8.6 is reduced to analysis of circuit shown in Fig. 8.7. This circuit consists of one nonlinear 2 m-pole, described by characteristic (8.2a), and one linear multipole, resistance matrix of which is defined at frequencies of all harmonics and is equal to $\tilde{\mathbf{Z}}(n\omega_0)$ $\forall\, n \in (-\infty, \infty)$.

During circuit analysis in Fig. 8.7 it is necessary to solve a system of $m(2N + 1)$-th order equations (N is the number of current harmonics held during the calculations). Consequently, solving equations of state for LAR in infinite array approximation is

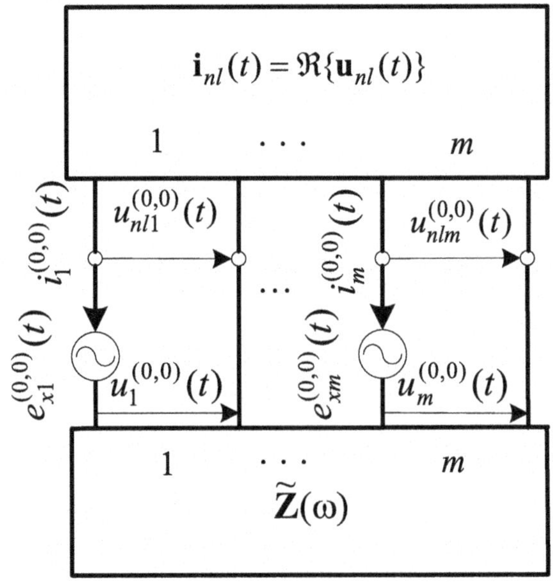

Fig. 8.7 "Equivalent" RRE in infinite array

reduced to solving system for one "equivalent" RRE in infinite array, circuit parameters of which are defined in this subsection. This makes it possible to develop an effective technique for calculating the rectennas with equal-amplitude excitation using available algorithms and programs for calculating single RREs. Obtained result is similar to concept of "single" cell known from the theory of linear antenna arrays [9].

8.1.3 External Parameters of Large-Aperture Rectennas

In item 8.1.2, when considering description of linear multipoles included in the rectenna circuit, it was indicated that at stage of solving equations of state it is reasonable to characterize them by matrices of eigenstates and mutual resistances. However, at determining the external parameters of the rectenna, different description is more convenient. Thus, at this stage, it is more appropriate to characterize the multipole LM-1, which is equivalent to system of radiators, by scattering matrix $\mathbf{S}_R(\omega)$, column of orthonormalized directivity diagram (DD) $\vec{e}(\omega, \theta, \varphi)$, column of partial DD $\vec{g}(\omega, \theta, \varphi)$, and orthogonalizing incident wave matrix, that is, to use descriptions adopted in antenna matrix theory [10]. This is due to fact that, as mentioned earlier, at determining external parameters of the rectenna for system of radiators, parameters characterizing its relating with external space should also be known. Such parameters are DD $\vec{e}(\omega, \theta, \varphi)$ and $\vec{g}(\omega, \theta, \varphi)$.

Matrix $\mathbf{S}_R(\omega)$ relates amplitudes of incident $\mathbf{a}''(\omega)$ and reflected $\mathbf{b}''(\omega)$ waves at inputs of radiator system (sections $\beta - \beta$, Fig. 8.8a) and amplitudes of convergent $u_c\prime(\omega_0)$ and divergent $u_d\prime(\omega)$ spherical waves in free space channels (sections $\delta - \delta$)

$$\begin{pmatrix} \mathbf{b}''(\omega) \\ \mathbf{u}'_d(\omega) \end{pmatrix} = \begin{pmatrix} \mathbf{S}_{\beta\beta} \ \mathbf{S}_{\beta\delta} \\ \mathbf{S}_{\delta\beta} \ \mathbf{S}_{\delta\delta} \end{pmatrix} \begin{pmatrix} \mathbf{a}''(\omega) \\ \mathbf{u}'_c(\omega_0) \end{pmatrix}. \tag{8.22}$$

Vector $\mathbf{u}'_c(\omega_0)$ is related to electric field strength $\vec{E}(\omega_0, \theta, \varphi)$ exciting plane wave coming from direction of (θ, φ), as follows:

$$\begin{aligned} \mathbf{u}'_c(\omega_0) &= -\frac{j\lambda_0}{\sqrt{2W_0}}\vec{E}(\omega_0, \theta, \varphi)\vec{e}(\omega_0, \theta, \varphi) \\ &= -\frac{j\lambda_0}{\sqrt{2W_0}}\vec{E}(\omega_0, \theta, \varphi)\mathbf{A}^T(\omega_0)\vec{g}(\omega_0, \theta, \varphi), \end{aligned} \tag{8.23}$$

where W_0 is the wave impedance of free space. In (8.23) it is taken into account that equation between $\vec{\mathbf{e}}(\omega_0, \theta, \varphi)$, $\mathbf{g}(\omega_0, \theta, \varphi)$ and $\mathbf{A}(\omega)$ is determined similarly to [10] by equation

$$\vec{\mathbf{e}}(\omega, \theta, \varphi) = \mathbf{A}^T(\omega)\vec{\mathbf{g}}(\omega, \theta, \varphi). \tag{8.24}$$

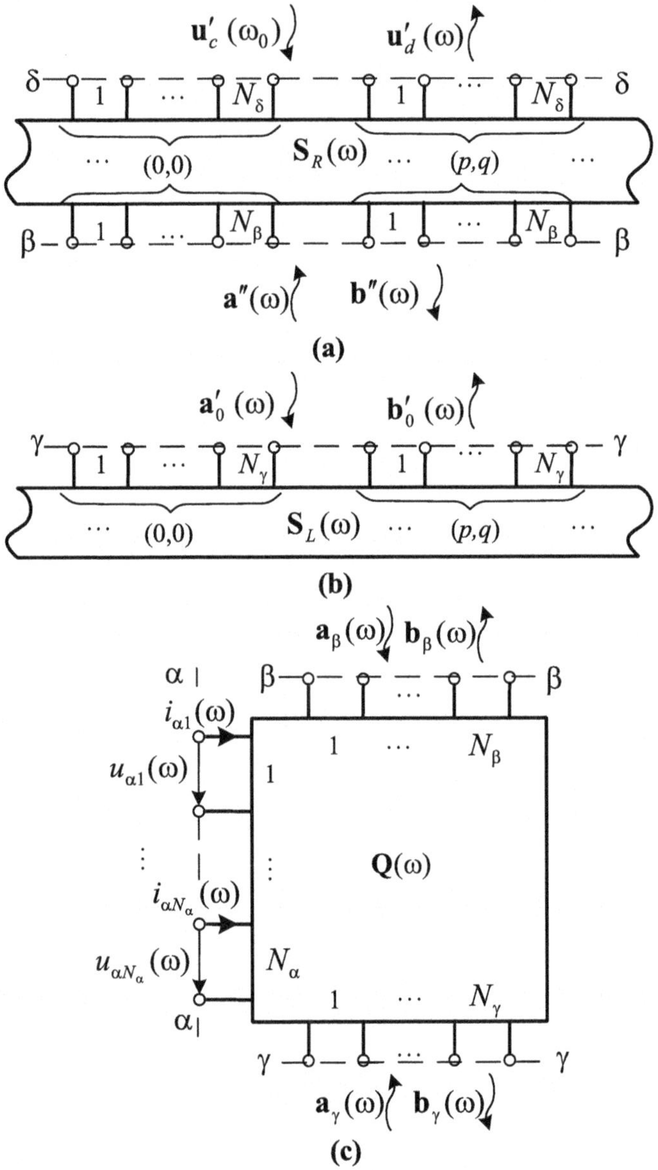

Fig. 8.8 Representation of linear subcircuits of the rectenna at stage of external parameters determination

The LM-2 multipole, describing the energy receivers connected to the rectenna, is characterized by the scattering matrix $\mathbf{S}_L(\omega)$, linking complex amplitudes of incident $\mathbf{a}'_0(\omega)$ and reflected waves $\mathbf{b}'_0(\omega)$ at its inputs (sections $\gamma - \gamma$, Fig. 8.8b),

$$\mathbf{b}'_0(\omega) = \mathbf{S}_L(\omega)\mathbf{a}'_0(\omega). \tag{8.25}$$

Linear multipole LM-3 will be described by the mixed matrix $\mathbf{Q}(\omega)$, which connects (see Fig. 8.8c) incident and reflected waves in the sections $\beta - \beta$ and $\delta - \delta$ with normalized currents $\mathbf{i}_\alpha(\omega)$ and voltages $\mathbf{u}_\alpha(\omega)$ in the section $\alpha - \alpha$, connecting LM-3 with nonlinear multipole:

$$\begin{pmatrix} \mathbf{u}_\alpha(\omega) \\ \mathbf{b}_\beta(\omega) \\ \mathbf{b}_\gamma(\omega) \end{pmatrix} = \begin{pmatrix} \mathbf{Q}_{\alpha\alpha}(\omega) & \mathbf{Q}_{\alpha\beta}(\omega) & \mathbf{Q}_{\alpha\gamma}(\omega) \\ \mathbf{Q}_{\beta\alpha}(\omega) & \mathbf{Q}_{\beta\beta}(\omega) & \mathbf{Q}_{\beta\gamma}(\omega) \\ \mathbf{Q}_{\gamma\alpha}(\omega) & \mathbf{Q}_{\gamma\beta}(\omega) & \mathbf{Q}_{\gamma\gamma}(\omega) \end{pmatrix} \begin{pmatrix} \mathbf{i}_\alpha(\omega) \\ \mathbf{a}_\beta(\omega) \\ \mathbf{a}_\gamma(\omega) \end{pmatrix} \tag{8.26}$$

Blocks of matrix $\mathbf{Q}(\omega)$ can be calculated if scattering matrix of the LM-3 multipole is known $\mathbf{S}(\omega)$ by the equations given in [8].

Normalized currents $\mathbf{i}_\alpha(\omega)$ and voltages $\mathbf{u}_\alpha(\omega)$ are related to vectors of complex amplitudes of currents $\mathbf{I}_\alpha(\omega)$ and voltage $\mathbf{U}_\alpha(\omega)$ by the known equations:

$$\mathbf{i}_\alpha(\omega) = 0.5\{\mathbf{Z}_w\}^{1/2}\mathbf{I}_\alpha(\omega)$$
$$\mathbf{u}_\alpha(\omega) = 0.5\{\mathbf{Z}_w\}^{-1/2}\mathbf{U}_\alpha(\omega)$$

where and$\{\mathbf{Z}_w\}^{1/2}\{\mathbf{Z}_w\}^{-1/2}$ are diagonal matrices whose elements are numbers $\sqrt{Z_{wl}(\omega)}$ for $\{\mathbf{Z}_w\}^{1/2}$ and $(1/\sqrt{Z_{wl}(\omega)}$ for $\{\mathbf{Z}_w\}^{-1/2})$; $Z_{wl}(\omega)$—impedance (at frequency ω) of transmission line connected in section $\alpha - \alpha$ to l-th input of nonlinear multipole. Further assume that impedances of all lines is the same, i.e. $Z_{wl}(\omega) = Z_w(\omega) \quad (l = 1, \ldots, N_\alpha)$. At such description of linear multipoles for calculation of external parameters of the rectenna following data are initial: an operating frequency ω_0; vector of input influences $\mathbf{u}'_c(\omega_0)$; parameters of linear multipole devices included in the circuit; vectors of complex current amplitudes $\mathbf{I}_n^{(p,q)}$, obtained by solving state equations and determining vector of normalized currents $\mathbf{i}_\alpha(\omega)$ at frequency of n-th harmonic in cross-sections $\alpha - \alpha$ of all RRE of the array:

$$\sqrt{Z_w}\mathbf{i}_\alpha(n\omega_0)$$
$$= \left(\ldots, \mathbf{I}_n^{(-1,-q)}, \mathbf{I}_n^{(0,-q)}, \mathbf{I}_n^{(1,-q)}, \ldots, \mathbf{I}_n^{(-1,0)}, \mathbf{I}_n^{(0,0)}, \mathbf{I}_n^{(1,0)}, \ldots, \mathbf{I}_n^{(-1,q)}, \mathbf{I}_n^{(0,q)}, \mathbf{I}_n^{(1,q)}, \ldots\right)^T. \tag{8.27}$$

The external parameters of the rectenna at frequency ω can be determined by calculating the vectors $\mathbf{a}'_0(\omega)$ and $\mathbf{u}'_d(\omega)$ at given input action $\mathbf{u}'_c(\omega_0)$ and a known (found from the results of solving equations of state) vector ($\mathbf{i}_\alpha(\omega)$ (hereinafter, where no special clarification is required, any of frequencies $n\omega_0$ for $n = 0, 1, \ldots$). At describing parameters of linear subcircuit, we will use mixed matrix $\mathbf{Q}^A(\omega)$, whose definition and equations for calculating its blocks through the parameters LM-1, LM-2 and LM-3 are given in [11]. Equations that describe relating between vectors $\mathbf{a}'_0(\omega)$, $\mathbf{u}'_d(\omega)$ from one hand, and vectors $\mathbf{u}'_d(\omega_0)$ and $\mathbf{i}_\alpha(\omega)$, from the other hand, have the form

$$\mathbf{a}'_0(\omega) = \mathbf{Q}^A_{\gamma\alpha}(\omega)\mathbf{i}_\alpha(\omega_0) + \begin{cases} \mathbf{Q}^A_{\gamma\delta}(\omega_0)\mathbf{u}'_c(\omega_0) & \text{if } \ \omega = \omega_0 \\ 0 & \text{if } \ \omega \neq \omega_0, \end{cases} \tag{8.28}$$

$$\mathbf{u}'_d(\omega) = \tilde{\mathcal{Q}}_{\delta\alpha}(\omega)\mathbf{i}_\alpha(\omega_0) + \begin{cases} \tilde{\mathcal{Q}}_{\delta\delta}(\omega_0)\mathbf{u}'_c(\omega_0) & \text{if } \ \omega = \omega_0 \\ 0 & \text{if } \ \omega \neq \omega_0. \end{cases} \tag{8.29}$$

Obtained expressions represent system of output equations and describe reactions in cross sections $\gamma - \gamma$ and $\delta - \delta$ generalized rectenna circuit (Fig. 8.2), in which NM is represented by equivalent current sources $\mathbf{i}_\alpha(\omega)$. Their value is determined from the solution of equations of state at given external actions $\mathbf{u}'_c(\omega_0)$.

Need to be noted that Eqs. (8.28) and (8.29) are formally linear equations linking vector of output parameters and vectors $\mathbf{u}'_c(\omega_0)$ and $\mathbf{i}_\alpha(\omega)$. In fact, however, nonlinear dependence $\mathbf{i}_\alpha(\omega)$ on $\mathbf{u}'_c(\omega_0)$, described by equations of state, leads to nonlinear dependence of vector of output parameters $\{\mathbf{a}'_0(\omega), \mathbf{u}'_d(\omega)\}^T$ on vector of input influences $\mathbf{u}'_c(\omega_0)$. In turn, this leads to nonlinear dependence of all external parameters of rectenna on magnitude of input action. Therefore, when characterizing rectenna by value of one or another external parameter, it is necessary to specify at what level of external influence given value was obtained. In addition, presence of nonlinear rectifier diodes leads to appearance of new spectral components in spectrum of currents ($\mathbf{i}_\alpha(\omega)$ (at the inputs $\alpha - \alpha$) of linear subcircuit) with frequencies ω, which do not coincide with the frequencies of input influences ω_0. At these frequencies $\mathbf{u}'_c(\omega \neq \omega_0) \equiv 0$ and (8.28), (8.29) are simplified

$$\mathbf{a}'_0(\omega) = \mathbf{Q}^A_{\gamma\alpha}(\omega)\mathbf{i}_\alpha(\omega_0), \tag{8.30}$$

$$\mathbf{u}'_d(\omega) = \tilde{\mathcal{Q}}_{\delta\alpha}(\omega)\mathbf{i}_\alpha(\omega_0). \tag{8.31}$$

Next, let's determine external key parameters of the rectenna: current and voltage of constant component in load; DC power absorbed in load multipole; rectenna efficiency; field over-radiated by rectenna at fundamental frequency and frequencies of higher harmonics. They are calculated as follows.

Within aperture of the rectenna, a single-type DC connection of the RRE is assumed. Given this assumption, current and voltage of the DC component in array load are determined. For series acquisition scheme, we have

$$U_0 = N_0 U_{ORRE}, \quad I_0 = I_{ORRE}. \tag{8.32}$$

For parallel acquisition circuit, the voltage and current

$$U_0 = U_{ORRE}, \quad I_0 = N_0 I_{ORRE}. \tag{8.33}$$

Here U_{ORRE}, I_{ORRE}—voltage and current in load of one RRE; U_0, I_0—voltage and current in rectenna load, with number of RREs equal to N_0.

It should be noted that the same-type (parallel or series) connection of DC RRE at equal-amplitude excitation of LAR is simultaneously optimal in sense that efficiency of such gathering circuits is equal to one (provided that the loads of RRE at fundamental frequency and frequencies of higher harmonics are equal). This follows from the identical characteristics of the RRE rectenna as DC (voltage) generators. At the same time, load impedances, at which the maximum efficiency of element, and hence the rectenna as a whole, is provided, will be different for series and parallel acquisition circuits. Options for structures of optimal collection circuits are not limited to a single-type connection of all RRE rectenna. For example, all series–parallel schemes of hierarchical type, in which at any level of hierarchy the numbers of elements to be combined (in series or parallel) are the same, will also have an efficiency equal to one.

DC power absorbed in the load multipole is calculated by the formula

$$P_0 = \mathbf{a}'^{*T}_0(\omega)\big(\mathbf{E} - \mathbf{S}^{*T}_L(\omega)\mathbf{S}_L(\omega)\big)\mathbf{a}'_0(\omega)\Big|_{\omega\to 0}. \tag{8.34}$$

Here the vector $\mathbf{a}'_0(\omega)\big|_{\omega\to 0}$ is defined from Eq. (8.30), $\mathbf{a}'_0(\omega)\big|_{\omega\to 0} = \mathbf{Q}^A_{\gamma\alpha}(n\omega_0)\mathbf{i}_\alpha(n\omega_0)\big|_{n=0}$; $\mathbf{E}$ is single matrix.

Efficiency of the rectenna is defined as ratio of the power P_0 to the maximum value of power P_M that can be extracted from incident field by radiator system at its optimal load:

$$\begin{aligned}\eta = \frac{P_0}{P_M} &= \frac{P_0}{\mathbf{u}'^{*T}_c(\omega_0)\mathbf{u}'_c(\omega_0)} \\ &= \frac{2W_0 P_0}{\lambda_0^2 \big|\vec{E}\big|^2 \vec{\mathbf{g}}^{*T}(\omega_0, \theta, \varphi)\mathbf{A}^*(\omega_0)\mathbf{A}^T(\omega_0)\vec{\mathbf{g}}(\omega_0, \theta, \varphi)}\end{aligned} \tag{8.35}$$

When the rectenna is excited by an electromagnetic wave with frequency ω_0, secondary field $\vec{E}_S$ is formed in surrounding space. At fundamental frequency, scattered field in wave zone can be represented as two components [10]:

$$\begin{aligned}\vec{E}_S(\omega_0, \theta, \varphi) &= \vec{E}_\Sigma(\omega_0, \theta, \varphi) + \vec{E}_C(\omega_0, \theta, \varphi) \\ &= \sqrt{2W_0}\mathbf{a}''^T(\omega_0)\vec{\mathbf{g}}(\omega_0, \theta, \varphi)\frac{e^{-jk_0 r}}{r} + \vec{E}_C(\omega_0, \theta, \varphi)\end{aligned} \tag{8.36}$$

Where $\vec{E}_\Sigma(\omega_0, \theta, \varphi)$ is secondary field caused by flow of currents on terminals of radiators of the rectenna; $\vec{E}_C(\omega_0, \theta, \varphi)$ is scattering field on matched radiators and elements of the rectenna structure; $\mathbf{a}''(\omega_0)$ is vector of incident waves on inputs of radiators of the rectenna, defined as

$$\mathbf{a}''(\omega) = \left(\mathbf{E} - \mathbf{S}_{\beta\beta}\tilde{\boldsymbol{Q}}_{\beta\beta}\right)^{-1}\left(\mathbf{S}_{\beta\beta}\tilde{\boldsymbol{Q}}_{\beta\alpha}\mathbf{i}_{\alpha}(\omega) + \begin{cases} \mathbf{S}_{\beta\gamma}\mathbf{u}_{c}^{;}(\omega_0) & \text{if } \omega = \omega_0 \\ 0 & \text{if } \omega = \omega_0 \end{cases}\right) \tag{8.37}$$

In (8.37) matrices $\tilde{\boldsymbol{Q}}_{ij}$ are calculated through the parameters of linear multipoles by expressions:

$$\tilde{\boldsymbol{Q}}_{ij} = \mathbf{Q}_{ij} + \mathbf{Q}_{i\gamma}\left(\mathbf{E} - \mathbf{S}_L\mathbf{Q}_{\gamma\gamma}\right)^{-1}\mathbf{S}_L\mathbf{Q}_{\gamma j}, \quad i = \alpha, \beta; \quad j = \alpha, \beta \tag{8.38}$$

At the frequencies of higher harmonics, rectenna radiators are excited from the input terminals, so the scattered field at each of frequencies $\omega_n = n\omega_0$ has only one component $\vec{E}_{\Sigma}(\omega_n, \theta, \varphi)$. The rectenna scattering field in far zone at fundamental frequency and frequencies of higher harmonics is defined by the expression

$$\vec{E}_S(\omega_n, \theta, \phi) = \sqrt{2W_0}\mathbf{a}{\prime\prime}^{T}(\omega_n)\vec{\mathbf{g}}(\omega_n, \theta, \varphi)\frac{\mathrm{e}^{-jnk_0 r}}{r} \tag{8.39}$$

Since in an equal-amplitude excited rectenna parameters of all RREs are identical, expression (8.39) can be represented in the form

$$\vec{E}_S(\omega_n, \theta, \varphi) = \sqrt{2W_0}\mathbf{a}_0^{\prime\prime T}(\omega_n)\vec{\mathbf{g}}_0(\omega_n, \theta, \varphi)F(\omega_n, \theta, \varphi)\frac{\mathrm{e}^{-jnk_0 r}}{r} \tag{8.40}$$

where $\mathbf{a}''_0(\omega_n)$ and $\vec{\mathbf{g}}_0(\omega_n, \theta, \varphi)$ is amplitude of incident wave and partial DD of the RRE array radiator at frequency of n-th harmonic; $F(\omega_n, \theta, \varphi)$—array multiplier at the same frequency.

If the parameters of the RRE radiator as part of array are determined, calculation of field $\vec{E}_S(\omega_n, \theta, \varphi)$ is not difficult in principle.

8.2 Analysis of Microstrip Radiating Structures of Periodic Large-Aperture Rectennas

Method and improved algorithm underlying program developed by the authors for analysis of periodic arrays of narrow microstrip radiators of complex geometry are considered. This method makes it possible to determine parameters of LARs, taking into account influence of DC power acquisition circuit conductors both at fundamental frequency and at frequencies of higher harmonics of fundamental frequency. It is based on well-known method of integral equations. Universal way to describe geometry of radiators of arbitrary configuration, including case of wires crossing the boundaries of the periodicity cell, is proposed, and new methods of accelerating calculations when solving system of integral equations at frequencies of higher harmonics are developed. Reliability of developed analysis algorithm is proved by comparing calculation results with known data of other authors.

In conclusion of this subsection we present materials on study of influence of power collection scheme and side radiation on efficiency of LARs.

8.2.1 Description of the Radiating Structure Configuration

Consider a LAR implemented on basis of flat periodic array of narrow microstrip radiators of complex configuration. The array geometry of the radiating system is shown in Fig. 8.1. By radiator system we will understand a set of connected in certain way rectilinear segments of ribbon conductors located within boundaries of one cell of periodicity of array. Conductors have zero thickness and their width matching conditions

$$2b_s << \lambda \text{ and } 2b_s << L_{\min}, \tag{8.41}$$

where $2b_s$ is the width of conductor, λ—wavelength of operating frequency in propagation environment, $L_{\min}$ is length of smallest straight segment of conductor in the antenna structure.

Radiators within one cell can be located in several parallel planes, coinciding with interfaces of layers of multilayer dielectric substrate placed over an infinite perfectly conducting screen (Fig. 8.9). Layers of the substrate differ in material parameters $\tilde{\varepsilon}_i$, and μ_i (i is the number of the layer). Values $\tilde{\varepsilon}_i$ can be complex, which takes into account losses in substrate dielectrics.

To describe geometry of radiators, we introduce concepts of branches and nodes. A branch is a homogeneous rectilinear segment of conductor (in Fig. 8.10, branches are indicated by numbers from 1 to 12). Homogeneity of branch implies absence of junctions for connecting loads or excitation sources, and, as a result, it is possible

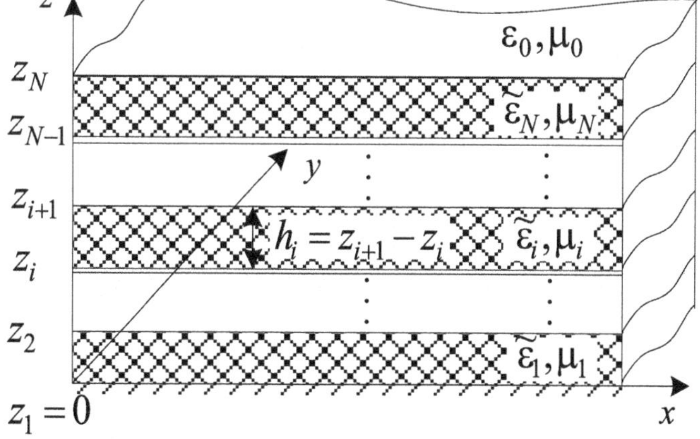

Fig. 8.9 Structure of rectenna system substrate

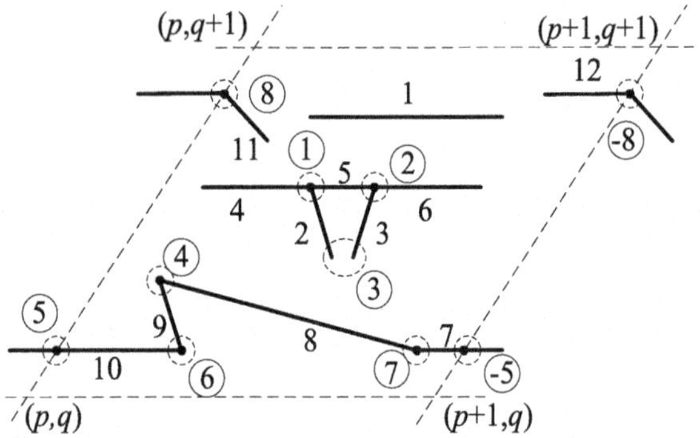

Fig. 8.10 Determining the nodes and branches of radiating structure of the rectenna

to describe the current flowing in it as a continuous function. Radiator node is a connection point of several branches (in Fig. 8.10, the nodes are circled and marked with numbers).

Let us define three types of radiator nodes:

- internal nodes—points of direct connection (by closing) of two or more branches (nodes 1, 2, 3, 4, 6, 7 in Fig. 8.10). Kirchhoff's law for currents and condition of zero divergence of latter, indicating the absence of charge accumulation in these points, must be fulfilled in all internal nodes;
- external nodes—connection points of two or more branches, in the gap between which external excitation sources or concentrated loads are included (node 3 in Fig. 8.10). Obviously, the nodes of this type are the input terminals of radiators, through which the radiating system is connected to the rest (linear and nonlinear) elements of the rectenna. The external nodes are modeled as infinitely narrow gaps;
- nodes formed by intersection of the radiator branches with the array periodicity cell boundaries. They occur when the radiating structure contains continuous conductors along the aperture (in the framework of the accepted model—infinite) (nodes 5 and -5 in Fig. 8.10) or when cell boundaries are defined in such way that the branches of one radiator appear in neighboring cells (nodes 8 and -8 in Fig. 8.10).

To describe the geometry of the radiator, we introduce rectenna coordinate system (x, y, z) with origin in the central cell of the array (cell with numbers 0,0). The coordinate plane $x0y$ is aligned with the screen plane. Coordinates (x_{pq}, y_{pq}, z_{pq}) of an arbitrary point M_{pq} located in the cell (p, q) can be expressed through the coordinates (x_{00}, y_{00}, z_{00}) of point M_{00} located similarly in the cell $(0, 0)$ and array parameters d_x, d_l, α_r as follows:

$$x_{pq} = x_{00} + p \cdot d_x + q \cdot d_l \cdot \cos\alpha_r,$$
$$y_{pq} = y_{00} + q \cdot d_l \cdot \sin\alpha_r, z_{pq} = z_{00}. \tag{8.42}$$

Consequently, characteristic of single periodicity cell provides a complete picture of geometry of the rectenna array as a whole.

In view of the above, we can propose following procedure for describing the geometry of the analyzed radiating structure. We preliminarily number all nodes and branches of radiators located within the same periodicity cell by sequence of numbers of natural number series. During numbering we assume that following conditions are satisfied:

(a) if three or more branches are connected in node, the smallest number has the one through which the total current flows;
(b) all free ends of the branches are considered to be connected to zero node;
(c) pair of nodes formed by the intersection of radiator conductors with opposite walls of the Flocked channel are assigned the same numbers with different signs.

Otherwise, the order of numbering is arbitrary.

Setting the coordinates of all nodes and free ends of branches, as well as specifying the node numbers corresponding to the beginning and end of each branch, uniquely determines the configuration of radiators. Additional information is provided by the width of the conductors that form them and the sequence of external node numbers. The dimensions and shape of the periodicity cell are determined by the values of parameters d_x, d_l, α_r respectively, and the geometry of the substrate by indicating z-coordinates of lower boundaries of all its layers (the region corresponding to free half-space above the radiating structure is considered as the N-th layer).

Obviously, the listed parameters form the minimum required amount of initial data, which allows us to adequately represent the analyzed radiating structure and, as will be shown below, to build an effective algorithm for its analysis.

In order to simplify further description, in addition to already described rectenna coordinate system (x, y, z) it is convenient to introduce local systems connected directly with each i radiator branch, the coordinate axes ξ_i and η_i of which are oriented along the axis and across the conductor of the corresponding branch (Fig. 8.11).

Then the position of any arbitrary point on the surface of the radiator conductor can be determined as follows:

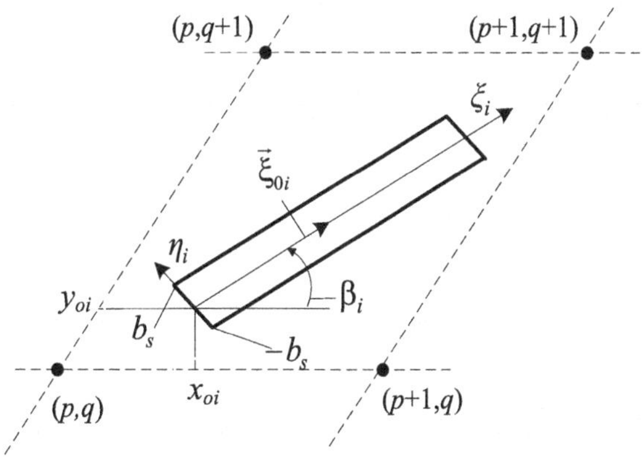

Fig. 8.11 Defining local coordinate systems

$$x_i = x_{oi} + \xi_i \cdot \cos\beta_i - \eta_i \cdot \cos\beta_i,$$
$$y_i = y_{oi} + \xi_i \cdot \sin\beta_i + \eta_i \cdot \cos\beta_i, z = z_o, \tag{8.43}$$

Where x_o,y_o,z_o are the coordinates of the starting point of the local system, defined as the coordinates of the beginning of the i-th radiator branch, β_i is the angle between the coordinate axes x and ξ_i.

In addition, to describe fields in the space above the array, we use the spherical coordinate system (r, θ, φ), the beginning of which coincides with the beginning of the rectenna coordinate system in cell (0, 0). The unit vector $\vec{r}_0$, characterizing direction (θ, φ) in the spherical coordinate system with respect to the axes of the rectenna system, is defined as follows:

$$\vec{r}_0 = S_x\vec{x}_0 + S_y\vec{y}_0 + S_z\vec{z}_0, \tag{8.44}$$

where

$$S_x = \sin\theta \cdot \cos\varphi, S_y = \sin\theta \cdot \sin\varphi, S_z = \cos\theta. \tag{8.45}$$

Proposed method of describing the geometry fully enough characterizes the radiating structure of any configuration with minimum set of input data and, as will be shown below, allows to build efficient algorithms for their analysis.

8.2.2 Integral Equations of Narrow Radiators of Arbitrary Configuration in the Infinite Array Composition

We obtain system of integral equations with respect to currents on the antenna array elements, configuration of which is described in Sect. 8.2.1. Taking into account conditions (8.41), it is natural to assume that longitudinal component of the vector of surface electric current density $\vec{J}_s$ in radiators of the considered type is significantly greater than the transverse one. Therefore, latter is not taken into account hereinafter. In case of narrow ribbon conductors the surface current density $\vec{J}_s(\xi, \eta)$ can be represented as a product of two functions, each of which depends only on one coordinate (ξ or η), i.e.:

$$\vec{J}_s(\xi, \eta) = J(\xi) \cdot f(\eta) \cdot \vec{\xi}_0, \tag{8.46}$$

where $\vec{\xi}_0$—single vector oriented along the conductor axis, $J(\xi)$—function describing the current density distribution along the conductor; $f(\eta)$—function describing the current distribution in the conductor cross section. In all the following equations, the indices indicating that the local coordinate system belongs to one or another branch will be omitted, if it does not lead to misunderstandings.

In the quasi-static approximation for a metal strip of width $2b_s << \lambda$ the function $f(\eta)$ is

$$f(\eta) = \frac{1}{\pi\sqrt{b_s^2 - \eta^2}}. \tag{8.47}$$

In this case ($2b_s << \lambda$ and $2b_s << L_{\min}$) we can assume that the transverse distribution of current density does not change along the conductor. If we introduce an integral characteristic $\vec{I}(\xi)$ called the total current of the radiator

$$\vec{I}(\xi) = \int_{-b_s}^{b_s} \vec{J}_s(\xi, \eta)d\eta, \tag{8.48}$$

then by substituting (8.46) into (8.48), we obtain

$$\vec{I}(\xi) = \Delta_s J(\xi)\vec{\xi}_0 = \Delta_s \vec{J}(\xi), \tag{8.49}$$

where $\Delta_s = \int_{-b_s}^{b_s} f(\eta)d\eta$.

Obviously, this corresponds to the replacement of the real surface current distribution by some equivalent total current $\vec{I}(\xi)$ flowing along the conductor axes, and system of integral equations can be obtained using procedure similar to that used in theory of thin-wire antennas [12]. Its starting point is the fulfillment of boundary condition on the surface of the conductors.

Following the above procedure, we equate to zero the tangential component of the full electric field strength vector at all points $\vec{r} = \vec{r}(x, y, z)$ on the conductor surface of the radiators, except for points corresponding to dissections at the outer nodes:

$$\left[\vec{n}(\vec{r}), \vec{E}(r)\right] + \left[\vec{n}(\vec{r}), \vec{E}^{is}(\vec{r})\right] = 0. \tag{8.50}$$

Here $\vec{n}(\vec{r})$ is single vector directed along the normal to the conductor surface at point $\vec{r}$; $\vec{E}(r)$ and $\vec{E}^{is}(\vec{r})$ are vectors of electric field strength created at point $\vec{r}$ by all radiators of the array and external sources of excitation, respectively.

Let the current distribution $\vec{J}_{p'q'}(\xi')$ on the radiators located within some cell of the investigated (p', q') radiating structure be established. Field generated by this current in an arbitrary point of space $\vec{r}$ can be defined as follows [13]:

$$\vec{E}^{(p',q')}(\vec{r}) = \frac{1}{j\omega\tilde{\varepsilon}_n}(\text{grad div} + k_n^2) \int_L \hat{G}\left(\vec{r}, \vec{r'}\right)\vec{J}_{p'q'}(\xi')d\xi', \tag{8.51}$$

where $\hat{G}\left(\vec{r},\vec{r'}\right)$ is the tensor Green's function (TGF) of a layered homogeneous lossy medium [14]; $\vec{r}, \vec{r'}$—radius vectors of observation and integration points, respectively; $\omega = 2\pi f$—circular frequency; $\tilde{\varepsilon}_n = \varepsilon_a(1 - jtg\delta_n)$; $\varepsilon_a = \varepsilon_0\varepsilon_n$; ε_n—relative dielectric permittivity of the n-th substrate layer, in which observation point is located; $tg\delta_n$—angle tangent of dielectric losses of the n-th layer material; $k_n = 2\pi/\lambda_n$—wave numbers; λ_n—frequency wave length f in environment with material parameters $\tilde{\varepsilon}_n, \mu_n$. Integration in (8.51) is performed along all radiator branches located in the cell with numbers $\left(p', q'\right)$.

Calculation of electric field created by the currents of all radiators of the array, on surface of the radiator located in the cell (p, q), consists in summing (8.51) over all indices p', q' in infinite limits. Taking into account (8.49), after summation we obtain:

$$\vec{E}(\vec{r}) = \frac{1}{j\omega\tilde{\varepsilon}_n\Delta_s}\left(\text{grad div} + k_n^2\right)\sum_{p'=-\infty}^{\infty}\sum_{q'=-\infty}^{\infty}\int_{L'}\hat{G}\left(\vec{r},\vec{r'}\right)\vec{I}_{p'q'}\left(\xi'\right)d\xi'. \tag{8.52}$$

Consider excitation of the array by plane electromagnetic wave arriving from direction (θ_0, φ_0) and having amplitude of the electric field strength in free space E_m^{is}. Then at point $\vec{r} = \vec{r}(x, y, z)$, located in n-th layer of the substrate, we have

$$\vec{E}^{is}(\vec{r}) = \vec{e}_0\dot{E}_m^{is}(n)e^{jk_n\left(x\cdot S_x^n + y\cdot S_y^n + z\cdot S_z^n\right)}, \tag{8.53}$$

where

$$\begin{aligned}\vec{e}_0 = {}& \vec{x}_0\left(-\cos\psi_0\cdot\cos\theta_0^n\cdot\cos\varphi_0 - \sin\psi_0\cdot\sin\varphi_0\right)\\ &+\vec{y}_0\left(-\cos\psi_0\cdot\cos\theta_0^n\cdot\sin\varphi_0 + \sin\psi_0\cdot\cos\varphi_0\right)\\ &+\mathbf{z}_0\left(\cos\psi_0\cdot\cos\theta_0^n\right)-,\end{aligned} \tag{8.54}$$

ψ_0 is polarization angle, defined as the angle between the vector $\vec{E}^{is}(\vec{r})$ and plane of incidence of excitation wave (Fig. 8.12);

$$\theta_0^n = 90^\circ - \arcsin\left(\sqrt{\frac{\tilde{\varepsilon}_{n+1}\tilde{\mu}_{n+1}}{\tilde{\varepsilon}_n\tilde{\mu}_n}}\sin\theta_0^n\right)-, \tag{8.55}$$

is angle of incidence of electromagnetic wave on lower boundary of the n-th layer of the substrate;

$$\dot{E}_m^{is}(n) = \dot{E}_m^{is}(n+1)\cdot\chi_{n+1,n}- \tag{8.56}$$

amplitude of the external field in the n-th layer of substrate; $\chi_{n+1,n}$—coefficient of wave propagation through the interface between the n-th and $(n + 1)$-th layers;

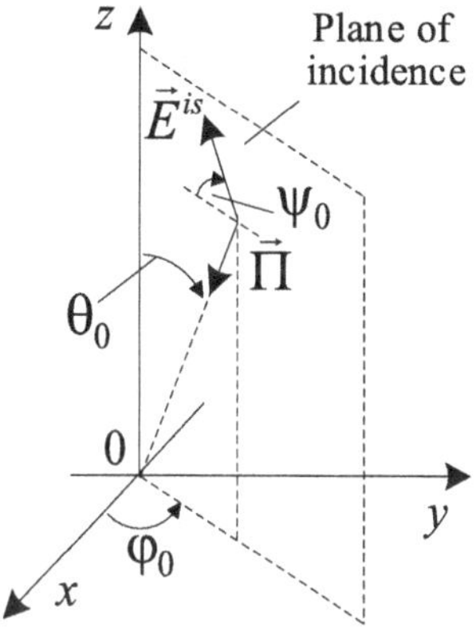

Fig. 8.12 Excitation of the array by plane wave

$$S_x^n = \sin\theta_0^n \cos\varphi_0;\ S_y^n = \sin\theta_0^n \sin\varphi_0;\ S_z^n = \cos\theta_0^n$$

Substituting (8.52) and (8.53) into (8.50), we obtain an infinite integral equations (IE) system with respect to unknown currents $\vec{I}_{p'q'}(\xi')$:

$$\frac{1}{j\omega\tilde{\varepsilon}_n}\sum_{p'=-\infty}^{\infty}\sum_{q'=-\infty}^{\infty}\left[\vec{n}(\vec{r}) \times \left(\text{grad div} + k_n^2\right)\int\limits_L \hat{G}\left(\vec{r}, \vec{r}'\right)\frac{\vec{I}_{p'q'}(\xi')}{\Delta_s}d\xi'\right]$$
$$= -\left[\vec{n}(\vec{r}) \times \mathbf{e}_0\right]E_m^{is}(n)\cdot e^{jk_n\left(x\cdot S_x^n + y\cdot S_y^n + z\cdot S_z^n\right)},\ -\infty < p < \infty,\ -\infty < q < \infty. \tag{8.57}$$

8.2.3 *Solving Integral Equations for Current*

To solve IE (8.57) we use Galerkin method, according to which we present required current $\vec{I}_{p'q'}(\xi')$ expansion in a system of basis functions $\left\{\vec{T}_{p'q'}^k(\xi')\right\}$:

$$\vec{I}_{p'q'}(\xi') = \sum_{k=1}^{K} I_{p'q'}^k \vec{T}_{p'q'}^k(\xi'), \tag{8.58}$$

where $I_{p'q'}^k$—unknown coefficients; K—number of basis functions (BF) sufficient to approximate with a given accuracy the current distribution of radiators located in cell (p', q').

Substituting (8.58) into equations of system (8.57), we form scalar products in their left and right parts by multiplication by weight functions $\vec{T}_{pq}^l(\xi)$, $l = \overline{1, K}$ and integration by coordinate ξ along all conductors of radiators located within cell (p, q). As a result, each Eq. (8.57) is converted into a system of linear algebraic equations of following type:

$$\sum_{p'=-\infty}^{\infty}\sum_{q'=-\infty}^{\infty}\sum_{k=1}^{K} Z_{lk}(p,p',q,q')I_{p'q'}^{k} = V_{pq}^{l} \quad \forall 1 \le l \le K. \tag{8.59}$$

where

$$Z_{lk}(p,p',q,q') = \frac{1}{j\omega\tilde{\varepsilon}_n}\int_L \vec{T}_{pq}^{l}(\xi) \cdot \left[\vec{n}(\xi)\times\left(\text{grad div}+k_n^2\right)\int_L \hat{G}(\xi,\xi')\frac{\vec{T}_{p'q'}^{k}(\xi')}{\Delta_s}d\xi'\right] - \tag{8.60}$$

are elements of square matrix **Z** of dimension $K \times K$, having the meaning of the eigen (at $k = l$) and mutual (at $k \neq l$) impedances between currents characterized by the functions $\vec{T}_{p'q'}^{k}(\xi')\vec{T}_{pq}^{l}(\xi)$;

$$V_{pq}^{l} = -\int_L \vec{T}_{pq}^{l}(\xi)\cdot[\vec{n}(\xi)\times\vec{E}^{is}(\xi)]d\xi - \tag{8.61}$$

elements of vector **V** of dimension K, having sense of stresses.

Considering further periodicity of the array and fact that identical BF systems are used in approximating required currents on all its radiators, it is easy to show that matrix **Z** of system of Eq. (8.59) is an infinite-dimensional block-template, i.e.

$$Z_{kl}(p,p',q,q') = Z_{kl}(p-p',q-q'). \tag{8.62}$$

It is known that systems of algebraic equations whose matrices depend on the difference of indices $(p-p')$, $(q-q')$ are conveniently solved using the apparatus of the Fourier transform. Thus, after multiplying both parts of Eq. (8.59) by $e^{j(\alpha' p+\beta' q)}$ and summing over p and q in infinite limits, we obtain a simple system of linear equations:

$$\sum_{k=1}^{K}\tilde{Z}_{lk}(\alpha',\beta')\tilde{I}_k(\alpha',\beta') = \tilde{V}_l(\alpha',\beta')\,\forall 1 \le l \le K, \tag{8.63}$$

known and unknown coefficients of which have the meaning of spectral densities and are defined as follows:

$$\tilde{I}_k(\alpha',\beta') = \sum_{p'=-\infty}^{\infty}\sum_{q'=-\infty}^{\infty} I_{p'q'}^{k} e^{j(\alpha' p'+\beta' q')}, \tag{8.64}$$

$$\tilde{V}_l(\alpha', \beta') = \sum_{p=-\infty}^{\infty} \sum_{q=-\infty}^{\infty} V_{pq}^l e^{j(\alpha' p + \beta' q)}, \tag{8.65}$$

$$\tilde{Z}_{kl}(\alpha', \beta') = \sum_{r=-\infty}^{\infty} \sum_{s=-\infty}^{\infty} Z_{kl}(r, s) e^{j(\alpha' r + \beta' s)}, \tag{8.66}$$

where α' and β' are Fourier transform variables; $r = p - p'$, $s = q - q'$.

After solving Eq. (8.63) with respect to $\tilde{I}_k(\alpha', \beta')$, values $I_{p'q'}^k$ are defined as Fourier series coefficients from the equation

$$I_{p'q'}^k = \frac{1}{4\pi^2} \int_0^{2\pi} \int_0^{2\pi} \tilde{I}_k(\alpha', \beta') e^{j(\alpha' p + \alpha' q)} d\alpha' d\alpha'. \tag{8.67}$$

If the excitation of the radiating structure is carried out by plane electromagnetic wave, the problem is significantly simplified. In this case, the strength of the external electric field at an arbitrary point of space is defined as $\vec{r} = \vec{r}(x, y, z)$

$$\vec{E}_{pq}^{is} = E_{00}^{is} e^{j(\alpha p + \beta q)}, \tag{8.68}$$

where

$$\alpha = k d_p S_x; \quad \beta = k d_p (S_x \cos \alpha_r + S_y \sin \alpha_r). \tag{8.69}$$

Therefore, currents of radiators located in the cells (p', q') and (0, 0), are related by a similar equation

$$I_{p'q'}^k = I_{00}^k e^{j(\alpha p + \beta q)}. \tag{8.70}$$

Then, taking into account Poisson equations [15], expressions (8.64) and (8.65) will take the form

$$\tilde{I}_k(\alpha', \beta') = 4\pi^2 I_{00}^k \cdot \delta(\alpha' - \alpha) \cdot \delta(\beta' - \beta), \tag{8.71}$$

$$\tilde{V}_l(\alpha', \beta') = 4\pi^2 V_{00}^l \cdot \delta(\alpha' - \alpha) \cdot \delta(\beta' - \beta) \tag{8.72}$$

and, respectively,

$$\sum_{k=1}^{K} \tilde{Z}_{lk}(\alpha', \beta') I_{00}^k \delta(\alpha' - \alpha) \delta(\beta' - \beta) = V_{00}^l \delta(\alpha' - \alpha) \delta(\beta' - \beta). \tag{8.73}$$

Here $\delta(x' - x)$ is Dirac delta function.

Fig. 8.13 Points of intersection of periodicity cell boundaries by structure conductors

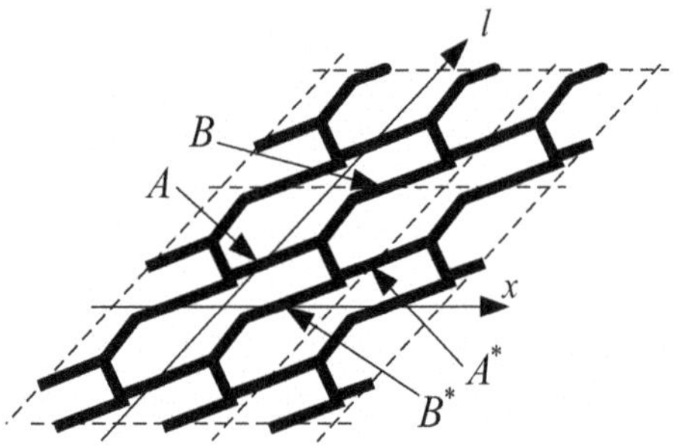

By integrity (8.73) on α' and β', we obtain

$$\sum_{k=1}^{K} \tilde{Z}_{lk}(\alpha, \beta) I_{00}^{k} = V_{00}^{l} \, \forall \, l = \overline{1, K}. \tag{8.74}$$

It should also be noted that in the case where the radiator conductors cross the opposite boundaries of periodicity cells (points *A*-*A* *and *B*-*B* * in Fig. 8.13), the uniqueness of the solution of system (8.73) will be provided only in the case when in these points currents match the conditions

$$I_{00}^{A^*} = I_{00}^{A} e^{j\alpha} \tag{8.75}$$

for points formed by the intersection of conductors with boundaries parallel to the axis l;

$$I_{00}^{B^*} = I_{00}^{B} e^{j\beta} \tag{8.76}$$

for points formed by the intersection of conductors with boundaries parallel to the axis x.

Thus, solution of system (8.74) together with conditions (8.75) and (8.76) makes it possible to determine distribution of currents along the conductors of the zero array cell, and then—along the conductors of any cell of the radiating structure in accordance with expression (8.70).

8.2.4 Characteristics of the Radiating Structure

System of Eq. (8.74) can be formally considered as an equation of a single radiator (group of radiators) corresponding to one periodicity cell. By determining the current along the conductors of this radiator, we can calculate its external characteristics as well. In this case, the matrix elements **Z** are calculated taking into account the mutual equation between the radiators of the structure, therefore, the calculated external characteristics take this equation into account as well.

At stage of calculating the external parameters of the radiating structure matrix **Z**, **I** and **V** Eq. (8.74) it is advisable to present in block form:

$$\begin{bmatrix} \mathbf{Z}_{11} & \mathbf{Z}_{12} \\ \mathbf{Z}_{21} & \mathbf{Z}_{22} \end{bmatrix} \begin{bmatrix} \mathbf{I}_1 \\ \mathbf{I}_2 \end{bmatrix} = \begin{bmatrix} \mathbf{V}_1 \\ \mathbf{V}_2 \end{bmatrix}, \tag{8.77}$$

and in such a way that the values characterizing the radiator's external nodes are grouped in $\mathbf{V}_2$, $\mathbf{Z}_{22}$, $\mathbf{I}_2$, blocks.

In order to determine the matrix of eigen and mutual resistances with respect to the inputs of the radiating structure, assume that voltage generators with amplitudes V_2^n, are connected to its external nodes, and there are no other excitation sources ($V_1^n = 0$) (V_1^n and V_2^n—elements of vectors $\mathbf{V}_1$ and $\mathbf{V}_2$). Then, from Eq. (8.77) it is easy to obtain a system of linear equations describing the radiator from its inputs (external nodes):

$$\hat{\mathbf{Z}} \cdot \mathbf{I}_2 = \mathbf{V}_2 \tag{8.78}$$

where $\hat{\mathbf{Z}} = \mathbf{Z}_{22} - \mathbf{Z}_{21}\mathbf{Z}_{11}^{-1}\mathbf{Z}_{12}$ is matrix of eigen and mutual resistances of the radiator with respect to the inputs.

By marking

$$\mathbf{H} = \mathbf{Z}_{11}^{-1} \cdot \mathbf{Z}_{12} \tag{8.79}$$

the expression for the matrix $\hat{\mathbf{Z}}$ can be written in a more compact form:

$$\hat{\mathbf{Z}} = \mathbf{Z}_{22} - \mathbf{Z}_{21}\mathbf{H}. \tag{8.80}$$

To calculate EMF induced at radiator inputs by field of excitation plane wave (8.53), it is sufficient to recalculate values obtained according to (8.61) V_1^n to the corresponding radiator inputs by the equation

$$e_x^{(0,0)} = V_2^i - \sum_{n=1}^{N_1} H_{i,n} V_1^n. \tag{8.81}$$

where $e_x^{(0,0)}$—EMF at i input of the radiator; N_1—Vector dimension $\mathbf{V}_1$, equal to the number of basis functions approximating the current distribution along branches and at internal nodes of the radiating structure; $H_{i,n}$—matrix element **H**, defined according to expression (8.79).

Let us consider the directivity characteristics of the radiator surrounded by other elements of the array. The expression for the field created by the radiator currents in the far field will be in matrix form [10].

$$\vec{E}(r, \theta, \varphi) = \frac{Z_0}{2\lambda}\mathbf{I}^T \vec{\Phi}(\theta, \varphi)\frac{e^{-jkr}}{r}, \tag{8.82}$$

where Z_0 is the wave impedance of environment, $\mathbf{I}^T$ is matrix-string of expansion coefficients of radiator current I_{00}^n, $\vec{\Phi}(\theta, \varphi)$ is vector of directional functions of currents characterized by distributions $\vec{T}_{00}^n(\xi)$.

If the vector of amplitudes of input currents is given $\mathbf{I}_2$, the DD of the radiator in the array can be calculated as follows:

$$\vec{f}(\theta, \varphi) = \frac{Z_0}{2\lambda}\left[\mathbf{I}_1^T[\mathbf{E}], \mathbf{I}_1^T[-\mathbf{H}]\right]\vec{\Phi}(\theta, \varphi), \tag{8.83}$$

where $\mathbf{E}$ is the single matrix.

The obtained equation describes directivity characteristic of the radiator located in zero cell of the array and excited by the vector of currents $\mathbf{I}_2$, i.e., it is its partial directivity characteristic.

8.2.5 Improved Algorithm for Solving System of Integral Equations

One of factors determining efficiency of solving the system of integral equations of type (8.57) by Galerkin's method is the choice of basis type on which the decomposition (8.59) of current distribution is performed. When constructing universal algorithms allowing to calculate characteristics of radiators of arbitrary configuration, it is most expedient to use as the basis the so-called subdomain functions [12] determined at separate sections of the radiator (subdomains). The application of functions of such type, preserving continuity at connection points of two or more conductors, allows to ensure the fulfillment of Kirchhoff's law in nodes without changing the type of IE system. If, however, to approximate desired current distribution, we use a decomposition by full domain functions, then the system of equations obtained in accordance with the above procedure should additionally include the equations establishing the equations between the branch currents entering each node of the radiator (the equations of the first Kirchhoff law). Obviously, in this case, the system of equations will take a different form for each new configuration of the radiator, and therefore, the principle of universality of the developed algorithm will not be observed.

The advantage of the basis of subdomains is also the possibility to select, already at the stage of description of the model of the analyzed radiator, subsequences of basis functions characterizing currents of open branches, as well as internal and external nodes, which, as will be shown below, will allow to organize more optimally the procedure of solution of the initial IE system and calculation of external parameters of the radiating structure.

Piecewise-trigonometric functions of the above type are the most preferable. Their use allows to accelerate the process of solving the IE system in comparison to case of piecewise-triangular and piecewise-constant bases, respectively, by 3–5 and 6–8 times [12]. In addition, if piecewise-sinusoidal functions are chosen as a basis, the condition of zeroing of the current distribution is automatically ensured at the free ends of the radiator conductors. Note also that the use of piecewise-sine functions allows us to obtain simple analytical expressions for the spectral densities of the eigen and mutual resistances and to reduce the volume and time of the calculations.

Let us define n-th piecewise sinusoidal basis function as follows (to shorten the record in this and following subsections, we assume that all linear dimensions are normalized by multiplication by the free space wave number):

$$\vec{T}^n(\xi) = \vec{T}^n_{00}(\xi) = \frac{\sin(\xi_n - L_n)}{\sin L_n}\vec{\xi}_0^{(n)} + \frac{\sin(L_{n+1} - \xi_{n+1})}{\sin L_{n+1}}\vec{\xi}_0^{(n+1)}, \tag{8.84}$$

where $\vec{\xi}_0^{(n)}, \vec{\xi}_0^{(n+1)}$ are unit vectors oriented along the radiator's axis, whose common origin is the center of the BF (common point of neighboring segments on which the function $\vec{T}^n$ is defined); ξ_n, ξ_{n+1} are coordinates counted in the local coordinate system associated with n the BF in the direction of vectors, $\mathbf{x}_0^{(n)}$ and $\mathbf{x}_0^{(n+1)}$ respectively; L_n, L_{n+1} are the lengths of segments on which the basis current function n is defined.

It is obvious that in the most general case for the complete formation of the impedance matrix $\mathbf{Z}$ of Eq. (8.74) it is enough to determine the mutual resistance between two elementary currents, which are characterized by functions in the form of sine segments (one of the terms in 8.83). For this we substitute expressions for $\vec{T}^k(\xi')$ and $\vec{T}^l(\xi')$ in (8.60) and (8.66) and perform necessary operations of integration and summation, as a result we obtain the following expression for the spectral density of mutual resistance of the indicated current elements:

$$\begin{aligned}\tilde{Z}_{kl}(\alpha, \beta) = & \frac{j4Z_c^i}{d_l d_x \sin\alpha_r \sin L_k \sin L_l} \\ & \times \sum_{\nu=-\infty}^{\infty}\sum_{\mu=-\infty}^{\infty}\left\{\left[\cos(\varphi_k - \varphi_l) - \delta_k^{\nu\mu}\delta_l^{\nu\mu}\right]\cdot Q_{\nu\mu} - \gamma_i^{\nu\mu}\delta_k^{\nu\mu}\delta_l^{\nu\mu}F_{\nu\mu}\right\} \\ & \times \frac{e^{-j\delta_k^{\nu\mu}L_k} - \cos(L_k) + j\delta_k^{\nu\mu}\sin L_k}{1 - \left(\delta_k^{\nu\mu}\right)^2}\cdot\frac{e^{j\delta_l^{\nu\mu}L_l} - \cos(L_l) - j\delta_l^{\nu\mu}\sin L_l}{1 - \left(\delta_l^{\nu\mu}\right)^2} \\ & \times \frac{\sin\left(\frac{\Delta_k^{\nu\mu}\Delta_s}{2}\right)}{\Delta_k^{\nu\mu}\Delta_s}\cdot\frac{\sin\left(\frac{\Delta_l^{\nu\mu}\Delta_s}{2}\right)}{\Delta_l^{\nu\mu}\Delta_s}\cdot e^{j\left[\chi_1^{\nu}(x_k - x_l) + \chi_2^{\nu\mu}(y_k - y_l)\right]},\end{aligned} \tag{8.85}$$

where Z_c^i is the characteristic resistance of the material of the i-th layer of substrate in which the observation point is located; L_k, L_l are the lengths of segments on which functions $\vec{T}^k(\xi)$ and $\vec{T}^l(\xi)$ are defined; φ_k,φ_l are angles between the axis x of the rectenna coordinate system and the vectors $\vec{\xi}_0^k$ and $\vec{\xi}_0^l$ respectively; $Q_{\nu\mu}$, $F_{\nu\mu}$ are factors whose values are determined by the geometry and material parameters of the substrate;x_k, y_k, x_l, y_l are the coordinates of the points taken as the origin of k-th and l-th base functions;

$$\chi_1^{\nu} = S_x + \frac{2\pi\nu}{d_l};\ \chi_2^{\nu\mu} = S_y - \frac{2\pi\nu}{d_l tg\alpha_r} + \frac{2\pi\mu}{d_x \sin\alpha_r} \tag{8.86}$$

propagation coefficients in the array plane;

$$\gamma_i^{\nu\mu} = \sqrt{\left(\chi_1^{\nu}\right)^2 + \left(\chi_2^{\nu\mu}\right) - 1} \tag{8.87}$$

is the propagation coefficient in direction normal to the array plane in the i-th layer of the substrate;

$$\delta_m^{\nu\mu} = \chi_1^{\nu}\cos\varphi_m + \chi_2^{\nu\mu}\sin\varphi_m,\ m = k, l, \tag{8.88}$$

$$\Delta_m^{\nu\mu} = \chi_1^{\nu}\sin\varphi_m - \chi_2^{\nu\mu}\cos\varphi_m,\ m = k, l. \tag{8.89}$$

However, directly by (8.85) it is reasonable to calculate only the mutual resistances between the currents at the connection points of radiator branches, i.e. at the nodes. To calculate the mutual resistance between elementary currents determined at the branches, it is reasonable to consider the mutual resistance between the currents shown in Fig. 8.14b, c. It should be taken into account that the spectral density of mutual resistance can be represented as

$$\tilde{Z}_{kl}(\alpha,\beta) = -\tilde{Z}_{kl}^{11}(\alpha,\beta) - \tilde{Z}_{kl}^{22}(\alpha,\beta) + \tilde{Z}_{kl}^{12}(\alpha,\beta) + \tilde{Z}_{kl}^{21}(\alpha,\beta). \tag{8.90}$$

Terms $\tilde{Z}_{kl}^{mn}(\alpha,\beta)$, $m = 1, 2$, $n = 1, 2$ in (8.89) are spectral densities of eigen (at $m = n$) and mutual (at $m \neq n$) resistances between corresponding current elements $\vec{T}^k(\xi)$ and $\vec{T}^l(\xi)$ are calculated by (8.85). Performing summation, we obtain

$$\begin{aligned}\tilde{Z}_{kl}(\alpha,\beta) &= \frac{-j4Z_c^i}{d_l d_x \sin\alpha_r \sin L_k \sin L_l}\\ &\times \sum_{\nu=-\infty}^{\infty}\sum_{\mu=-\infty}^{\infty}\left\{\left[\cos(\varphi_k-\varphi_l) - \delta_k^{\nu\mu}\delta_l^{\nu\mu}\right]\cdot Q_{\nu\mu} - \gamma_i^{\nu\mu}\delta_k^{\nu\mu}\delta_l^{\nu\mu}F_{\nu\mu}\right\}\\ &\times \frac{\cos\left(\delta_k^{\nu\mu}L_k\right) - \cos(L_k)}{1-\left(\delta_k^{\nu\mu}\right)^2}\cdot\frac{\cos\left(\delta_l^{\nu\mu}L_l\right) - \cos(L_l)}{1-\left(\delta_l^{\nu\mu}\right)^2}\\ &\times \frac{\sin\left(\frac{\Delta_k^{\nu\mu}\Delta_s}{2}\right)}{\Delta_k^{\nu\mu}\Delta_s}\cdot\frac{\sin\left(\frac{\Delta_l^{\nu\mu}\Delta_s}{2}\right)}{\Delta_l^{\nu\mu}\Delta_s}\cdot e^{j\left[\chi_1^{\nu}(x_k-x_l)+\chi_2^{\nu\mu}(y_k-y_l)\right]}.\end{aligned} \tag{8.91}$$

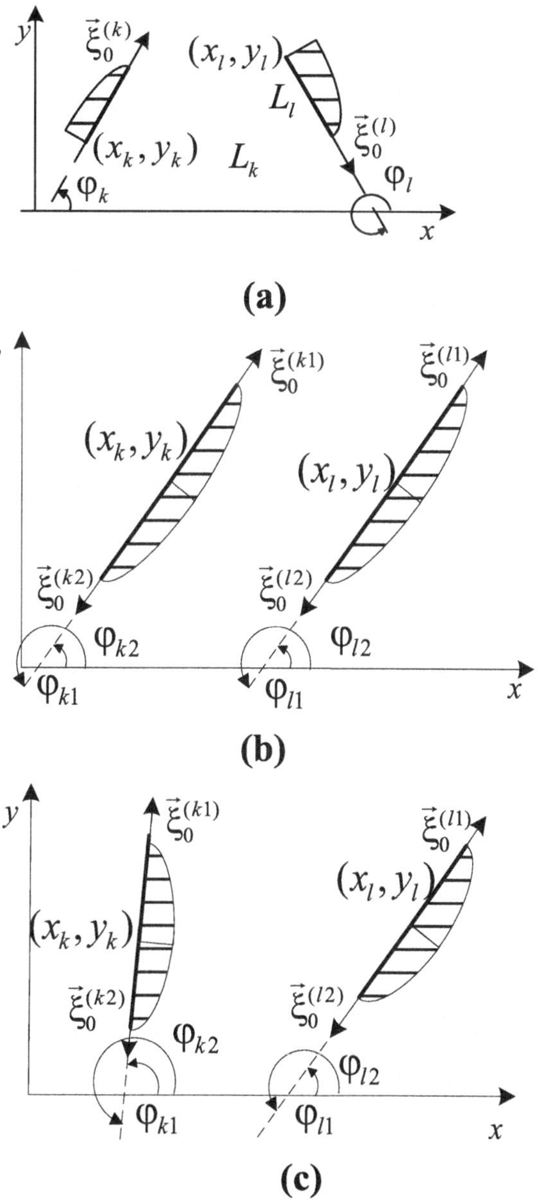

Fig. 8.14 Variants of reciprocal arrangement of the basic current functions

In the special case, when currents $\vec{T}^k(\xi)$ and $\vec{T}^l(\xi)$ are determined on identical parallel segments (Fig. 8.12b), expression (8.91) takes the form

$$\tilde{Z}_{kl}(\alpha, \beta) = \frac{-j16Z_c^i}{d_l d_x \sin \alpha_r \sin^2 L}$$

$$\sum_{\nu=-\infty}^{\infty}\sum_{\mu=-\infty}^{\infty}\{[1-\delta_{\nu\mu}^2]\cdot Q_{\nu\mu}-\gamma_i^{\nu\mu}\delta_{\nu\mu}^2 F_{\nu\mu}\}$$
$$\times\left[\frac{\cos(\delta_{\nu\mu}L)-\cos L}{1-\delta_{\nu\mu}^2}\right]^2\left[\frac{\sin\left(\frac{\Delta_{\nu\mu}\Delta_s}{2}\right)}{\Delta_{\nu\mu}\Delta_s}\right]^2\cdot e^{j\left[\chi_1^{\nu}(x_k-x_l)+\chi_2^{\nu\mu}(y_k-y_l)\right]} \tag{8.92}$$

If it is necessary to determine characteristics of radiator on the array of values α and β (i.e., for different directions of arrival of the excitation plane wave), the volume and, therefore, the computation time can be reduced by partially calculating the sums in expressions (8.85), (8.91) and (8.92). In the case of array with an arbitrary cell, algorithm for calculating the elements of the matrix of generalized impedances on the array of values α and β has features, the essence of which is the need to calculate the value of the summation region by a variable μ for each certain value of the variable ν.

Sums in (8.85), (8.91), and (8.92) can be represented as several partial sums, which in turn are combined into two subgroups:

$$S_i(\theta_0,\varphi_0)=\sum_{\nu=-\infty}^{\infty}\sum_{\mu=-M_{\min}}^{M_{\min}}\ldots+\sum_{\nu=-N_{\min}}^{N_{\min}}\sum_{\mu=-\infty}^{-M_{\min}-1}\ldots$$
$$+\sum_{\nu=-N_{\min}}^{N_{\min}}\sum_{\mu=-M_{\min}-1}^{\infty}\ldots \tag{8.93}$$

and

$$S_2(\theta_0,\varphi_0)=\sum_{\nu=-\infty}^{-N_{\min}-1}\sum_{\mu=M_{\min}+1}^{\infty}\ldots+\sum_{\nu=N_{\min}+1}^{\infty}\sum_{\mu=M_{\min}+1}^{\infty}\ldots$$
$$+\sum_{\nu=-\infty}^{-N_{\min}-1}\sum_{\mu=-\infty}^{-M_{\min}-1}\ldots+\sum_{\nu=N_{\min}+1}^{\infty}\sum_{\mu=-\infty}^{-M_{\min}-1}\ldots \tag{8.94}$$

Expressions under the signs of sums are omitted for brevity.

Thus, in $S_2(\theta_0,\varphi_0)$ groups terms with indices $|\nu|>N_{\min}$ and $|\mu|>M_{\min}$, and in $S_1(\theta_0,\varphi_0)$ groups all the others. Since dependence of elements of the matrix of generalized impedances $\tilde{Z}_{kl}(\alpha,\beta)$ on θ_0 and φ_0 is defined through the parameters χ_1^{ν} and $\chi_2^{\nu\mu}$(see expressions (8.86)), values of $N_{\min}$ and $M_{\min}$ can be chosen such that the equations are valid:

$$1<<\frac{2\pi\nu}{d_l}, \tag{8.95}$$

$$1<<\left|\frac{2\pi\nu}{d_l\mathrm{tg}\alpha_r}+\frac{2\pi\mu}{d_x\sin\alpha_r}\right|. \tag{8.96}$$

When writing (8.95) and (8.96), it is taken into account that the maximum value of the parameters S_x and S_y cannot be more than one. Moreover, condition (8.96) can be satisfied for any μ and ν, if $d_x \sin \alpha_r \neq d_l tg\alpha_r$. Otherwise, the right-hand side of this inequality may be zero. Consequently, the elements with indices $\nu = \left(d_l / d_x \cos \alpha_r\right)$ must also be included in the sum of $S_1(\theta_0, \varphi_0)$.

Thus, if conditions (8.95) and (8.96) are matching, change in values of S_x and S_y has no significant effect on the value of parameters χ_1^{ν} and $\chi_2^{\nu\mu}$. Therefore, the sum $S_2(\theta_0, \varphi_0)$ is enough to calculate only once at $\theta_0 = \varphi_0 = 0$, and approximate values of the impedance matrix elements determine by the formula

$$\tilde{Z}_{kl}(\alpha, \beta) = S_1(\theta_0, \varphi_0) + S_2(0, 0).$$

Numerical experiments to assess accuracy of such an algorithm for calculating the elements of the impedance matrix showed that the calculation error does not exceed 1%, if the values $N_{\min}$ and $M_{\min}$ constitute 20% of number of terms in μ and V, which must be taken into account to accurately determine the elements $\tilde{Z}_{kl}(\alpha, \beta)$. In this case, calculation time can be reduced by a factor of two or three.

8.2.6 Recurrence Algorithm for Determining the Tensor Green's Function

One of the essential points when calculating microstrip antennas is the determination of the tensor Green's function (TGF) of the multilayer dielectric medium by which their substrate is modeled. The TGF elements can be represented in different ways [14], in particular, as the corresponding components of the electric and/or magnetic vector potentials of some local current distribution in one of the dielectric layers.

Consider the problem of electromagnetic field excitation in a medium, the geometry of which is shown in Fig. 8.9. Let in i layer in point $M\left(x', y', z'\right)$ elementary electric dipole of length dl is placed, characterized by momentum $\vec{p} = J \cdot dl \cdot \vec{p}_0$, where J is the dipole current, $\vec{p}_0$ is a single vector characterizing the direction of the dipole axis. At infinity, the zero emission conditions are fulfilled, and conditions of continuity of the electromagnetic field vectors are fulfilled at the boundaries of the layers:

$$\vec{B}_{\perp}^{1} = 0; \vec{E}_{||}^{1} = 0 \ \text{at}\, z = 0 \tag{8.97}$$

and.

$$\vec{B}_{\perp}^{n-1} = \vec{B}_{\perp}^{n}; \vec{H}_{||}^{n-1} = \vec{H}_{||}^{n}; \vec{E}_{||}^{n-1} = \vec{E}_{||}^{n} \ \text{at}\, z = z_n \cdots 2 \leq n \leq N \tag{8.98}$$

where the indices $\perp$ and $||$ respectively denote normal and tangential components of the vectors.

We will characterize field in the considered environment by the electric vector potential $\vec{A}$, which matches to system of wave equations

$$\Delta\vec{A}^n + k_n^2\vec{A}^n = \begin{cases} 0, & \text{if } n \neq i; \\ -\vec{J}\cdot\delta(x-x')\cdot\delta(y-y')\cdot\delta(z-z'), & \text{if } n = i, \end{cases} \tag{8.99}$$

where $k_n = \omega\sqrt{\tilde{\varepsilon}_n\mu_n}$—wave number, ω—circular frequency.

It is evident from the geometry of the problem that components of current J_x and J_y generate two components of the vector potential: A_x, A_z and A_y, A_z, respectively, whereas the current component J_z generates only one component of the vector potential A_z. Therefore, in the most general form, the desired TGF can be represented by the following matrix [14]:

$$\hat{G} = \begin{bmatrix} G_{xx} & 0 & 0 \\ 0 & G_{yy} & 0 \\ G_{zx} & G_{zy} & G_{zz} \end{bmatrix}. \tag{8.100}$$

Elements G_{lk} of this matrix represent l-th components of vector potential $\vec{A}$, due to the k-th components of current $\vec{J}$. Two pairs of tensor elements G_{xx}, G_{zx} and G_{yy}, G_{zy} are of direct interest for the planar array of microstrip radiators. Latter have an identical form, since the medium is unbounded in plane $x0y$. Thus, to solve the problem it is sufficient to determine the components of the vector potential due to the current J_x.

Given the above, the system of wave Eq. (8.99) can be written as a system of partial differential equations with respect to the unknown components of the vector potential and A_xA_z:

$$\begin{aligned} &\frac{\partial^2 A_x^n}{\partial x^2} + \frac{\partial^2 A_x^n}{\partial y^2} + \frac{\partial^2 A_x^n}{\partial z^2} + k_n^2 A_x^n = \begin{cases} 0, & n \neq i; \\ -J_x\cdot\delta(x-x')\cdot\delta(y-y')\cdot\delta(z-z'), & n \neq i; \end{cases} \\ &\frac{\partial^2 A_z^n}{\partial x^2} + \frac{\partial^2 A_z^n}{\partial y^2} + \frac{\partial^2 A_z^n}{\partial z^2} + k_n^2 A_z^n = 0, \qquad n = \overline{1, N}. \end{aligned} \tag{8.101}$$

To represent boundary conditions (8.97) and (8.98) in terms of vector potential, we use the equations known from electrodynamics [13]:

$$\begin{cases} \vec{H}^n = \text{rot}\vec{A}^n \\ \vec{E}^n = \frac{1}{j\omega\tilde{\varepsilon}_n}\left(\text{grad div} + k_n^2\right)\vec{A}^n; \quad n = \overline{1, N} \end{cases}, \tag{8.102}$$

where $\vec{H}^n$ and $\vec{E}^n$ are magnetic and electric field strength vectors in the n-th layer of the dielectric structure.

In this case $\vec{A}^n = A_x^n\vec{x}_0 + A_z^n\vec{z}_0$, therefore (8.102) will take the form

$$\begin{cases} \vec{H}^n = \frac{\partial A_z^n}{\partial y} \cdot \vec{x}_0 + \left(\frac{\partial A_x^n}{\partial z} - \frac{\partial A_z^n}{\partial x} \right) \cdot \vec{y}_0 - \frac{\partial A_x^n}{\partial y} \cdot \vec{z}_0, \quad n = \overline{1, N} \\ \vec{E}^n = \frac{1}{j\omega\tilde{\varepsilon}_n} \left[\left(\frac{\partial}{\partial x} \cdot \vec{x}_0 + \frac{\partial}{\partial y} \cdot \vec{y}_0 + \frac{\partial}{\partial z} \cdot \vec{z}_0 \right) \left(\frac{\partial A_x^n}{\partial x} + \frac{\partial A_z^n}{\partial z} \right) \right. \\ \left. \qquad + k_n^2 (A_x^n \cdot \vec{x}_0 + A_z^n \cdot \vec{z}_0) \right], \end{cases} \quad (8.103)$$

where $\vec{x}_0, \vec{y}_0, \vec{z}_0$, are unit vectors of x, y, z axes of entered rectenna coordinate system.

Using boundary conditions (8.97), (8.98) and Eq. (8.103), we obtain the following continuity conditions for the vector potential components at the boundaries of the structure layers:

$$\begin{aligned} &\tilde{\mu}_1 \frac{\partial A_x^1}{\partial y} = 0; \ \frac{\partial A_x^1}{\partial z} = \frac{\partial A_z^1}{\partial x}; \ \frac{\partial A_z^1}{\partial y} = 0; \\ &\frac{\partial}{\partial x} \left(\frac{\partial A_x^1}{\partial x} + \frac{\partial A_z^1}{\partial z} \right) + k_1^2 A_x^1 \\ &= 0; \ \left(\frac{\partial}{\partial y} \left(\frac{\partial A_x^1}{\partial x} + \frac{\partial A_z^1}{\partial z} \right) \right) = 0 \text{ if } z = 0; \end{aligned} \quad (8.104)$$

and

$$\begin{aligned} &\tilde{\mu}_{n-1} \frac{\partial A_x^{n-1}}{\partial y} = \tilde{\mu}_n \frac{\partial A_x^n}{\partial y}; \ \frac{\partial A_x^{n-1}}{\partial z} - \frac{\partial A_z^{n-1}}{\partial x} = \frac{\partial A_x^n}{\partial z} - \frac{\partial A_z^n}{\partial x}; \ \frac{\partial A_z^{n-1}}{\partial y} = \frac{\partial A_z^n}{\partial y}; \\ &\frac{1}{j\omega\tilde{\varepsilon}_{n-1}} \left[\frac{\partial}{\partial x} \left(\frac{\partial A_x^{n-1}}{\partial x} + \frac{\partial A_z^{n-1}}{\partial z} \right) + k_{n-1}^2 A_x^{n-1} \right] \\ &= \frac{1}{j\omega\tilde{\varepsilon}_n} \left[\frac{\partial}{\partial x} \left(\frac{\partial A_x^n}{\partial x} + \frac{\partial A_z^n}{\partial z} \right) + k_n^2 A_x^n \right]; \\ &\frac{1}{j\omega\tilde{\varepsilon}_{n-1}} \left[\frac{\partial}{\partial y} \left(\frac{\partial A_x^{n-1}}{\partial x} + \frac{\partial A_z^{n-1}}{\partial z} \right) \right] \\ &= \frac{1}{j\omega\tilde{\varepsilon}_n} \left[\frac{\partial}{\partial y} \left(\frac{\partial A_x^n}{\partial x} + \frac{\partial A_z^n}{\partial z} \right) \right] \text{ if } \ z = z_n, \quad n = \overline{2, N}. \end{aligned} \quad (8.105)$$

Thus, the field generated by current J_x in the layered dielectric can be determined as a result of solving the Cauchy problem for system of differential Eq. (8.101) if conditions (8.104) and (8.105) are fulfilled. We will search for the above solution in the spectral domain by presenting the components of vector potential in the form of Fourier integral expansion in a homogeneous cross section of the structure (plane $x0y$):

$$\overline{\overline{A}}_\alpha^n = \frac{1}{2\pi} \int_{-\infty}^{+\infty}\int A_\alpha^n(x, y, z) e^{(-j\chi_1 x - j\chi_2 y)} dx dy, \quad (8.106)$$

where the index α can take the value of x, y or z, and the parameters χ_1 and χ_2 have meaning of propagation constants in the cross section. Heterogeneity of the structure in this case will be taken into account in the characteristic part of the solution related to the coordinate z.

As a result of the twofold Fourier transform, boundary value problems for partial differential equations are reduced to simpler boundary value problems for ordinary differential equations of the following type:

$$\begin{cases} \frac{d^2\overline{\overline{A}}_x^n}{dz^2} - \gamma_n^2 \overline{\overline{A}}_x^n = \begin{cases} 0 & , if\ n \neq i; \\ P_x \cdot \delta(z - z') & , if\ n = i; \end{cases} \\ \frac{d^2\overline{\overline{A}}_z^n}{dz^2} - \gamma_n^2 \overline{\overline{A}}_z^n = 0 \quad , if\ n = \overline{1, N} \end{cases} \tag{8.107}$$

with boundary conditions

$$\overline{\overline{A}}_x^1 = 0; \frac{\overline{\overline{A}}_x^1}{dz} = 0; \overline{\overline{A}}_z^1 = 0; \frac{\overline{\overline{A}}_z^1}{dz} = 0; if\ z = 0, \tag{8.108}$$

and

$$\begin{aligned} \tilde{\mu}_{n-1} \overline{\overline{A}}_x^{n-1} &= \tilde{\mu}_n \overline{\overline{A}}_x^n; \frac{\overline{\overline{A}}_x^{n-1}}{dz} = \frac{\overline{\overline{A}}_x^n}{dz}; \overline{\overline{A}}_z^{n-1} \\ &= \overline{\overline{A}}_z^n; \frac{1}{\tilde{\varepsilon}_{n-1}} \left(\frac{d\overline{\overline{A}}_z^{n-1}}{dz} + j\chi_1 \overline{\overline{A}}_z^{n-1} \right) \\ &= \frac{1}{\tilde{\varepsilon}_n} \left(\frac{d\overline{\overline{A}}_z^n}{dz} + j\chi_1 \overline{\overline{A}}_z^n \right) if\ z = z_n, n = \overline{2, N}. \end{aligned} \tag{8.109}$$

In Eq. (8.107) propagation constant along the axis has meaning $\gamma_n^2 = \chi_1^2 + \chi_2^2 - k_n^2$ of z, and $P_x = -\frac{J_x}{2\pi} e^{(-j\chi_1 x' - j\chi_2 y')}$.

Solutions of equations of system (8.107) are well known and have the following form:

$$\begin{cases} \overline{\overline{A}}_x^n = B_1^n e^{\gamma_n z} + B_2^n e^{-\gamma_n z}, & n \neq i; \\ \overline{\overline{A}}_x^i = B_1^i e^{\gamma_i z} + B_2^i e^{-\gamma_i z} - \frac{P_x}{2\gamma_i} e^{-\gamma_i (z - z') m}, & n = i; \\ \overline{\overline{A}}_z^n = C_1^n e^{\gamma_n z} + C_2^n e^{-\gamma_n z}, & 1 \leq n \leq N, \end{cases} \tag{8.110}$$

Where $m = \frac{(z - z')}{|z - z'|}$.

Taking into account zero emission conditions at infinity, the solutions for the layer $n = N$ contain only one summand:

$$\begin{cases} \overset{=N}{A}_z = B_2^N \mathrm{e}^{-\gamma_n z}, \\ \overset{=N}{A}_z = C_2^N \mathrm{e}^{-\gamma_n z}. \end{cases} \tag{8.111}$$

Unknown coefficients $B_{\alpha_1}^n$ and $C_{\alpha_1}^n$ $(1 \le n \le N,\ \alpha_1 = 1, 2)$ are determined based on the given boundary conditions by substituting expressions (8.110) and (8.111) into Eqs. (8.108) and (8.109). As a result, we obtain the following system of equations:

$$\begin{aligned}
&B_1^1 + B_2^1 = 0;\\
&\gamma_1 B_1^1 + \gamma_1 B_2^1 = j\left(1 + \frac{\chi_1}{\chi_2}\right);\\
&\gamma_1\left(B_1^1 \mathrm{e}^{\gamma_1 z_2} + B_2^1 \mathrm{e}^{-\gamma_1 z_2}\right) = \gamma_2\left(B_1^2 \mathrm{e}^{\gamma_2 z_2} + B_2^2 \mathrm{e}^{-\gamma_2 z_2}\right);\\
&\mu_{i-1}\left(B_1^{i-1} \mathrm{e}^{\gamma_{i-1} z_i} + B_2^{i-1} \mathrm{e}^{-\gamma_{i-1} z_i}\right) = \mu_i\left(B_1^i \mathrm{e}^{\gamma_i z_i} + B_2^i \mathrm{e}^{-\gamma_i z_i} - \frac{P_x}{2\gamma_i} e^{-\gamma_i z'} \mathrm{e}^{\gamma_i z_i}\right);\\
&\gamma_{i-1}\left(B_1^{i-1} \mathrm{e}^{\gamma_{i-1} z_i} + B_2^{i-1} \mathrm{e}^{-\gamma_{i-1} z_i}\right) = \gamma_i\left(B_1^i \mathrm{e}^{\gamma_i z_i} + B_2^i \mathrm{e}^{-\gamma_i z_i} - \frac{P_x}{2\gamma_i} e^{-\gamma_i z'} \mathrm{e}^{\gamma_i z_i}\right);\\
&\mu_i\left(B_1^i \mathrm{e}^{\gamma_i z_{i+1}} + B_2^i \mathrm{e}^{-\gamma_i z_{i+1}} - \frac{P_x}{2\gamma_i} e^{-\gamma_i z'} \mathrm{e}^{\gamma_i z_{i+1}}\right) = \mu_{i+1}\left(B_1^{i+1} \mathrm{e}^{\gamma_{i+1} z_{i+1}} + B_2^{i+1} \mathrm{e}^{-\gamma_{i+1} z_{i+1}}\right);\\
&\gamma_i\left(B_1^i \mathrm{e}^{\gamma_i z_{i+1}} + B_2^i \mathrm{e}^{-\gamma_i z_{i+1}} - \frac{P_x}{2\gamma_i} e^{-\gamma_i z'} \mathrm{e}^{\gamma_i z_{i+1}}\right) = \gamma_{i+1}\left(B_1^{i+1} \mathrm{e}^{\gamma_{i+1} z_{i+1}} + B_2^{i+1} \mathrm{e}^{-\gamma_{i+1} z_{i+1}}\right);\\
&\mu_{N-2}\left(B_1^{N-2} \mathrm{e}^{\gamma_{N-2} z_{N-1}} + B_2^{N-2} \mathrm{e}^{-\gamma_{N-2} z_{N-1}}\right) = \mu_{N-1}\left(B_1^{N-1} \mathrm{e}^{\gamma_{N-1} z_{N-1}} + B_2^{N-1} \mathrm{e}^{-\gamma_{N-1} z_{N-1}}\right);\\
&\gamma_{N-2}\left(B_1^{N-2} \mathrm{e}^{\gamma_{N-2} z_{N-1}} + B_2^{N-2} \mathrm{e}^{-\gamma_{N-2} z_{N-1}}\right) = \gamma_{N-1}\left(B_1^{N-1} \mathrm{e}^{\gamma_{N-1} z_{N-1}} + B_2^{N-1} \mathrm{e}^{-\gamma_{N-1} z_{N-1}}\right);\\
&\mu_{N-1}\left(B_1^{N-1} \mathrm{e}^{\gamma_{N-1} z_N} + B_2^{N-1} \mathrm{e}^{-\gamma_{N-1} z_N}\right)\\
&= \mu_N B_1^N \mathrm{e}^{-\gamma_N z_N};\ \mu_{N-1}\left(B_1^{N-1} \mathrm{e}^{\gamma_{N-1} z_N} + B_2^{N-1} \mathrm{e}^{-\gamma_{N-1} z_N}\right)\\
&= \mu_N B_1^N \mathrm{e}^{-\gamma_N z_N};\\
&\gamma_{N-1}\left(B_1^{N-1} \mathrm{e}^{\gamma_{N-1} z_N} - B_2^{N-1} \mathrm{e}^{-\gamma_{N-1} z_N}\right) = -\gamma_N B_2^N \mathrm{e}^{-\gamma_N z_N};\\
&C_1^1 + C_2^1 = -j/\chi_2;\ \gamma_1 C_1^1 + \gamma_1 C_2^1 = 0;\\
&\frac{1}{\tilde{\varepsilon}_1}\left(\gamma_1 C_1^1 \mathrm{e}^{\gamma_1 z_2} - \gamma_1 C_2^1 \mathrm{e}^{-\gamma_1 z_2} + j\chi_1 \left.\overset{=1}{A}_x\right|_{z=z_2}\right)\\
&= \frac{1}{\tilde{\varepsilon}_2}\left(\gamma_2 C_1^2 \mathrm{e}^{\gamma_2 z_2} - \gamma_2 C_2^2 \mathrm{e}^{-\gamma_2 z_2} + j\chi_1 \left.\overset{=2}{A}_x\right|_{z=z_2}\right);\\
&C_1^{N-1} \mathrm{e}^{\gamma_{N-1} z_N} + C_2^{N-1} \mathrm{e}^{-\gamma_{N-1} z_N} = C_2^N \mathrm{e}^{-\gamma_N z_N};\\
&\frac{1}{\tilde{\varepsilon}_{N-1}}\left(\gamma_{N-1} C_1^{N-1} \mathrm{e}^{\gamma_{N-1} z_N} - \gamma_{N-1} C_2^{N-1} \mathrm{e}^{-\gamma_{N-1} z_N} + j\chi_1 \left.\overset{=N-1}{A}_x\right|_{z=z_N}\right)\\
&= \frac{1}{\tilde{\varepsilon}_N}\left(-\gamma_N C_2^N \mathrm{e}^{-\gamma_N z_N} + j\chi_1 \left.\overset{=N}{A}_x\right|_{z=z_N}\right).
\end{aligned} \tag{8.112}$$

From consideration of the structure of the written down system it is obvious that it is most rational to first determine the coefficients of $B^n_{\alpha_1}$ $(\alpha_1 = 1, 2)$, by solving the first $2(N - 1)$ equations, and then, by substituting the values obtained in the remaining equations of the system, find the coefficients of $C^n_{\alpha_1}$ $(\alpha_1 = 1, 2)$. Besides, the diagonal character of the matrix of the written system allows us to use in determining the vector of unknown coefficients a recurrence algorithm by which the original problem is reduced to solving a sequence of systems of two equations with two unknowns. The physical interpretation of such an algorithm consists in recalculating the parameters of all above- and below-lying substrate layers to the layer in which the point current source in question is located.

Obtained as a result of solving the system of Eq. (8.112) values of coefficients $B^n_{\alpha_1}$ and $C^n_{\alpha_1}$ $(\alpha_1 = 1, 2)$ allow us to determine the necessary to calculate the elements of the impedance matrix $\mathbf{Z}$ in (8.85), (8.91) and (8.92) coefficients Q_{in} and F_{in} in form

$$Q_{in}(\chi_1, \chi_2, z_n) = \begin{cases} -Q_1 \frac{\mu_i}{\mu_n} \coprod\limits_{k=1}^{i-1} T_k, \ 1 \le n \le i-1; \\ -Q_1, \ n = i; \\ -Q_1 \frac{\mu_i}{\mu_n} \coprod\limits_{k=n}^{i-1} T_k, \ i+1 \le n \le N, \end{cases} \tag{8.113}$$

$$\begin{aligned} F_{in}(\chi_1, \chi_2, z_n) = & -d_n \sum_{k=1}^{n-1} (1 + \xi_n) S_k \prod_{l=k+1}^{n-1} R_l \mathrm{e}^{-\gamma_l h_l} \\ & - \mathrm{e}(-\gamma_n h_n)(1 + \varphi_n) \sum_{k=n}^{N-1} S_k d_{k+1} \prod_{l=n+1}^{k} t_l. \end{aligned} \tag{8.114}$$

The values included in these ratios are given in [16].

8.2.7 Numerical Investigations of Complex Geometry Radiators

On the basis of the described method, a set of application programs (SAP) for analysis of electrodynamic characteristics of periodic structures of microstrip radiators of complex geometry was developed. To evaluate its reliability and efficiency, a number of test calculations were performed, the results of which were compared with the data given in the works of other authors. As the simplest test example, the calculation of the input resistance of symmetrical wire vibrators in an infinite antenna array, located in free space above the screen, was considered. Figure 8.15 shows the dependences of the values of active and reactive components of the input resistance on the phasing direction of the array (the solid line—in the plane E, the dashed line—in the plane H), given in [17].

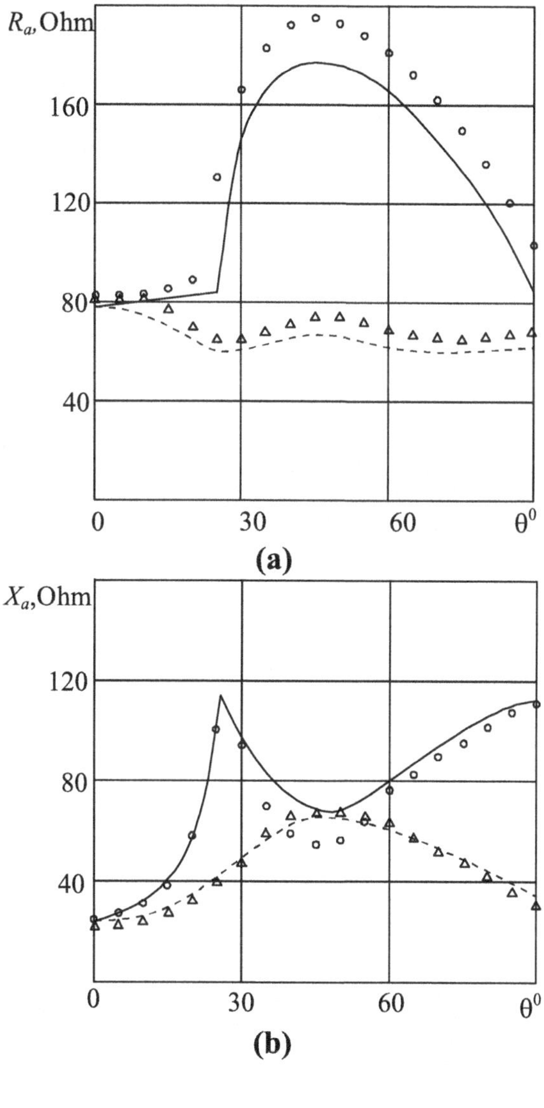

Fig. 8.15 Calculation results of the simplest test case

The array had the following geometrical parameters: $d_l = 0.5\lambda$, $d_x = 0.5\lambda$, $\alpha_r = 90°$, $h = 0.5\lambda$, $L = 0.25\lambda$ where h—height of arrangement of its elements above the screen;L —length of the vibrators arm. Accordingly by circles and triangles the values of similar parameters determined using the developed set of applied programs are marked. As it can be seen from Fig. 8.15, the character of variation of the considered dependences turns out to be identical. When quantifying the difference in the results, which does not exceed 5%, it is necessary to take into account, in particular, the difference in the models used to calculate the characteristics of the ARs under study.

For example, such features: recalculation of the vibrator radius into the equivalent ribbon conductor width, calculation of the free space Green's function by the obtained recurrence equations for the case of a two-layer substrate, with the corresponding free space parameters set as the material parameters of the layers. As another test example, the problem of the reflection coefficient of the infinite array of symmetrical printed vibrators located on the dielectric layer above the screen was solved. Figure 8.16a shows the dependences of the reflection coefficient modulus on angle of incidence of the excitation plane wave in the E and H planes (solid and dashed lines, respectively) of antenna array of vibrators with the arm length $L = 0.19\lambda$.

The vibrators are located in nodes of a periodic array with rectenna cell shape whose sides are sized $d_l = 0.5\lambda$ and $d_x = 0.5\lambda$ on nonmagnetic lossless substrate with thickness $h = 0.26\lambda$ with relative permittivity $\varepsilon = 2.55$. Data obtained on the basis of developed SAP are marked on the corresponding graphs by circles and rectangles. In Fig. 8.16b similar dependences for antenna array (AA) with other geometrical parameters $d_l = 0.5155\lambda$, $d_x = 0.5\lambda$, $\alpha_r = 60°$, $h = 0.19\lambda$, $\varepsilon = 2.55$,

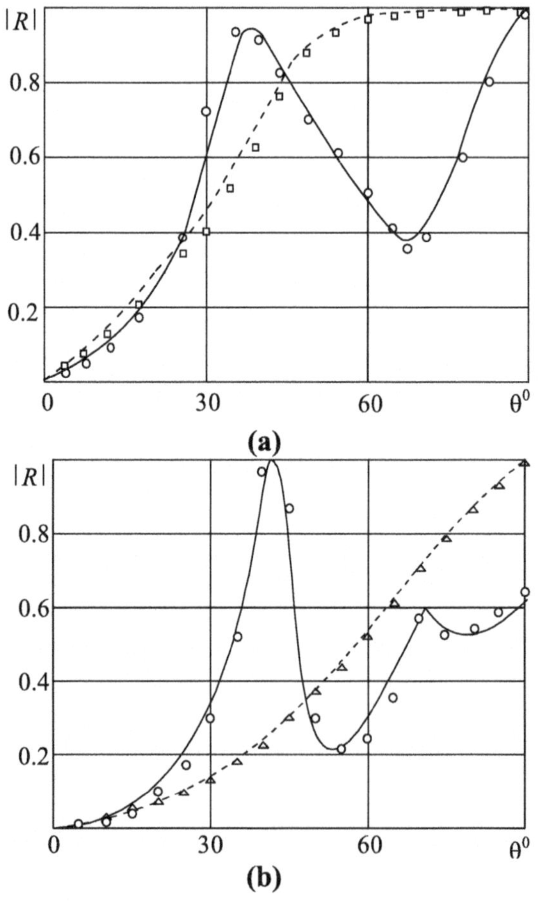

Fig. 8.16 Results of calculating reflection coefficient of antenna array

Table 8.1 Input impedances of printed symmetrical vibrators in an infinite antenna array

AA number	Z_{in}(data [18])	Z_{in}(calculation)
1	$11{,}0 + j2$	10,61 – j3,21
2	$73{,}0 + j4$	69,18 – j1,29

$L = 0.195\lambda$ are shown. In addition, in Table 8.1 values of input impedance of vibrators in above described antenna arrays under the condition of their excitation by a plane wave coming from the direction $\theta_0 = \varphi_0 = 0°$.

Analysis of given data shows that the results of calculation of relative parameters (reflection coefficient) almost completely coincide with the data from [18], and the error of calculation of absolute values (input resistance) does not exceed 3%. To evaluate the reliability of the algorithm in solving the problems of radiating structure analysis of complex geometry, we used data from [17], which considers a system of small-sized coiled vibrators located in free space above the screen. Values of input resistances of elements of such array, given in the mentioned work and obtained on the basis of the developed SAP, differ by 3–5%. Correctness of functioning of the algorithm when calculating the characteristics of periodic structures containing conductors continuous along the aperture was checked by calculating the amplitude-phase distribution of current in the arrays; the geometry of periodicity cells is shown in Figs. 8.17 and 8.18.

Table 8.2 shows the values of current amplitudes and phases in the array-type structure (Fig. 8.17) at normal incidence of a plane wave on it. Considered two cases of vector polarization $\vec{E}$: "horizontal" (along conductor $A - E$) and "vertical" (along conductor $F - P$).

Fig. 8.17 Geometry of cell of periodic array-type structure

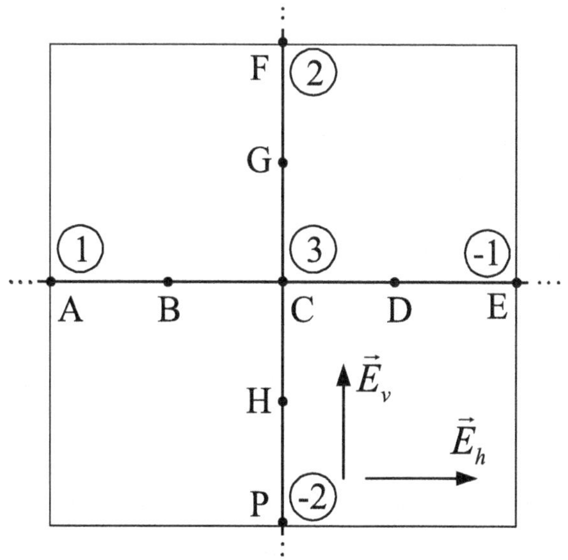

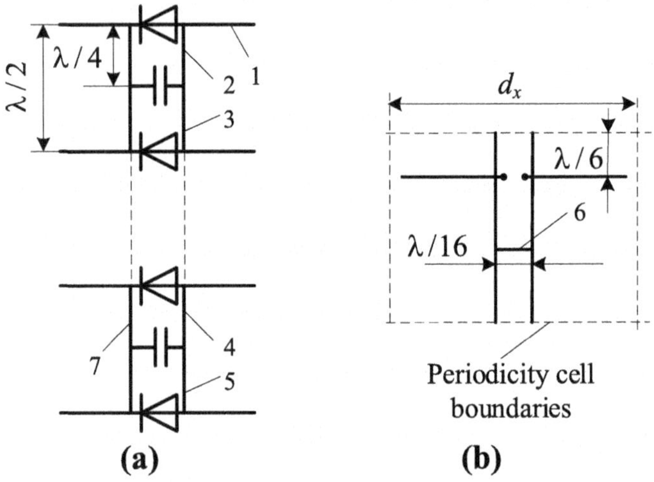

Fig. 8.18 Investigations of periodicity cell of rectenna

Table 8.2 Current distribution in a periodic array structure

	"Horizontal" polarization		"Vertical" polarization	
	$\|I\|$, mA	φ_A^0	$\|I\|$, mA	φ_A°
A	0.73743	15	0.0	– 165
B	0.73743	– 165	0.0	15
C_1	0.0	15	0.73743	15
C_2	0.0	15	0.0	– 165
C_3	0.73743	– 165	0.0	15
D	0.73743	– 165	0.0	15
E	0.73743	– 165	0.0	15
F	0.0	15	0.73743	15
G	0.0	– 165	0.73743	– 165
H	0.0	15	0.73743	– 165
P	0.0	– 165	0.73743	– 165

From data given in the table it can be seen that for the cell size $\lambda/2$ phase shift in the points corresponding to crossing of conductor by the periodicity cell boundary is the value of $\prod$. When polarization plane of wave is rotated by 90°, the current distribution that takes place in the first case in the section of $A - E$, is set in section $F - P$ of, and vice versa. For point C, there are three values of current expansion coefficients in given basis corresponding to the three basis functions approximating the current at junction point of four branches.

Table 8.3 Results of calculations of rectenna element parameters

Excluding the impact of conductors of the collection scheme		Taking into account the impact of conductors of the collection scheme	
Z_{in}, Ohm	η_r, %	Z_{in}, Ohm	η_r, %
$273.1 + j0.3$	83.0	$350.3 + j24.5$	80.0

With the help of the developed program we also investigated influence of number of harmonics and conductors of DC power collection circuits on parameters of the RRE of LAR. One string of this array is shown in Fig. 8.18a. Array consists of rectilinear resonant vibrators 1 of length $2L$ located above the screen at height h, loaded with Schottky diodes. Vibrators are connected in parallel to DC power acquisition circuit conductors 2, 3, 4, 5. To decouple the diodes from the load resistance at operating frequency through the gaps $\lambda/4$ capacitances with resistance $1/\omega C \rightarrow 0$ are included in acquisition circuit. It was assumed that all lines of periodic array are spaced $\lambda/2$ and the periodicity cell is rectenna (Fig. 8.18b). In calculations, the capacitance was modeled by the short-circuit jumper 6 (Fig. 8.18b). The numerical experiment consisted in following. First, input impedance of resonant tuned vibrator and rectification efficiency of rectifier element were calculated, assuming that conductors 2, 3, 4, 5 in Fig. 8.18b are absent. Then the same calculations were carried out, taking into account influence of DC power acquisition circuit conductors. In calculations of the efficiency it was assumed that Schottky diode with an admissible power $P_d = 0.5$ W is used on the array. Results of the calculations at the frequency $f = 2.45$ GHz are presented in Table 8.3.

These results were obtained at the vibrator arm length $kl = 1.49$ and height $kh = 0.825$ at $\varepsilon_r = 1$ and $P_{in} = 0.1$ W. From the comparison of the above data, it follows that the collection circuit conductors significantly affect both the input impedance of radiator, and parameters of the rectenna as a whole.

The calculated values of the vibrator input impedance $Z_{in}(nf)$ in the infinite-rectenna array at the fundamental frequency and frequencies of higher harmonics are given in Table 8.4.

These data were used to calculate the rectification efficiency of the rectifier element (Fig. 8.19) without considering the influence of harmonics (curve 1) and considering the influence of three, five and seven harmonics (curves 2, 3, 4, respectively).

Numerical experiment carried out using developed SAP has confirmed that 5–7 higher harmonics must be taken into account for the correct calculation of the energy characteristics of the rectenna. Effectiveness of using developed SAP when designing rectennas with radiators of complex configuration was studied by the example of

Table 8.4 Dependence of linear vibrator input impedance on harmonic number

n	1	2	3	4	5	6	7
Z_{in}, Ohm	273.1 + j0.3	129.9 + j136.4	50.4 + j109.9	25.0 + j79.0	14.6 + j61.5	9.53 + j50.16	6.7 + j42.3

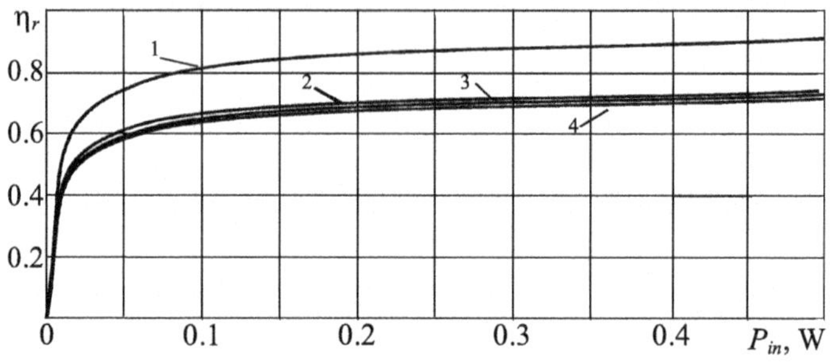

Fig. 8.19 Rectenna element efficiency

calculation of a two-half-period rectenna element based on a loop vibrator located on a two-layer substrate (Fig. 8.20). Thickness of lower layer with $\varepsilon_r = 1$ $kh_1 = 0.35919$, thickness of upper layer $\varepsilon_r = 1$ $kh_1 = 0.103$ (see Fig. 8.9), vibrator dimensions are $kl \times kb_1 = 1.67 \times 0.257$, $kb = 0.514$.

Table 8.5. presents $Z_{in}(nf)$ values calculated with respect to one pair of loop vibrator terminals for seven harmonics of 2.45 GHz frequency at six basis functions in the current distribution approximation and twenty spatial harmonics.

Fig. 8.20 Periodicity cell with loop vibrator

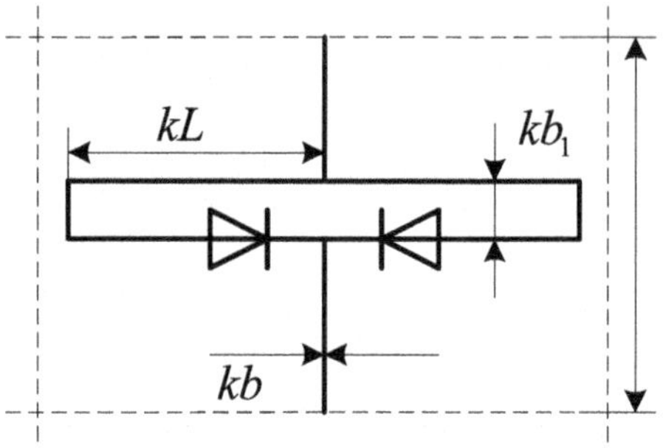

Table 8.5 Input impedances of the loop vibrator at harmonic frequencies

n	1	2	3	4	5	6	7
Z_{in}, Ohm	69.8 + j0.3	54.4 + j29.0	32.5 + j34.8	19.5 + j31.3	12.5 + j26.6	8.6 + j22.3	6.6 + j15.4

8.3 Numerical Investigations of Properties of RRE Infinite Rectenna Arrays

Let us study characteristics of RREs functioning as part of equal-amplitude excited LARs. Using above described methodology, calculations should be performed in two stages. The first stage involves determining matrix of eigen and mutual resistances of the RRE radiator at fundamental and harmonic frequencies and calculating RRE operation mode, and the second stage, using obtained data as inputs, involves calculating required energy characteristics. To carry out calculations we used developed SAP, which allows us to determine, in infinite array approximation, all electrodynamic parameters of ribbon strip vibrators of complex configuration located on multilayered substrate, and energy characteristics of the rectenna.

In the calculations, the Schottky diode is represented as a resistive nonlinearity, assuming that its reactive elements are part of linear part of the RRE circuit. Volt-ampere characteristic of nonlinearity was described by the expression

$$i = I_S\left(e^{\alpha u} - 1\right), \tag{8.115}$$

where i is the current via diode; I_S is saturation current; α is constant coefficient; u is junction voltage.

Table 8.6 shows the characteristics of typical Schottky diodes with different values of permissible input power P_d, forward current I_d and reverse voltage U_d.

First, we investigate single half-period RRE with ideal input and output filters (no losses or side radiations). This approximation allows us to consider the efficiency limits of the RRE in LARs. Let us also assume that the RRE is based on a resonant half-wave vibrator located above the screen on a substrate with relative permittivity ε_r and height k_0h.

An indirect confirmation of reliability of the calculation technique proposed in the previous subsection (it can be established only experimentally) are dependences of DC power P_{0e} and optimum load resistance $R_{L\mathrm{opt}}$ on step $d_x = d_l = d$ of array with square cell, which are shown in Figs. 8.21 and 8.22. In these figures, similar parameters of single RRE are dashed. Values depend significantly P_{0e} on the array step and, as it increases, asymptotically approach the DC power value of solitary RRE at any substrate height. Calculated dependences $P_{0e} = P_{0e}(d/\lambda_0)$ qualitatively repeat the dependence of antenna directivity element of periodic antenna array [18], and their course is primarily predetermined by course of radiator input impedance curves $R_a = R_a(d/\lambda_0)$. For the RRE case, this relationship can be detected by considering

Table 8.6 Schottky diode parameters

Diode type	P_d, W	I_d, mA	U_d, B	I_S, A	α
2A204A	0.2	–	–	$1.066 \cdot 10^{-5}$	8.55
3A208A	0.5	50	20	$2.972 \cdot 10^{-4}$	3.69

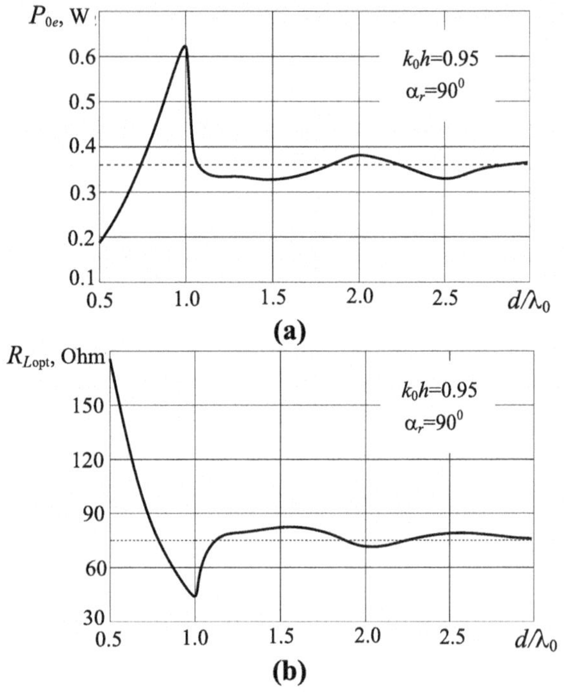

Fig. 8.21 Dependences of RRE parameters on array pitch at k_0h=0.95

the curves $P_{0e} = P_{0e}(d/\lambda_0)$ and $R_a = R_a(d/\lambda_0)$, since in the RRE with ideal filters $R_{Lopt} = 2R_a$.

Described results are obtained for case of Schottky diode with $P_d = 1$ W and dielectric substrate with $\varepsilon_r = 1$. For other values of ε_r qualitative picture remains the same, only the RRE efficiency deteriorates.

Calculations showed that in practical rectenna constructions the substrates with high value of ε_r, should not be used, because when it increases, the rectification efficiency decreases (Fig. 8.23). This is due to the fact that with the growth ε_r EMF, induced on vibrator terminals decreases (see Table 8.7), and, consequently, the losses on the internal resistance of the rectifier diode grow. Data of Table 8.7 correspond to the incident power flux density $\Pi = 0.089$ W/m^2.

Let us now consider dynamic characteristics of the RRE, i.e., the dependence of its energy parameters on level of input action. Calculations were performed for three versions of the RRE, in which Schottky diodes with different values of permissible power P_d were used (Table 8.5). Calculated values of rectification efficiency η of the RRE from the normalized power flux density $\overline{\Pi} = \Pi/\Pi_d$, falling on the rectenna, are shown in Fig. 8.24. Π_d is understood as value of power flux density, at which input power $P_{in} = P_d$ (see Table 8.8).

The dependencies are $\eta(\overline{\prod})$ well approximated by power function

$$\eta = \eta_{\max}\overline{\Pi}^{\mu}, \tag{8.116}$$

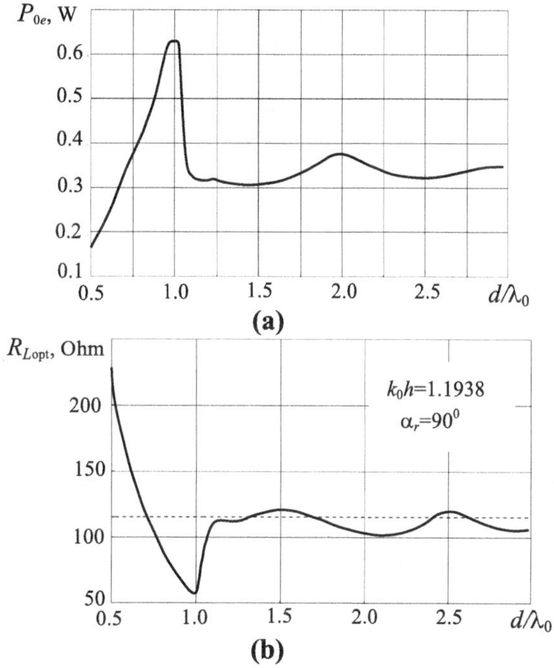

Fig. 8.22 Dependences of RRE parameters on array pitch at k_0h=1.1938

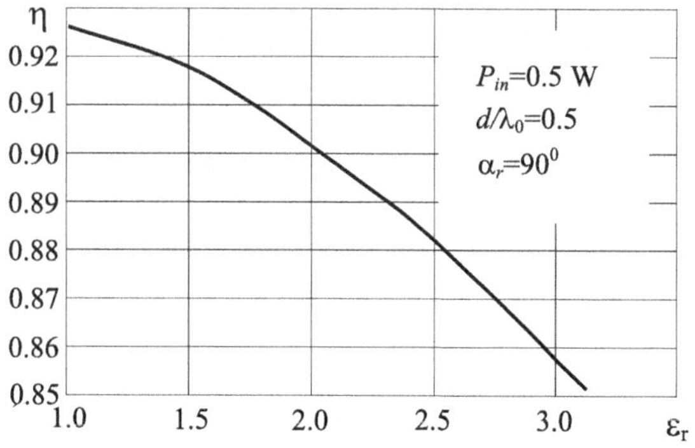

Fig. 8.23 Effect of dielectric constant on RRE efficiency

Table 8.7 Values of e_i at $d/\lambda_0 = 0.5$ and $k_0h = 1.1938$

ε_r	1	1.5	2	3
e_i, V	0.585	0.529	0.502	0.438

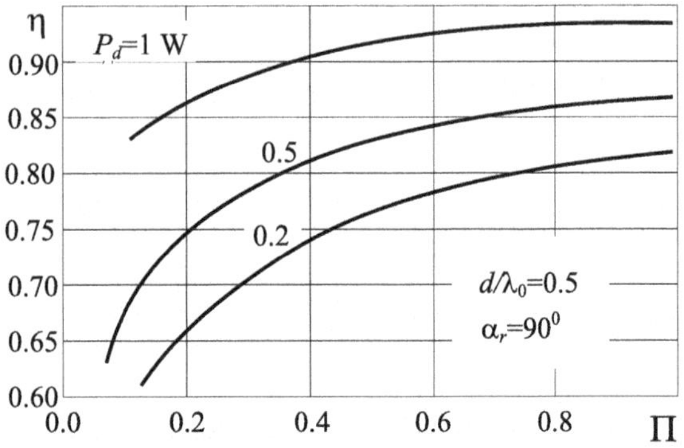

Fig. 8.24 Rectifying efficiency of RRE from the normalized power flux density

Table 8.8 Parameters of three RRE variants

Calculation option	P_d, W	Π_d, W/m ^{2}	η_{max},%	μ
1	0.2	52	83.7	0.142
2	0.5	131	89.7	0.121
3	1.0	275	94.8	0.053

where $\eta_{\max}$ is the value η at $\Pi = \Pi_d$, μ is the coefficient characterizing degree of nonlinearity of dependence (8.116). The values $\eta_{\max}$ and μ are given in Table 8.8.

From presented data it follows that rectification efficiency increases nonlinearly with increasing Π magnitude, with the most linear dependence $\eta\left(\overline{\Pi}\right)$ being if the Schottky diode with the highest value of P_d is used in the RRE.

Calculations have shown that approximation (8.116) is also applicable in cases of varying the array step and shape (see Fig. 8.25a). Some decrease in the rectification efficiency at increasing array step and changing its shape from $\alpha_r = 60°$ to $\alpha_r = 90°$ is explained by reduction of EMF e_i.

Another dynamic characteristic of the RRE $P_{in} = P_{in}(\Pi)$ is shown in Fig. 8.25b. It follows from graphs presented here that in case of transition from the square ($\alpha_r = 90°$) to the triangular ($\alpha_r = 60°$) form of cell or with decreasing its step value of power flux density grows, at which $P_{in} = P_d$. This result is natural, since decreasing the angle α_r or the array pitch along its axes leads to a decrease in the cell area, i.e. to denser placement of RRE in rectenna aperture. In turn, denser placement of RREs predetermines growth of microwave power, which can be extracted by the rectenna from the free space without burning diodes, and, consequently, the increase in the specific DC power $\overline{P}_S$, removed from a unit area of the aperture (see Table 8.9).

Table 8.9 shows values of the maximum possible value $\overline{P}_S$, which for an equal-amplitude excited aperture can be calculated by expression

$$\overline{P}_S = \Pi_d \cdot \Delta S_{ef} / \Delta S \Pi,$$

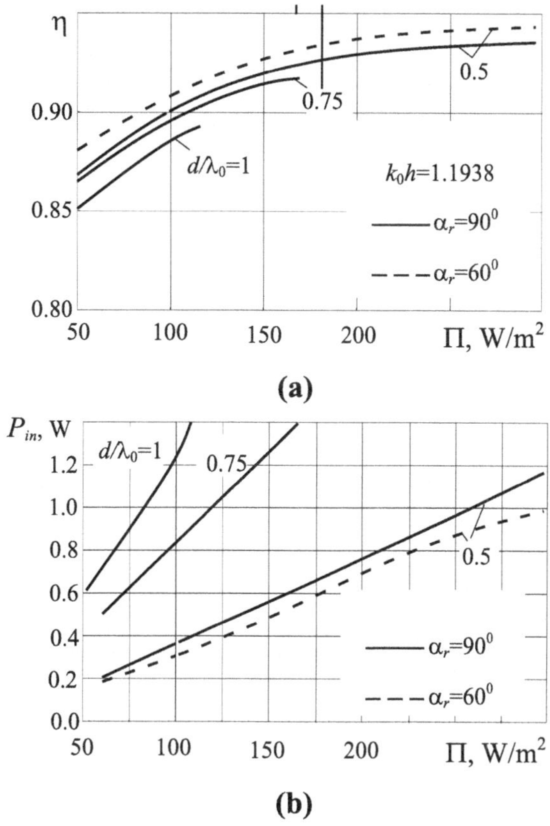

Fig. 8.25 Dynamic characteristics of the RRE

Table 8.9 Characteristics of RRE with different mesh sizes ($k_0h = 0.95$)

$d_x \times d_l$	$(0.5 \times 0.5)\lambda_0$	$(0.5 \times 0.4)\lambda_0$	$(0.5 \times 0.3)\lambda_0$	$(0.5 \times 0.25)\lambda_0$
R_a, Ohm	93.7	118.0	160.5	195.3
e_i, V	27.5	31.0	36.0	40.0
Π_d,W/m^2	269.0	339.6	448.4	546.0
$\overline{P}_S$, W/m^2	250.4	295.5	421.5	508.1

where ΔS_{ef} and ΔS are effective and geometric area of the array cell. Here we also present the data on the values of R_a, e_i, Π_d at which the values of specific power $\overline{P}_S$ were calculated.

It is easy to see that by changing the cell area one can change parameters R_a and e_i, i.e., regulate in wide range characteristics of RRE as an equivalent DC generator, which is very important in creating rectenna with high efficiency. Thus, compared with the solitary RRE in rectified array due to additional degrees of freedom (size and shape of the cell) there are more opportunities in varying parameters of both individual RRE and the rectenna as a whole.

Let us now consider factors affecting directivity characteristics of the rectenna at the fundamental frequency. Due to the fact that the rectenna is a nonphased antenna array, its directivity characteristics in the equidistant structure are completely determined by the properties of a single RRE. Directivity diagram of the rectenna at fundamental frequency will be understood as dependence of rectification efficiency of the RRE on angular coordinates with incident field flux density and DC load unchanged. Angular dependences of RRE efficiency calculated at $P_d = 1$ W, $\mu = 0.053$, $R_L = R_{L\text{opt}}$ in E and H planes are shown in Fig. 8.26a, b. Here curves 1 correspond to square array step of $0.5\lambda_0$, and curves 2 to λ_0. Similar to conventional antenna, directivity diagram of the rectenna antenna depends on pitch of the array. In significant range of angles (up to about $\pm$ 40° from the normal), weak changes in efficiency are observed, with a less noticeable decrease in efficiency characteristic of rectenna array with a smaller pitch. In general, based on the data obtained and using the terminology generally accepted in antenna technology, it is possible to characterize the rectenna arrays as weakly directional. A peculiarity of rectenna DD is its dependence on degree of nonlinearity of dynamic response (8.116). As can be seen from Fig. 8.26a, b, curves 3 (calculation conditions—$P_d = 0.2$ W, $\mu = 0.142$, $d/\lambda_0 = 0.5$, $R_L = R_{L\text{opt}}$), with increasing μ value from 0.053 to 0.142 the shape of DD in planes E and H changes strongly. Moreover, the greater the nonlinearity, the drop in the efficiency is sharper and the DD is narrower.

Qualitative changes in DD also occur when load resistance is varied. This conclusion follows from a joint consideration of curves 1 and 4 in Fig. 8.26a. Curve 4 is calculated under the same conditions as curve 1, but the value of R_L was taken as $0.5R_{L\text{opt}}$. In this case, the DD in the sector of angles up to 40° becomes smoother, but has two maxima in directions $\pm$ 20° at a lower absolute value of the rectification efficiency. The presence of higher harmonics also reduces the efficiency of the RRE (curve 5 in Fig. 8.26a). The DD, taking into account generation of higher harmonics, is calculated at $P_d = 1$ W, $R_L = R_{L\text{opt}}$ and the absence of the input LPF.

Values of parameters of three versions of the RRE without input filters are given in Table 8.10. Here, for comparison, parameters of the RRR with ideal filters are also given in parentheses. As $\xi(nf_0)$ in Table 8.10 meaning emission level of the n-th harmonic equal to the ratio of the power of emission of this harmonic $P_\Sigma(nf_0)$ to the input power P_{in}. From data given in Table 8.10 we can state the following:

- presence of higher harmonics significantly changes the optimum DC resistance and reduces the efficiency of the RRE;
- -at absence of input filtering circuits, there is very noticeable side radiation in the RRE, which deteriorates the electromagnetic compatibility of the rectenna.

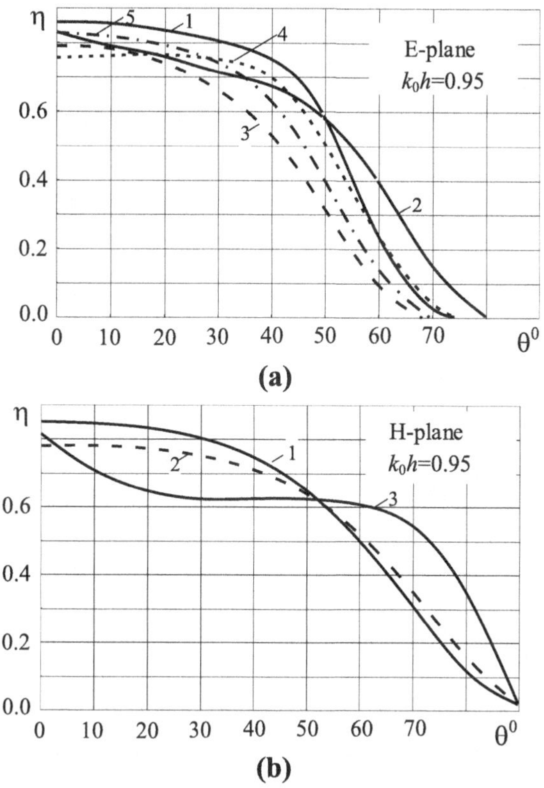

Fig. 8.26 Directivity diagrams of rectenna on rectifier efficiency

Table 8.10 RRE parameters without input LPFs

P_d, W	R_{Lopt}, Ohm	η	ξ($3f_0$), dB	ξ($5f_0$), dB	ξ($7f_0$), dB
0.2	127 (187)	0.772 (0.779)	– 17.0	– 24.2	– 36.8
0.5	120 (187)	0.756 (0.783)	– 16.8	– 24.0	– 36.6
1.0	117 (187)	0.807 (0.860)	– 17.3	– 24.7	– 37.9

8.4 Analysis of Large-Aperture Rectennas in Unequal-Amplitude Excitation

The analysis of LARs at unequal-amplitude excitation (Fig. 8.27) is a difficult task. The difficulty of analysis under such excitation is that, due to its nonuniformity, LAR cannot be described by a model in the form of an infinite periodic array, within which its analysis is reduced to the analysis of a single central, uniformly irradiated RRE. In this case, it is possible to create an effective methodology for analyzing LAR only by introducing a simplifying assumption, the essence of which lies in fact that the amplitude distribution in the LAR opening is a slowly changing function of coordinates. In case of regular arrangement of RREs, this allows to represent the rectenna aperture as a series of circular zones in which equally-amplitude excitable RREs with the same parameters are placed (Fig. 8.27), i.e. to approximate the amplitude distribution by a stepped function (Fig. 8.28). As a result, this allows us to consider the rectenna aperture as a set of non-interacting periodic arrays of RREs, and analyze each such array as a fragment of an infinite periodic antenna with nonlinear elements when determining its operating mode.

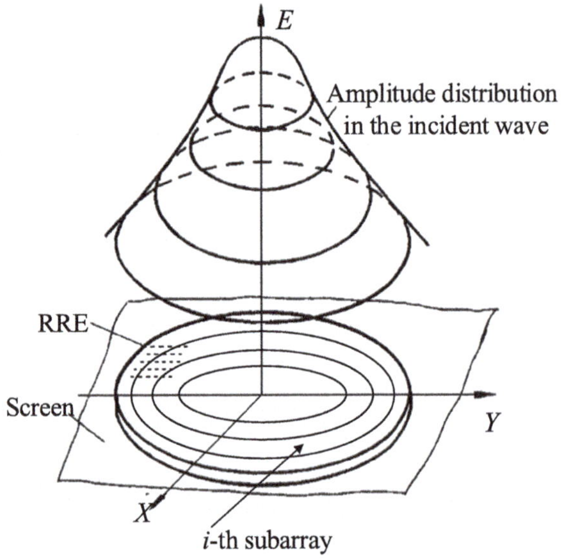

Fig. 8.27 Representation of the LAR aperture as a set of concentric zones

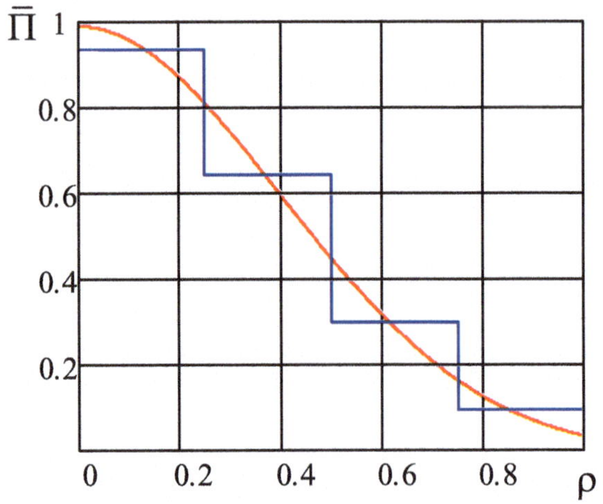

Fig. 8.28 Normalized power flux density distribution on the rectenna and its approximation (ρ—normalized rectenna radius)

The value of the approximating function in each zone should be determined so that the difference in the field amplitudes on the left and right boundaries of zone relative to the values of this value in the point of approximation was as minimal as possible.

8.5 Experimental Studies of Large-Aperture Rectennas

Objectives of experimental work were as follows:

- by comparing theoretical and experimental results, to establish the possibility of using the developed mathematical models for calculations of WPT systems;
- carry out experiments under uneven irradiation of the aperture of the rectenna, prove the existence of sufficiently large DC power collection losses for this case, evaluate the real efficiency of the WPT systems and the possibility of creating remote-controlled power sources on their basis;
- to work out one of possible variants of modular design of the rectenna, which allows technologically simple aperture buildup, and hence an increase in power in the rectenna load.

8.5.1 Description of Experimental Setup and Measurement Methods

Structural scheme of the setup is shown in Fig. 8.29.

Measurements were carried out in an anechoic chamber 1 of size 6 × 2.5 m, made of material "Ekran-320". In fact, the setup is an experimental system of energy transmission by a microwave beam. The transmitting part of the WPT system consists of microwave generator 2, control unit 3, which is placed outside the anechoic chamber, and transmitting antenna 4. The generator operates at frequency of 2.45 GHz. In used design the output power was not regulated, and its value, measured with wattmeter 5, was 550 ± 50 W. To change density of power flux, falling on rectenna 6 of microwave

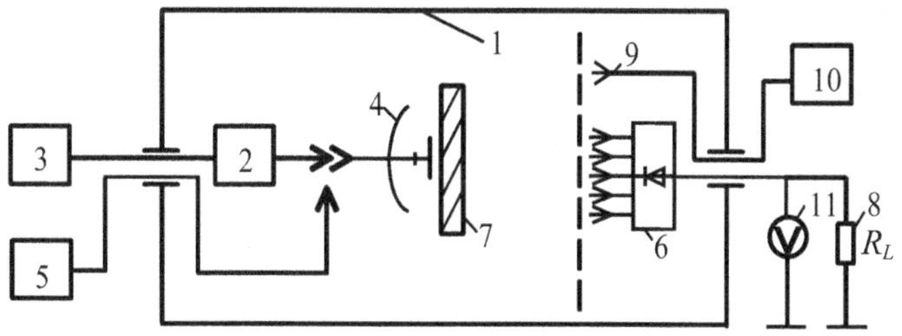

Fig. 8.29 Block diagram of laboratory prototype of the WTP system

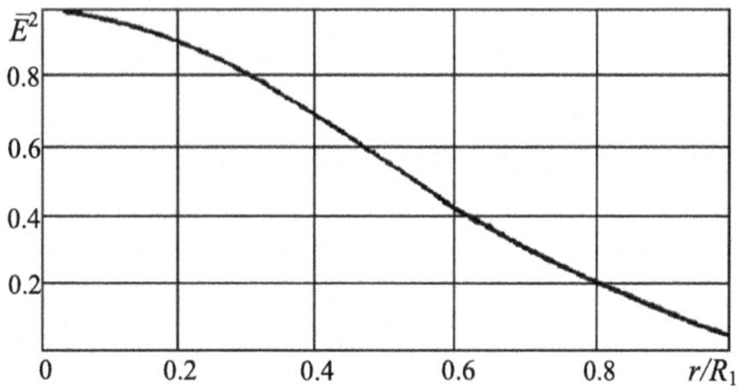

Fig. 8.30 Normalized amplitude distribution over the transmitting antenna

beam, an absorbing screen 7 of material "Ekran-320", installed near the transmitting antenna, was used. Value of attenuation in screen from 4.5 to 15 dB was changed by increasing its thickness (number of vertically mounted sheets of absorbing material).

Transmitting antenna contains mirror and irradiator. The mirror surface is parabolic with the ratio $f/2R_1 = 0.38$ where R_1 is the radius of the mirror ($2R_1 = 90\,\text{cm}$), f is the focal length. Illuminator is made in form of vibrator with reflector. The illuminator is powered by rigid coaxial feeder. To meet the conditions of axial symmetry of directivity diagram, the vibrator arms are tilted to the feeder. The illuminator is mounted so that its phase center, located between the vibrator and the reflector, coincides with the focus, i.e. the aperture of the transmitting antenna is excited in-phase. The normalized amplitude distribution over the mirror opening was calculated by the formula

$$\bar{E}^2 = f\left(f - \frac{r^2}{4f}\right)^2 \Big/ \left(f + \frac{r^2}{4f}\right)^3$$

and is shown in Fig. 8.30. This distribution was used as the initial distribution when calculating the intercept and rectification efficiencies.

Let us now consider the design of tested rectenna [19]. It consists of nine modules of the same type (three rows and three floors) and has a square shape with an overall size of 0.7 m. Scheme of one module is shown in Fig. 8.31. The module consists of 16 RREs (eight parallel connected strings, in each of which RREs are connected in series). Elements of the module are located in the nodes of the array with a triangular array and spacing $\lambda/2$. As the module element, the loop vibrator above the screen with 3A208A diodes, having an admissible power of 0.5 W, included in its cross sections, was chosen. The vibrator was tuned in resonance together with the diode capacitance. The peculiarity of the module is that the decoupling of the RRE rectifier circuits and DC circuits in it is realized by including the wires of the power collecting circuit in the points of zero potential of the Pistolcors vibrators. The module board was made of foil glass-textolite ($\varepsilon_r = 6$) 2 mm thick and was placed over the screen at height of 9 mm. The rectenna contained 144 elements. Structurally, the rectenna is designed so that the modules can be connected to each other by a direct current in series, parallel or series–parallel. The rectenna has a possibility to connect its own

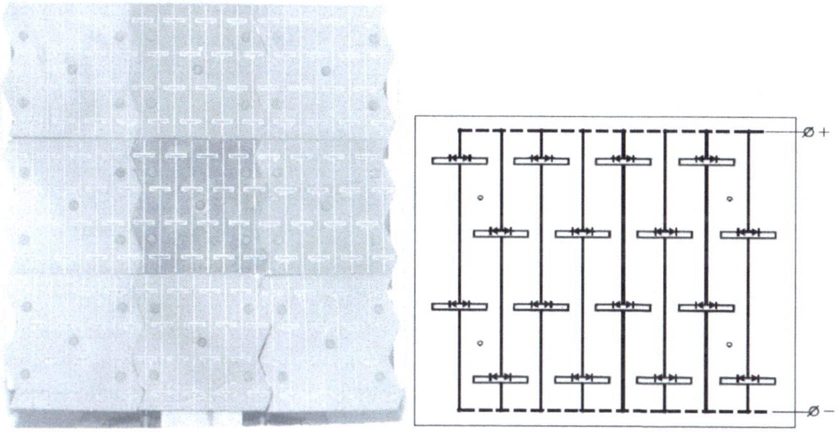

(a) (b)

Fig. 8.31 The experimental rectenna: **a** design; **b** schematic diagram of the base module

load resistance to each module. It is technologically possible to connect the DC load to any module line.

Let us consider the technique of measuring the efficiency of the experimental WPT system. This value is found as the ratio of the DC power in the rectenna load to the radiation power of the transmitting antenna, indicating the value of the power flux density in the rectenna opening, at which measurements were carried out.

Actual radiation power was determined as follows. The antenna of the energy flux density meter 9 was installed in place of the rectenna, and with the microwave generator switched on, the value Π in the center of the rectenna aperture was recorded on indicator unit 10 of the energy flux density meter. Required value $\prod_{max}$ and the field distribution over the beam cross section was set by selecting the distance between the apertures of the transmitting antenna and the rectenna, as well as by changing the thickness of the absorbing screen 7. Then, measurements of the attenuation coefficient in the absorbing screen α, power P_g at the output of the microwave generator and standing wave coefficient of the transmitting antenna (at frequency of 2.45 GHz it is equal to 1.3) were carried out in a standard way. When connecting the antenna with the generator used short sections of rigid coaxial lines, heat loss in which negligible, so the radiation power was calculated by the formula

$$P_{\sum} = P_g(1 - \rho^2)/\alpha,$$

where ρ is the modulus of reflection coefficient of the transmitting antenna.

DC power in load P_0 was calculated by measuring the voltage U_0 on the resistance R_L with a voltmeter 11. Results of measurements were processed by statistical method. The mean values of the measured value and the limits of intervals for finding the real values of the error at a probability of 0.95 were calculated. Due to the complexity of the experiment (considerable power of the generator) only ten measurements were carried out to determine each of the parameters.

8.5.2 Investigation of String and Module of the Rectenna Array Under Equal-Amplitude Excitation

Series of experiments described below was performed to check validity of the methodology proposed in chapter 8.1 for calculating LARs under uniform irradiation of their aperture. Experimental studies were preceded by adjusting the rectenna to maximum power in the load, this operation was performed by changing distance between the board and the screen of each module. It turned out that the screen distance of 9 mm was calculated correctly. This result is one of proofs of practical applicability of the proposed analysis technique.

As was shown in Sect. 8.1, a large-aperture equally-amplitude excitable rectenna can be calculated by calculating the parameters of a single RRE in an infinite array. Design of experimental sample of the rectenna did not make it possible to study the characteristics of single RRE in the array, but it was possible to measure the parameters of string consisting of two RREs.

This experiment was performed in following way. Variable load resistance and voltmeter were connected to one of lines of central module of the rectenna. All modules were connected to load by DC in series. Results of experimental investigations of DC power P_0 and U_0 voltage of the string from load resistance in a 144-element array are shown in Fig. 8.32a, b (marked by circles). Theoretical curves (solid lines), calculated in the infinite array approximation, are also shown here. In the calculations, typical IV characteristic of 3A208A diode was used. From above data, it follows that when value changes in Π wide dynamic range, there is a fairly good correspondence between the calculated and measured values.

Similar measurements were performed for central rectenna module under the condition of equal-amplitude irradiation of its aperture. Irregularity of incident field was controlled by energy flux density meter and was not more than $\pm$ 0.5 dB. Results of measurements and calculations are shown in Fig. 8.33a, b. Each of the modules was designed as periodic structure with periodicity cell boundaries crossed by collection circuit wires using the methods outlined in Sect. 8.2.

First, the geometric dimensions of loop vibrators at frequency of 2.45 GHz were found, taking into account Schottky diode capacitances C_c and C_0, and then energy parameters of the module were determined. These characteristics were specified taking into account the nonlinear dependence of the dynamic capacitance $C(u)$ on the input influence level:

$$C(u) = \begin{cases} \frac{C_0}{\left(1-\frac{u}{\varphi_C}\right)^{\nu}}, & \text{if } u \le U_{\max} \\ C_{\max}\{k(u - U_{\max}) + 1\}, & \text{if } u > U_{\max}. \end{cases}$$

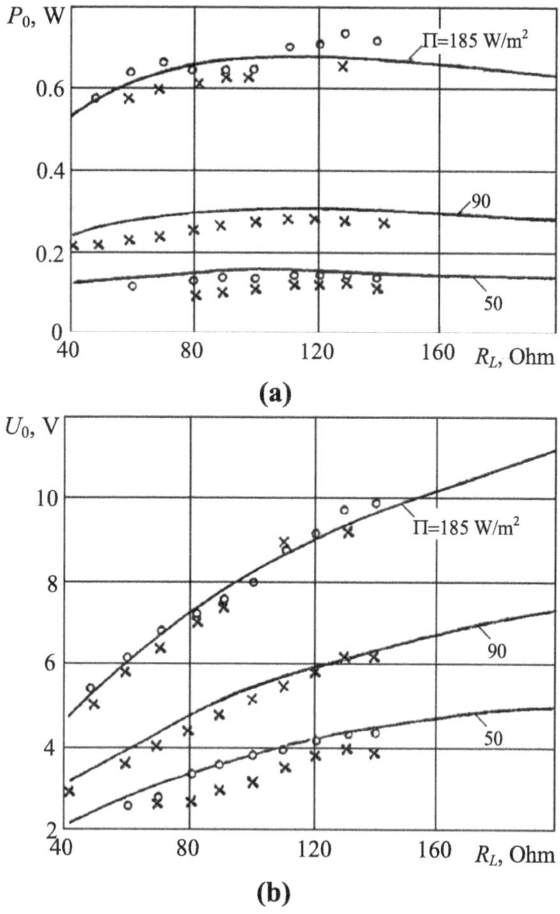

Fig. 8.32 Parameters of string of two RRE

Here $U_{max} = \varphi_c\left(1 - e^{\frac{1}{\nu}\ln\left(\frac{C_0}{C_{max}}\right)}\right)$; $k = \frac{\nu C_0}{\varphi_c C_{max}\left(1-\frac{U_{max}}{\varphi_k}\right)^{1+\nu}}$; u—input voltage; C_0—diode capacitance at $u = 0$; C_{max}—diode capacitance at $u = U_{max}$; U_{max}—voltage determined from condition of $(\varphi_c - U_{max})/ /U_{max} \ll 1 \neq 0$; φ_c—contact potential difference; $\nu = 0.5$.

Analyzing the obtained results, it is easy to conclude that taking into account dependence of capacitance of Schottky rectifier diodes on level of input influence significantly approximates calculation data to the experimental data and allows calculating the dynamic characteristics of the rectifier with a smaller error.

From practical point of view, results of measurements of the RRE string parameters as part of single module (marked with crosses in Fig. 8.32a, b) and a single module (marked with crosses in Fig. 8.33c) are of interest.

As a result of the experiments it was found that in these cases parameters of string and module do not differ much from similar parameters measured for the

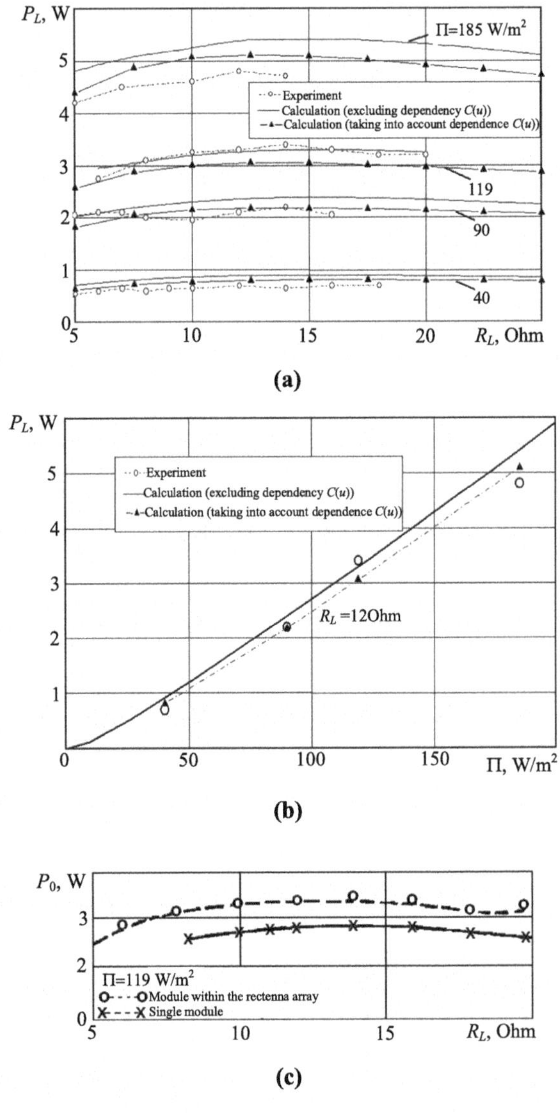

Fig. 8.33 Module parameters

string and module, which are in the rectenna. Obtained result suggests that stage of rectenna tuning for maximum rectification efficiency can be significantly simplified by determining, for example, optimal dimensions to the screen based on the results of one module test.

Let's note one more important result of the performed experiments related to the functionality of the rectenna module. At optimal load resistance and $\Pi = 185$ W/m^2 (see Figs. 8.32a, 8.33a) power in the module load is practically equal to eight times DC power of string, and optimal module resistance is exactly eight times less than the optimal string resistance. Above data indicate that the DC power acquisition circuit of the module is well decoupled from loop radiators and installation of output filters in the two-half-period RREs is not required. This significantly simplifies the module scheme and increases the manufacturability of the module. Considering that the configuration of the module allows a relatively easy build-up of receiving and rectifying aperture, we can recommend the developed design of the module as a basic one, on the basis of which various improvements of the rectenna can be made.

8.5.3 Measurement Results for Unequal-Amplitude Excitation of Rectennas

Main tasks of measurements in case of unequal-amplitude excitation of the rectenna were as follows:

- to find out possibility of practical use of technique described in Sect. 8.4 for calculating the energy characteristics of rectennas whose aperture is illuminated by a Gaussian beam;
- experimentally determine degree of repeatability of parameters of RRE systems when using series rectifier diodes from different batches in rectennas;
- demonstrate presence of significant losses during DC power collection in rectennas made of the same type RRE;
- evaluate the real efficiency of WPT systems with mirror antennas at different degrees of irradiance of aperture rectenna when changing the value of Fresnel number τ ($\tau = \sqrt{S_1 S_2}/\lambda_0 D = \pi R_1 R_2/\lambda_0 D$, where S_1, S_2—areas of circular apertures of transmitting antenna and rectenna; R_1, R_2– their radius, respectively; λ_0—operating wavelength; D—distance between apertures), which is equivalent to changing the distance between the transmitting antenna and rectenna.

To solve the above tasks the experiments were performed with two rectenna samples. Second sample differed from the first one only by the fact that it used rectifier diodes from a different batch. Measurements were performed for two distances ($D = 3$ m and $D = 5$ m). Calculations showed that with the previously specified dimensions of the rectenna, transmitting antenna, and excitation conditions of the latter, the distance $D = 3$ m corresponds to a Fresnel number $\tau = 1.535$, and the distance $D = 5$ m corresponds to $\tau = 0.95$. In the first case, the power flux density of the microwave beam decreases from the center to the edge of the aperture of the rectennas by 10 dB, and in the second—by 3.5 dB.

First, measurements were made at distance of 3 m for the first sample of the rectenna (the results given in Sect. 8.5.2 were obtained on this sample). Dependences of DC power and voltage at the rectenna output on the load resistance were studied. Results of experiments and calculations are shown in Fig. 8.34a, b. Circles show average values of measured values. Theoretical values of parameters (solid line), as before, were calculated using the averaged characteristics of diodes.

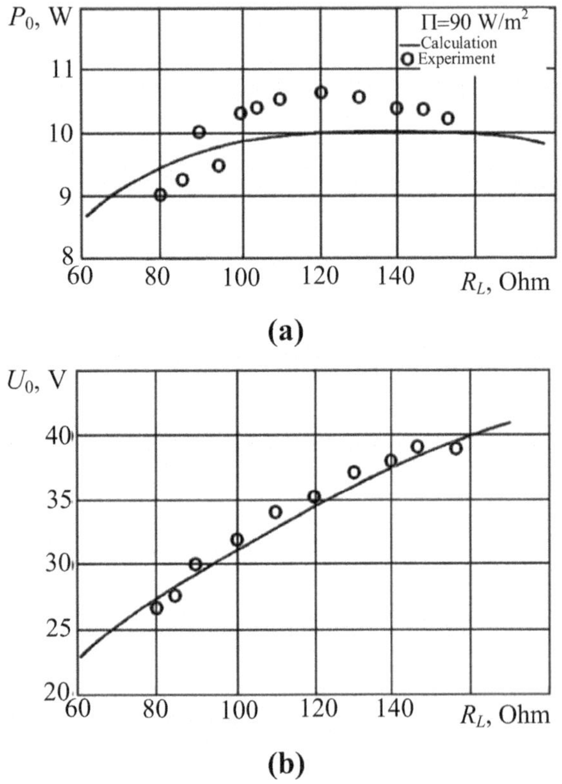

Fig. 8.34 Parameters of the rectenna

With known value of optimal load resistance, efficiency of laboratory sample WPT system was also measured. All necessary data related to this experiment are recorded in Table 8.11, where the following designations are taken: P_M—sum of powers measured at the outputs of nine modules when each of them is loaded to the optimum resistance; $\overline{\eta} = P_M / P_\Sigma$—efficiency of interception and rectification; $\eta_0 = P_0 / P_M$—efficiency of DC power collection.

Table 8.11 WPT system parameters at $\tau = 1.535$

Data	P_Σ,W	P_M,W	P_0,W at$R_L = R_{L\text{opt}}$	$R_{L\text{opt}}$, Ohm	$\overline{\eta}$, %	η_0, %	η_c, %
Calculation	–	–	10.0	30.0	40.0	–	–
Experiment (first sample rectenna)	7.9	12.0	10.55	120.0	43.0	87.9	37.8
Experiment (second sample rectenna)	27.9	13.9	12.0	110.0	49.8	86.2	43.0

Table 8.12 Data on ten measurements of the WTP system parameters at optimum load of the rectenna

D, m	Measurement number	1	2	3	4	5	6	7	8	9	10
3	P_0, W	12.4	10.5	12.0	13.2	12.4	9.8	12.8	12.0	12.0	11.6
	η_c, %	44.4	37.6	43.0	47.3	44.4	35.1	45.9	43.0	43.0	41.6
5	P_0, W	45.5	47.5	45.5	49.5	51.6	51.6	53.8	51.6	51.6	55.9
	η_c, %	40.2	42.0	40.2	43.7	45.6	45.6	47.5	45.6	45.6	49.4

Table 8.13 Total measurement data of the WPT system parameters

D, m	τ	Π_{max}, W/m^2	P_Σ, W	P_0, W	η_c, %	Δ_m, %	η_0, %
3	1.535	90	27.2	12 ± 2.3	43 ± 8.4	$\pm$ 19.5	86.2
5	0.950	185	113.2	50 ± 7.7	44.2 ± 6.8	$\pm$ 15.5	94.0

Limits of intervals, within which real measurement errors are located, were calculated from the data of 10 measurements, presented in Table 8.12 for the second sample of rectenna. As a result of processing the data of Table 8.12, the values of P_0, η and relative error of measurements Δ_m, which are presented in Table 8.13, were obtained. Here, the results of demonstration single measurements of collection efficiency and other data detailing the conditions of experiments at distances of 3 and 5 m are also given.

From the results given in Tables 8.11, 8.13 it is easy to make following generalization. With measurement errors realized in the laboratory sample of the WTP system, the theoretical values of the parameters are within the confidence intervals of the experimental characteristics of the WTP system in case of using both the first and the second sample rectenna. Based on this, it can be assumed that the proposed analysis techniques are quite applicable in practice. However, it should be pointed out that in order to achieve more accurate calculation data, it is necessary to use the measured characteristics of the batch of diodes that will be used in the rectenna, rather than the typical ones. Another important conclusion follows from the measurement data of the collection efficiency at different values of the Fresnel number. A decrease in the collection efficiency is one of the reasons that makes it impossible to increase the efficiency of the WPT system by increasing the irregularity of illumination of the aperture of the rectenna.

8.6 Analysis of EMC Parameters of Large-Aperture Rectennas

Let's consider the technique of analyzing EMC characteristics of LAR. From point of view of electromagnetic compatibility, two of its characteristics are of interest—absolute value of power of incident radiation at frequencies of higher harmonics and angular distribution of incident radiation field. Initial information for calculations is distribution of currents at frequencies of higher harmonics $I(nf_0)$. Calculations showed that when aperture of rectenna is irradiated with a Gaussian beam, amplitude distribution of currents at frequency of any harmonic is also described by the Gaussian function

$$I(nf_0) = I_{\max}(nf_0)e^{-\nu(nf_0)\tau^2(x_i^2+y_i^2)}. \tag{8.117}$$

Where $I_{\max}(nf_0)$ is the current amplitude of the n-th harmonic in center of aperture $\Pi = \Pi_{\max}$; $\nu(nf_0)$ is constant coefficient characteristic of the n-th harmonic and also depending on the degree of irradiation irregularity τ.

With accepted approximations, impedances of radiators and values of all RREs are rectenna at harmonic frequencies. Then, radiating power of the i-th RRE at frequency nf_0 can be determined from the equation

$$P_{\Sigma i}(nf_0) = \frac{1}{2}I_i^2(nf_0)R_{in}(nf_0) \tag{8.118}$$

Side radiation power of the entire rectenna

$$P_\Sigma(nf_0) = \sum_{i=1}^{M} P_{\Sigma i}(nf_0) = \frac{1}{2}R_{in}(nf_0)\sum_{i=1}^{M} I_i^2(nf_0). \tag{8.119}$$

For LAR ($M \to \infty, d \to 0$), taking into account approximation 8.117), the expression for power of incident radiation is rewritten as follows:

$$P_\Sigma(nf_0) = \frac{1}{2}R_{in}(nf_0)I_{\max}^2(nf_0) \\ \frac{1}{\Delta\overline{S}}\iint_{\overline{S}_2} e^{-2\nu(nf_0)\tau^2(x^2+y^2)}dxdy, \tag{8.120}$$

where $\Delta\overline{S}$ is the area of one array cell normalized to the radius squared of the rectenna.

For the special case of the square aperture

$$P_\Sigma(nf_0) = \frac{R_{in}(nf_0)I_{\max}^2(nf_0)}{2\Delta\overline{S}}\int_0^{\frac{1}{\sqrt{2}}}\int_0^{\frac{1}{\sqrt{2}}} e^{-2\nu(nf_0)\tau^2(x^2+y^2)}dxdy \\ = \frac{\pi R_{in}(nf_0)I_{\max}^2(nf_0)}{16\nu(nf_0)\tau^2\Delta\overline{S}}erf^2\left[\sqrt{\nu(nf_0)\tau^2}\right]. \tag{8.121}$$

In the case of circular aperture

$$P_{\Sigma}(nf_0) = \frac{R_{in}(nf_0)I^2_{\max}(nf_0)}{2\Delta\overline{S}} \int_0^1 \int_0^{\pi} e^{-2\nu(nf_0)\tau^2\rho^2} \rho d\rho d\varphi$$
$$= \frac{\pi R_{in}(nf_0)I^2_{\max}(nf_0)}{4\nu(nf_0)\tau^2\Delta\overline{S}}(1 - e^{-2\nu(nf_0)\tau^2}). \quad (8.122)$$

As can be seen from the presented expressions (8.121) and (8.127), the power of incident radiation depends not only on current amplitude $I_{\max}(nf_0)$, but also on properties of linear multipole connected to the diode (resistance $R_{\text{in}}(nf_0)$), irregularity of irradiation (parameter τ) and the geometric characteristics of aperture.

Let us now consider a technique for analyzing the angular distribution of the sidelobe radiation field. A variety of methods known from antenna theory can be used to determine the directivity diagram of a rectenna at frequencies of higher harmonics (see Sect. 8.2). The essence of our technique was as follows.

A flat periodic rectenna array of arbitrary shape with oblique-angle array periodic along the X and l axes has been considered as initial one. Period along the x-axis is d_x, along the l-axis is d_l, and angle between the axes is α_r. Array was assumed to be located over an infinite perfectly conducting screen in plane $Z = 0$, and consisted of N_l rows parallel to X axis, with M_p RREs in each p-row. Thus, the total number of RREs in the array

$$N = \sum_{p=1}^{N_l} M_p. \quad (8.123)$$

It was assumed that coordinates of the center of cell with number (1, 1) with respect to the chosen general coordinate system, as well as the coordinates and orientation of the RRE radiator with respect to the center of the cell, are known. The given data refer to geometric characteristics of rectenna and allow unambiguously determining location of the RRE in a cell with any arbitrary number (p, q).

To calculate side radiation field, data describing electrical parameters of the rectenna are also needed. The most complete characteristic of this type is current distribution on surface of the rectenna radiators. However, for strict calculation of the current distribution at frequencies of higher harmonics, it is necessary to solve the boundary problem of electrodynamics for a system of large electrical dimensions, which is a LAR, and taking into account the nonlinear characteristics of rectifier diodes, which is hardly feasible at present and in the near foreseeable future. In this connection, during the development of the methodology for the analysis of side radiation of rectenna, such simplifying assumptions were made:

1. The influence of edge effects in the array is neglected.
2. The RRE in a rectified array is assumed to have the same parameters as it would have in an infinite array of identical elements.

3. Amplitude-phase distribution of currents at harmonic frequencies is considered to be known, determined in the course of the analysis of the energy parameters of the rectenna.
4. Current distribution over surface of the rectenna radiators is already known from the solution of the problem of analysis of the energy parameters of the rectenna and is approximated by a system of piecewise sinusoidal basis functions of subdomains in the form

$$\vec{I}(\xi) = \sum_{n=1}^{M} \vec{I}_n(\xi), \tag{8.124}$$

where M is the number of basis functions that approximate the current distribution on surface of the RRE radiator; $\vec{I}_n(\xi)$—is the basis piecewise sine function defined on two neighboring segments of the radiator as follows:

$$\vec{I}_n(\xi) = I_n \frac{\sin k(l_{mn} - \xi)}{\sin kl_{mn}} \vec{\xi}_{mn},\ m = 1, 2. \tag{8.125}$$

In (8.125) l_{1n}, l_{2n} are lengths of first and second segments of radiator conductors, on which the n-th basis function is defined; $\vec{\xi}_{1n}$, $\vec{\xi}_{2n}$ are unit vectors characterizing the direction of the corresponding segments; ξ—coordinate along the first or second segments, counted from the point of their connection; $k = 2\pi/\lambda$—wave number for the i-th harmonic; I_n—amplitude factor of n-th basis function.

The second stage of calculation of power parameters of rectenna includes the procedure of calculation of currents flowing through rectifier diodes at basic frequency and at frequencies of higher harmonics formed in the rectification process. As a result, the amplitude-phase distribution of currents of higher harmonics over the rectifier aperture becomes known, which, in turn, determines the values of amplitude coefficients I_n of basic current functions in expression (8.125).

Thus, the incident radiation field of the rectenna in Fresnel zone can be found in following sequence. First, the field created by n-th basis current function is determined. Then, using the coefficients known from the solution of the rectenna excitation problem I_n, the field generated by one RRE is found. Sum of fields of all RREs gives the total field of the rectenna at the chosen frequency of incident radiation.

In general terms, the field created by the current (8.125) can be written in this way:

$$\begin{aligned} E_\theta(r, \theta, \varphi) = &-\frac{j30I}{\sin(kl_{mn})} \frac{e^{-jkr}}{r} \\ &e^{jk\vec{r}_0\vec{r}} \frac{e^{jkl_{mn}\left(\vec{\xi}\vec{r}\right)} - j\left(\vec{\xi}\vec{r}\right)\sin kl_{mn} - \cos kl_{mn}}{1 - \left(\vec{\xi}\vec{r}\right)^2} \\ &\times \quad (\cos\gamma_1 \cos\theta \cos\varphi + \cos\gamma_2 \cos\theta \sin\varphi - \cos\gamma_1 \sin\theta), \end{aligned} \tag{8.126}$$

$$E_{\varphi}(r,\theta,\varphi) = -\frac{j30I}{\sin(kl_{mn})}\frac{e^{-jkr}}{r}$$

$$e^{jk\vec{r}_0\vec{r}}\frac{e^{jkl_{mn}(\vec{r})} - j\left(\vec{\xi}\vec{r}\right)\sin kl_{mn} - \cos kl_{mn}}{1-\left(\vec{\xi}\vec{r}\right)^2}$$

$$\times \quad (\cos\gamma_1 \sin\varphi - \cos\gamma_2\cos\varphi). \tag{8.127}$$

In expressions (8.130) and (8.127) $\vec{r}_0$—radius-vector drawn from the origin to the junction point of two segments, on which the basis current function is defined; $\vec{r}$—unit vector in the direction of the observation point; $\cos\gamma_1$, $\cos\gamma_2$, $\cos\gamma_3$—directional cosines of the unit vector $\overline{\xi}$.

Presence of the screen is accounted for by introduction of mirror image of current function, field of which is determined taking into account the difference in phase and position of the mirror image current. Finally, the field of the total current with number n is defined as follows:

$$\left.\begin{aligned} E_{\theta n} &= E_{\theta n1} - E_{\theta n2} - E'_{\theta n1} + E'_{\theta n2} \\ E_{\varphi n} &= E_{\varphi n1} - E_{\varphi n2} - E'_{\varphi n1} + E'_{\varphi n2} \end{aligned}\right\}. \tag{8.128}$$

Here the symbol ($'$) marks the mirror image fields of the first and second segments.

Then, applying the superposition of fields principle, it is not difficult to find the field created by i-th RRE of the rectenna

$$E^{(i)}_{\theta,\varphi} = \sum_{n=1}^{M} E_{\theta,\varphi n}, \tag{8.129}$$

and then total side radiation field of the rectenna

$$E^{\Sigma}_{\theta,\varphi} = \sum_{n=1}^{M} E^{(i)}_{\theta,\varphi n}, \tag{8.130}$$

where M is the number of RRE in the rectenna.

The above expressions are valid for the case when the radiating structure of the rectenna consists of array of radiators of complex geometry located over a conducting screen in a medium with $\varepsilon_r = 1$. At $\varepsilon_r \neq 1$, calculation formulas will be different, but the order of calculation remains the same.

Let us consider examples of calculating the directivity diagrams of incident radiation. Figure 8.35a, b show the DD at the frequencies of the second and third harmonics of the experimental array, which is described in detail in Sect. 8.5.1. Calculations assumed that the array is irradiated in-phase.

In Fig. 8.35 curves 1 show the DD for equal-amplitude excitation. It can be seen that at frequency of the second harmonic the DD has one main maximum. As the

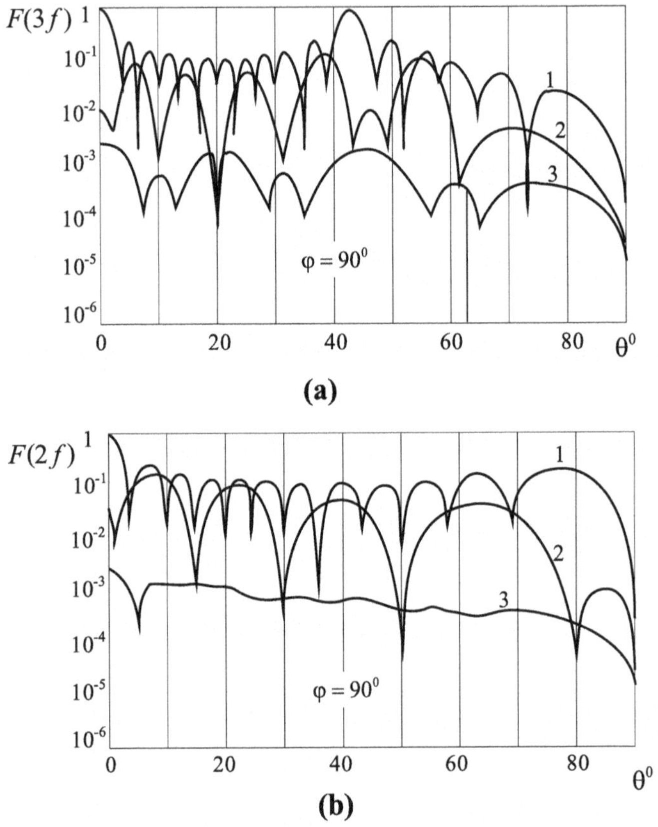

Fig. 8.35 Results of calculations of side radiation field of the rectenna

harmonic number increases, the condition of singleness of the main lobe is no longer satisfied and the DD has several diffraction maxima. Thus, at harmonic frequencies the rectenna functions as an ordinary active phased antenna array, which can concentrate a significant part of the power of incident radiation in certain directions, and thus significantly worsen the electromagnetic compatibility of radio facilities, located in zone of operation of RRE systems. Consequently, to ensure acceptable EMC characteristics of the rectenna, it is necessary to apply a number of technical solutions, which reduce the directivity of the rectenna at frequencies of incidental radiation.

Reduction of directivity of the rectenna can be achieved by defocusing the RRE at frequencies nf_0 in any way known from the practice of antenna theory. For example, by changing number of harmonic filter links in the RRE, by spatial shift of the rectenna modules relative to each other (conformal aperture), etc. General disadvantage of these methods is complexity of calculations and technical implementation of the rectenna. Below we propose a very simple way to reduce the directionality of side radiation of LAR. Its essence is that neighboring modules of the rectenna or neighboring lines of RREs within the module at frequencies nf_0 are excited in antiphase. Technically, this is realized by changing the direction of inclusion of Schottky diodes in cross sections of the RRE vibrators to opposite one.

Possible application of described method is confirmed by the results of calculations shown in Fig. 8.35. Here the DDs of rectenna of nine modules at their antiphase inclusion (curves 2) and antiphase inclusion of strings in modules (curves 3) are shown. Earlier the DDs of this rectenna were calculated for in-phase aperture excitation at harmonic frequencies (see Fig. 8.35, curves 1). From the above results of numerical experiments the following conclusions follow:

- for the 144-element antenna under consideration, the antiphase excitation of the modules at frequencies of the second and third harmonics can reduce maximum level of spurious radiation by only − 9.5 dB;
- more significant decrease in directivity of the rectenna at frequencies of higher harmonics is observed at antiphase excitation of neighboring arrays in each module (up to − 40...− 43 dB).

Effect of reducing level of side radiation when diodes are inversely phased in neighboring lines of a rectenna module was also confirmed experimentally. For this purpose, measurements of the harmonic level of modules with antiphase and in-phase connection of rectifier diodes were performed. Structure of measuring setup is shown in Fig. 8.36. Setup was built on the basis of the experimental WPT system, described in Sect. 8.5.1.

Measurements were carried out in an anechoic chamber 1. The investigated rectenna module 2 was loaded by direct current to the optimum load resistance. The beam falling on microwave energy module was formed by transmitting mirror antenna 3 with mirror opening radius of 0.45 m. Source of powerful microwave oscillations with a frequency of 2.45 GHz was magnetron generator 4, powered by power supply 5. Amplitudes of spectral components were measured using a spectrum analyzer 7 with a remote probe 6.

Main difficulty in measuring level of harmonics emitted by the rectenna module was that the magnetron generator 4 has fairly significant level of its own higher harmonics. Therefore, only the relative level of harmonics of the module with respect to level of harmonics of the generator was measured.

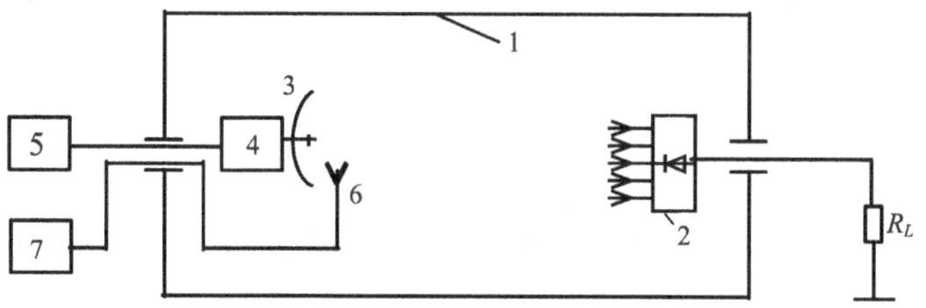

Fig. 8.36 Installation for measuring level of side radiation

Table 8.14 Results of side radiation level measurements

Number harmonics	Harmonic level, dB			
	Generator	Module with in-phase switching of diodes	Module with anti-phase switching of diodes	Winning
1	0	0	0	
2	–32.5	–26.3	–31.0	–4.7
3	–52.3	–43.4	–52.3	> −8.9

The measurement data presented in Table 8.14 show that use of the principle of antiphase excitation of Schottky diodes only in one rectenna module out of 16 RRE already allows obtaining a tangible gain in suppression of the second and third harmonics (respectively–4.7 dB and more than –8.9 dB).

References

1. Gretskih DV, Luchaninov AI, Vishniakova JV, Katrich VA, Nesterenko MV (2018) Electrodynamic model of a wireless power transmission system. In: Proceedings of international seminar/workshop on direct and inverse problems of electromagnetic and acoustic wave theory, pp 80–85
2. Gretskih DV, Luchaninov AI, Katrich VA, Nesterenko MV (2019) Electrodynamic approach to designing wireless power transfer systems (internal system processes). In: Proceedings of international conference on information and telecommunication technologies and radio electronics, pp 1–6
3. Luchaninov AI, Gretskih DV, Gomozov AV, Katrich VA, Nesterenko MV (2019), Electrodynamic approach to designing WPT systems with accounting for non-system interactions. In: Proceedings Ukraine conference on electrical and computer engineering, pp 107–111
4. Gretskih D, Luchaninov A, Katrich V, Nesterenko M, Gomozov A (2019) Extemal parameters of wireless power transmission systems. In: Proceedings of international seminar/workshop on direct and inverse problems of electromagnetic and acoustic wave theory, pp 117–121
5. Alieksieiev V, Gretskih D, Luchaninov A, Lykhograi V, Shcherbina A (2021) Applying the electrodynamic approach to modeling wireless power transmission systems. In: Proceedings of international seminar/workshop on direct and inverse problems of electromagnetic and acoustic wave theory, pp 111–115
6. Shinohara N (2014) Wireless power transfer via radiowaves. Wiley, NY
7. Nikoletseas S, Yang Y, Georgiadis A (2016) Wireless power transfer algorithms, technologies and applications in ad hoc communication networks. Springer International Publishing
8. Luchaninov AI, Shifrin YS (2007) Mathematical model of antenna with lumped nonlinear elements. Telecommun Radio Eng 66(9):763–803
9. Amitay N, Galindo V, Wu CP (1972) Theory and analysis of phased array antennas. Wiley, NY
10. Sazonov DM (2015) Multi-element antenna systems, matrix approach. Publishing House "Radiotekhnika", Moscow (in Russian)
11. Shifrin YS, Luchaninov AI (1997) Antennas with nonlinear elements. The Reference Manual on Antenna Equipment, Moscow (in Russian)
12. Mittra R (1973) Computer techniques for electromagnetic. Pergamon, NY
13. Moore J, Pizer R (1984) Moment methods in electromagnetics: techniques and applications (antenna series). Research Studies Press, NY, Letchworth
14. Dmitriev VV (1969) Electromagnetic fields in layered inhomogeneous media. Moscow State University, Moscow (in Russian)

15. Korn G, Korn T (1968) Handbook in mathematics. Nauka, Moscow (in Russian)
16. Shokalo VM, Luchaninov AI, Konovaltsev AA, Luchaninov SA, Omarov MA (2001) An algorithm for the analysis of the equidistant array of ribbon microstrip radiators of arbitrary geometry, adapted to the calculation of large-aperture antennas with nonlinear elements. 2. Solution of a system of integral equations. Radiotekhnika 118:3–7 (in Russian)
17. Bodrov BB, Markov GT (1977) Excitation of periodic antenna arrays. Collect Sci Method Art Appl Electrodyn 1:129–162 (in Russian)
18. Pozar D, Schaubert D (1984) Scan blindness in infinite phased arrays of printed dipoles. IEEE Trans Antennas Propag 32(6):602–610
19. Luchaninov AI, Shokalo VM, Konovaltsev AA, Rybalko AM, Shcherbina AA (1998) Theoretical and experimental studies of large-aperture rectenna arrays. Telecommun Radio Eng 52(6):87–92

Chapter 9
Ultrawideband Combined Vibrator-Slot Antenna Arrays

Ultrawideband technologies are growing significantly in the last decades. The efforts were aimed at creating both efficient single impulse antennas, such as IRA [1–3], and antenna arrays. For example, there are MIMO communication systems that need to develop ultra-wideband antenna arrays [4–6]. Ultrawideband (UWB) pulsed radiators are also an integral part of impulse georadars. The development and optimization of such devices is a difficult task, as its main purpose is to register low-power signals reflected from hidden objects. Therefore, the received signals must have minimal noise and be resistant to external interferences [7]. Another possible application of UWB georadars is a nondestructive measurement of thicknesses of layers of complex medium for the estimation of quality of road coverings. [8]. It is a very important application of the radars in humanitarian demining tasks where nanosecond pulses help to identify hidden explosive objects [9].

The latest trend in ultrawideband signal usage for Internet of Things applications has appeared as well. Ultrawideband radiators have also found applications in medicine. Thus, antenna arrays are developed for detecting cancers in the human chest [10, 11]. Such systems need efficient UWB radiators [12]. The above-mentioned applications of UWB radiators are needed antenna arrays as well. So, the development and optimization of ultra-wideband antenna arrays is a separate important area [13–15].

The dimensions of the antenna array element can be reduced by using a strong interaction between the radiators of electrical and magnetic types. An example of such an antenna is the Clavin radiator, which consists of a narrow rectangular slot and two short-circuited monopoles [17, 18]. An ultrawideband analog of such a radiator was presented in the works [19, 20]. As a magnetic radiator the ultrawideband Barnes slot is used [16], and two perfectly conducting cones are ultra-wideband monopoles. Due to the strong interaction of the cones with the slot, it was possible to improve the radiation in the low-frequency range and to increase the directivity of this radiator.

So, the chapter is devoted to build UWB antenna arrays with the monopole-slot radiator of Clavin type and to study its radiation characteristics.

M. V. Nesterenko et al., *Vibrator and Slot Antenna Arrays for Modern Applications*, Lecture Notes in Electrical Engineering 1244,
https://doi.org/10.1007/978-3-031-68346-6_9

9.1 Ultrawideband Vibrator-Slot Radiator

The transient electromagnetic field is radiated from a slot with the form given by the Eq. (9.1), which was presented in [16]

$$y(x) = \frac{\cos(\pi x)(1 - \cos(\pi x))}{4}, \tag{9.1}$$

where y(x) is the graph of the curve depending on the x coordinate. The slot is cut in a thin metal screen with sizes of 540 × 440 × 1 mm.

To operate in 2–10 GHz frequency range, the slot, shown in Fig. 9.1, should be placed in 140 × 17.6 mm rectangular with a gap of 0.5 mm in the center of the slot where the delta-generator with the internal resistance of 70 Ω is situated.

The approach that consists in improving the narrow band slot characteristics by monopoles [17, 18] was used in our previous works [19, 20] for widening the working frequency range. The more precise application of formula (9.1) caused the change of optimal geometrical parameters of cones and their positions in comparison with ones presented in [19, 20]. The view of the improved radiator is presented in Fig. 9.2.

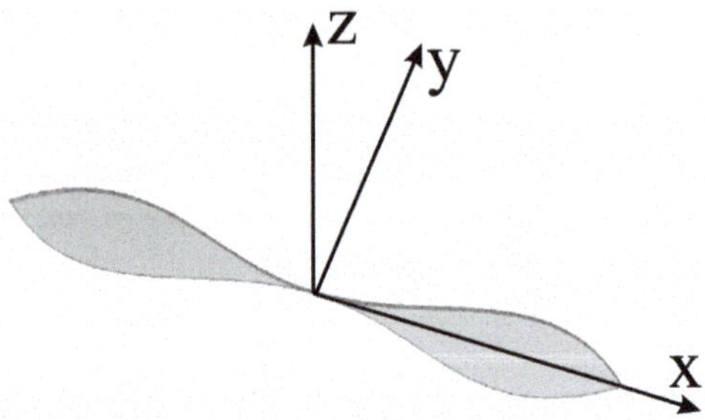

Fig. 9.1 Ultrawideband slot and coordinate system

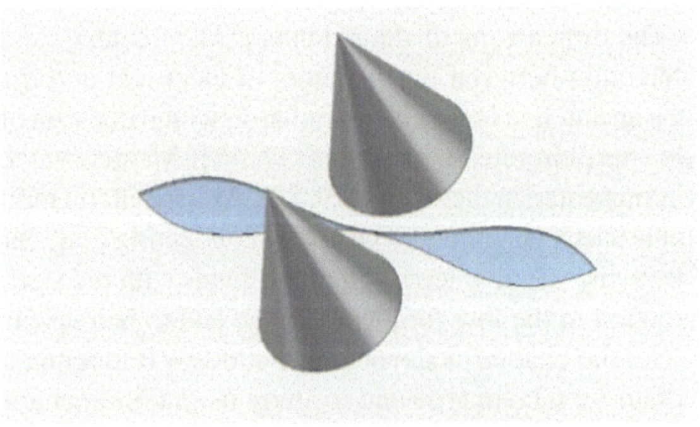

Fig. 9.2 Ultrawideband analogue of Clavin radiator

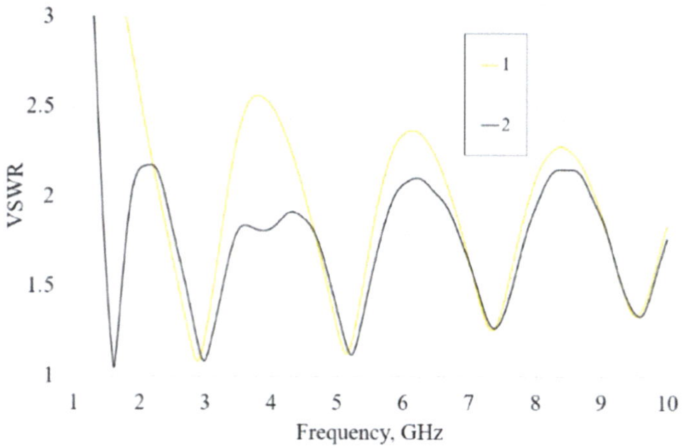

Fig. 9.3 VSWR versus frequency of ultrawideband slot (curve 1), ultrawideband analog of Clavin radiator shown in Fig. 9.2

Here, the optimal geometric parameters of a single radiator are 55.5 mm height and 20 mm base radius of conductive cones, 5 mm distance from the geometric center of the slot to the cone edge. The parameters of excitation are 0.08 ns exciting pulse duration and 1 V voltage amplitude of the source in form of delta-generator.

All numerical calculations of the electromagnetic problem and optimizations are performed using the finite difference in time-domain (FDTD) method. The calculated voltage standing wave ratio (VSWR) demonstrates better matching of the slot with cones (Fig. 9.2) in comparison to the single slot (Fig. 9.1) in a wide range of operating frequencies as seen in Fig. 9.3. The resonant properties of the cones allow to almost halve the lower frequency of the operating range of the radiator.

The radiation along the normal to the screen for these two radiators is compared in Fig. 9.4. The electrical field strength is calculated at a distance of 0.3 m, which corresponds to the far-field zone for the radiator. The use of cones increases the amplitude of the radiated electromagnetic pulse by one and half time and causes a delay in its appearance [20]. The delay is caused by the additional capacitance and inductance between the cones and the slot.

The time dependences of the electric field strength for different angles of observation are demonstrated in Figs. 9.5 and 9.6. We observe small distortions of the radiated pulse and the small difference in the amplitude at angles $\varphi = 0^\circ$ and $\varphi = 90^\circ$ only in Fig. 9.5 for the slot (curves 1 and 3). Considering the ultrawideband version of the Clavin radiator (curves 2 and 4) in the same figure, one can note that the electric field strengths have larger amplitudes compared to the ones radiated by a single slot. However, at an angle $\varphi = 90$ the amplitude becomes smaller. The conductive cones that obscure the field and generally increase the directivity of the radiation can explain it. One can mark, at the same time, a noticeable angular dependence of the shape of the radiated pulse can be used in impulse positioning systems [21, 22].

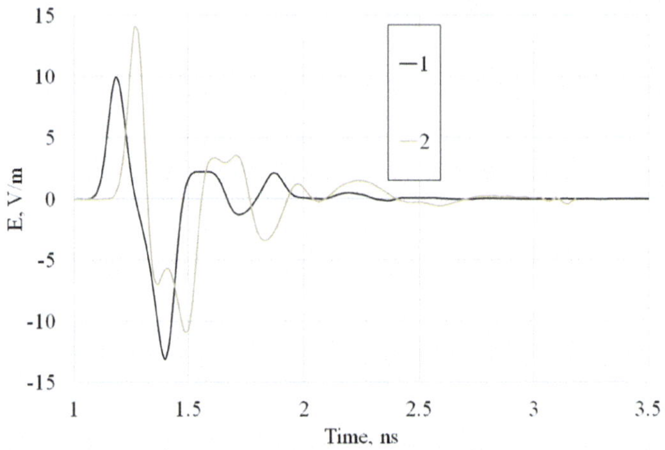

Fig. 9.4 Time dependences of the amplitude of the electrical component of the radiated field for the distance of observation $r = 0.3$ m for $\theta = 0°$ and $\varphi = 0°$, where 1—slot, 2—monopole-slot radiator in Fig. 9.2 (Phi component)

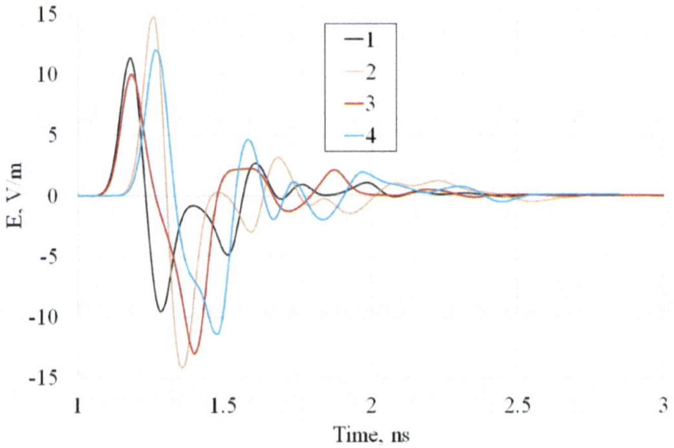

Fig. 9.5 Time dependences of the amplitude of electrical component of the radiated field for the distance of observation $r = 0.3$ m, where 1—slot, 2—Clavin ($\theta = 30°$ and $\varphi = 0°$) (Phi component), 3—slot, 4—Clavin ($\theta = 30°$ and $\varphi = 90°$) (Theta component)

However, when the angle of observation θ is increased to 60 degrees, as shown in Fig. 9.6, the Clavin radiator generates stronger oscillations in the pulse tail region than a single slot [23].

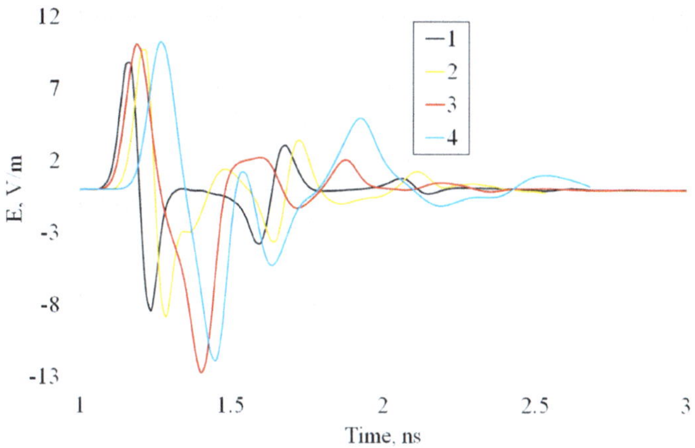

Fig. 9.6 Time dependences of the amplitude of the electrical component of the radiated field for the distance of observation $r = 0.3$ m, where 1—slot, 2—Clavin ($\theta = 60°$ and $\varphi = 0°$) (Phi component), 3—slot, 4—Clavin ($\theta = 60°$ and $\varphi = 90°$) (Theta component)

9.2 Vibrator-Slot Antenna Arrays

A single radiator described above is used to construct ultrawideband antenna arrays. Examples of antenna arrays considered are the combination of two Clavin radiators with a common H-plane (see Fig. 9.7), a combination of two radiators with a common E-plane (Fig. 9.8), and its combination of four radiators (Fig. 9.9). All calculations are performed at the observation distance $r = 0.6$ m in far zone.

Fig. 9.7 Antenna array with two UWB radiators with common H-plane

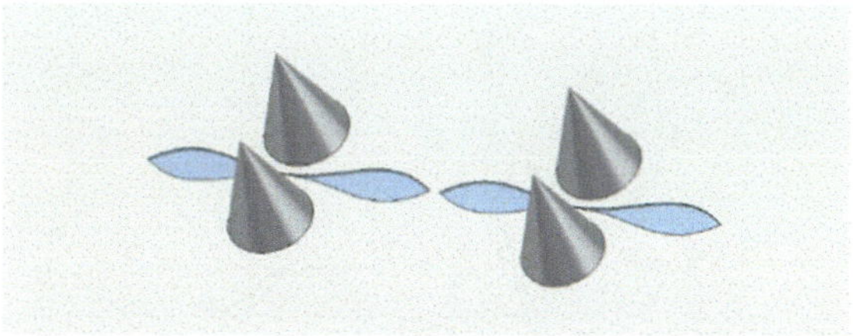

Fig. 9.8 Antenna array with two UWB radiators with common E-plane

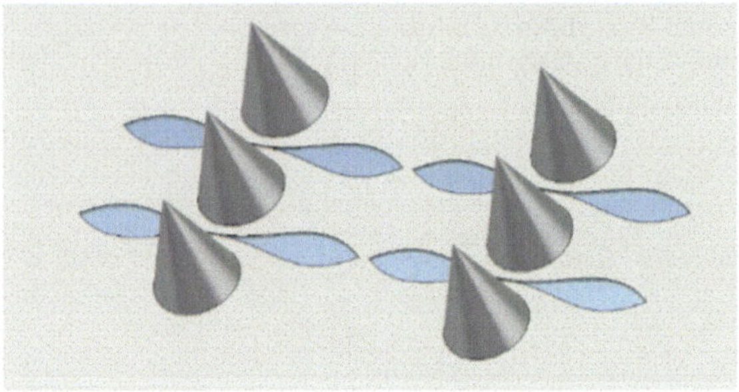

Fig. 9.9 The antenna array of four UWB radiators

First, it is interesting to analyze the character of the radiated electromagnetic field in the direction of normal to the plane of the screen. The time dependences for the Phi component of radiated field along normal are shown in Fig. 9.10.

It can be seen that if the number of radiators is doubled, curves 2 and 3, the field amplitude doubles. In the case of the antenna array of four radiators, it is possible to predict the proportional increase in the amplitude of the radiated field compared to other options. The peculiarities of the 2 × 2 antenna array can also include the fact that with increasing amplitude of the pulse front, the amplitude of its tail does not increase significantly, although the duration of attenuating oscillations is greater compared to a single radiator and AA shown in Figs. 9.7 and 9.8.

An important characteristic for ultrawideband radiators is the dependence on the angle of observation. Thus, Figs. 9.11, 9.12, 9.13 and 9.14 demonstrate different cases for different field components and angles. It is noticeable that at different angles of observation, other than normal, the antenna array 2 × 2 does form the

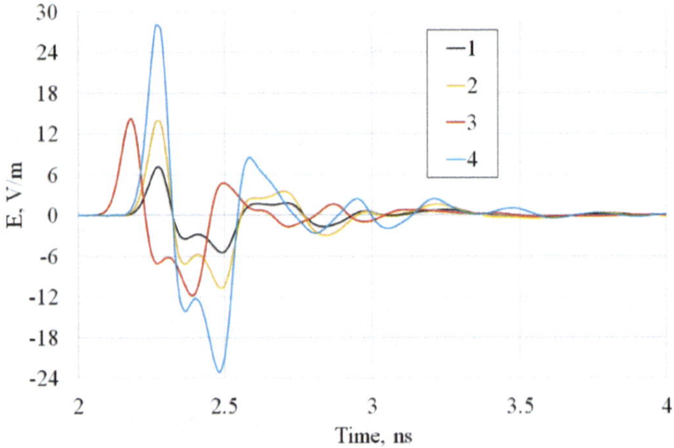

Fig. 9.10 Time dependences of the amplitude of the electrical component of the radiated field for the distance of observation $r = 0.6$ m, where 1—one radiator, 2—AA in H-plane, 3—AA in E-plane, 4—2 × 2 AA ($\theta = 0°$ and $\varphi = 0°$) (Phi component)

same maximum amplitude as other AA. For example, Fig. 9.11 shows that for the AA with four elements, the pulse is splitted into two, which follow one another. This phenomenon can also be observed for some angles and other AAs: in Fig. 9.12 for 2-AA in H-plane and 2 × 2 AA in Fig. 9.13 for all AAs. On the other hand, Figs. 9.12 and 9.14 show a strong change in the shape of the radiated pulse for AA. It points to the improved directive properties of the AAs and AA 2 × 2 in comparison with other AAs.

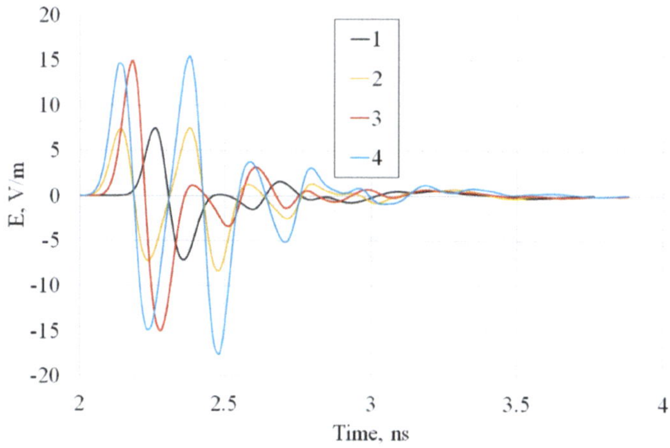

Fig. 9.11 Time dependences of the amplitude of the electrical component of the radiated field for the distance of observation $r = 0.6$ m, where 1—one radiator, 2—AA in H-plane, 3 –AA in E-plane, 4—2 × 2 AA ($\theta = 30°$ and $\varphi = 0°$) (Phi component)

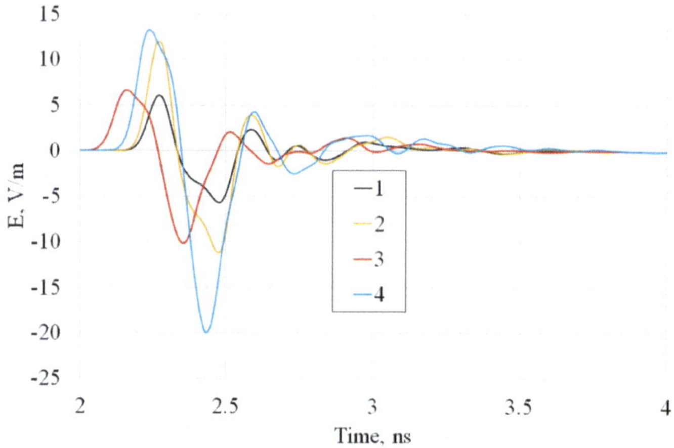

Fig. 9.12 Time dependences of the amplitude of the electrical component of the radiated field for the distance of observation $r = 0.6$ m, where 1—one radiator, 2—AA in H-plane, 3—AA in E-plane, 4—2 × 2 AA ($\theta = 30°$ and $\varphi = 90°$) (Theta component)

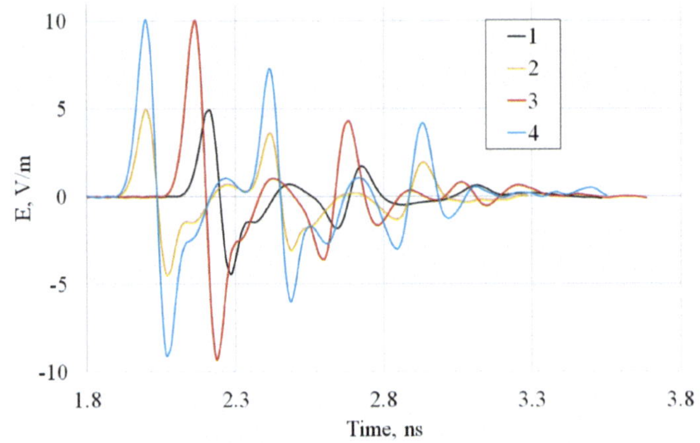

Fig. 9.13 Time dependences of the amplitude of the electrical component of the radiated field for the distance of observation $r = 0.6$ m, where 1—one radiator, 2—AA in H-plane, 3—AA in E-plane, 4—2 × 2 AA ($\theta = 60°$ and $\varphi = 0°$) (Phi component)

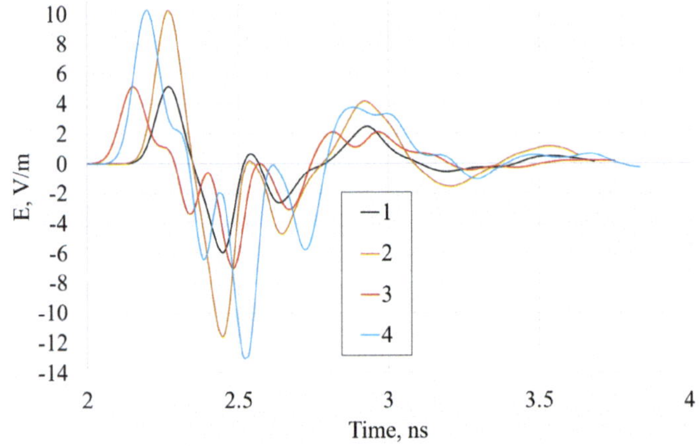

Fig. 9.14 Time dependences of the amplitude of the electrical component of the radiated field for the distance of observation $r = 0.6$ m, where 1—one radiator, 2—A in H-plane, 3—AA in E-plane, 4—2 × 2 AA ($\theta = 60°$ and $\varphi = 90°$) (Theta component)

The possibility of using an ultrawideband analog of the Clavin radiator to construct antenna arrays was shown. The angular dependence of the main radiating characteristics shows the ability to direct ultrawideband waves effectively without strong changes of their impulse time form. A more strong change of radiated impulse form of the antenna arrays considered can improve the parameters of the impulse positioning system [21].

References

1. Baum C (1989) Radiation of impulse-like transient fields. Department of Electronics and Communication Engineering, University of New Mexico, Albuquerque, NM, USA, Sensor Simul. Notes 321:1–29
2. Baum C et al (2004) JOLT: a highly directive, very intensive, impulse like radiator. Proc IEEE 92(7):1096–1109
3. Miletic M, Olcan D (2022) Impulse radiating antenna with six feeding arms and a tapered Balun. IEEE Trans Antennas Propag 70(8):6414–6422
4. Khanl M, Shafique F, Capobianco A, Autizi E, Imran I (2013) Compact UWB-MIMO antenna array with a novel decoupling structure. In: Proceedings of 10th international Bhurban conference on applied sciences and technology. Islamabad, Pakistan, pp 347–350
5. Sipal D, Abegaonkar M, Koul S (2017) Easily extendable compact planar UWB MIMO antenna array. Ant Wirel Prop Lett 16:2328–2331
6. Yang X, Salmi J, Molisch A, Qiu S, Sangodoyin S, Wang B (2012) Trapezoidal monopole antenna and array for UWB-MIMO applications. In: Proceedings of international conference on microwave and millimeter wave technology, pp 1–4
7. Orlenko OA, Pochanin GP, Korzh VG (2020) Radiation of electromagnetic field pulses by active and passive UWB slot antennas. In: Proceedings of Ukrainian Microwave Week: 1056–1060
8. Batrakov DO, Antyufeyeva MS, Batrakova AG, Urdzik SN (2020) The effect of secondary reflections on the quality of layers thickness assessment using UWB GPR signals. In: Proceedings of Ukrainian Microwave Week, pp 1132–1135
9. Pochanin G, Capineri L, Bechtel T, Ruban V, Falorni P, Crawford F, Ogurtsova T, Bossi L (2020) Radar systems for landmine detection. In: Proceedings of IEEE Ukrainian Microwave Week, pp 1118–1122
10. Foroutan F, Nikolova N (2019) UWB active antenna for microwave breast imaging sensing arrays. IEEE Ant Wireless Propag Lett 18(10):1951–1955
11. Sugitani T, Kubota S, Toya A, Kikkawa T (2012) Compact planar UWB antenna array for breast cancer detection. Proc Int Symp Ant Propag 1–2
12. Akhmedov R (2020) Neural radio in DS-UWB IoT applications. In: Proceedings of Ukrainian Microwave Week, pp 1073–1078
13. Huang B, Xu Y (2010) Analysis and design of a novel UWB antenna array. In: Proceedings of international conference on microwave and millimeter wave technology, pp 313–316
14. Wu H, Kishk A, Chen N (2005) A linear antenna array for UWB applications. In: Proceedings of antannas and propagation society international symposium, pp 594–597
15. Revna A, Balderas L, Panduro M (2018) 4D antenna array of UWB vivaldi elements with low side lobes and harmonic suppression. In: Proceedings of international symposium on antennas and propagation and USNC-URSI radio science meeting, pp 1505–1506
16. Barnes MA (2000) Ultra-wideband magnetic antenna. US patent 6091374, issued July 18
17. Penkin Yu, Semenikhin V, Yatsuk L (1987) Investigation of the internal and external characteristics of radiators such as a Clavin radiator. Radio Eng 83:3–10
18. Berdnik SL, Blinova NK, Katrich VA, Nesterenko MV, Penkin YM (2015) Spherical antenna with a Clavin radiator. In: Proceedings of XXth IEEE international seminar/workshop on direct and inverse problems of electromagnetic and acoustic wave theory, pp 75–77
19. Dumin O, Fomin P, Plakhtii V, Nesterenko M (2020) Ultrawideband combined monopole-slot radiator of Clavin type. In: Proceedings of XXVth international seminar/workshop direct and inverse problems of electromagnetic and acoustic wave theory. Tbilisi, Georgia, pp 32–36
20. Nesterenko MV, Katrich VA, Penkin YM, Berdnik SL, Dumin OM (2020) Combined vibrator-slot radiators in antenna arrays. Combined vibrator-slot structures: theory and applications. Lecture notes in electrical engineering. Springer, Cham 689:257–275
21. Dumin O, Plakhtii V, Persanov I, Shuaishuai C (2020) Positioning system using classification of ultra short electromagnetic pulse forms by ANN. In: Proceedings of 15th international conference on advanced trends in radioelectronics, telecommunications and computer engineering, Lviv-Slavske, Ukraine

22. Persanov I, Plakhtii V, Pryshchenko O, Dumin O, Fomin P (2020) Noise immunity of UWB positioning system on ANN. In: Proceedings of Ukrainian microwave week. Kharkiv, Ukraine, pp 1083–1088
23. Fomin P, Dumin O, Plakhtii V, Nesterenko M (2021) UWB antenna arrays with the monopole-slot radiator of Clavin type. In: Proceedings of 2021 IEEE IIIrd Ukrainian conference electrical and computer engineering. Lviv, Ukraine, pp 258–261

Appendix A

Green's Functions of the Various Electrodynamic Volumes for Antenna Arrays

I. Electrical Dyadic Green's Functions

1. The unbounded space with the permittivity and the permeability of the medium ε_1 and $\mu_1 (k_1 = k\sqrt{\varepsilon_1 \mu_1})$.

$$\hat{G}^e(\vec{r}, \vec{r}\,') = \hat{\text{I}} \frac{e^{-ik_1|\vec{r}-\vec{r}\,'|}}{|\vec{r} - \vec{r}\,'|}. \tag{A.1}$$

2. The half-space over the perfectly conducting plane with the permittivity and the permeability of the medium ε_1 and μ_1 (Fig. A.1).

$$\begin{aligned}
\hat{G}^e(\vec{r}, \vec{r}\,') = \hat{I} \frac{e^{-ik_1 R}}{R} - (\vec{e}_x \otimes \vec{e}_{x'}) \frac{e^{-ik_1 R_1}}{R_1} - (\vec{e}_y \otimes \vec{e}_{y'}) \frac{e^{-ik_1 R_1}}{R_1} \\
+ (\vec{e}_z \otimes \vec{e}_{z'}) \frac{e^{-ik_1 R_1}}{R_1}, \\
R = \sqrt{(x - x')^2 + (y - y')^2 + (z - z')^2}, \\
R_1 = \sqrt{(x - x')^2 + (y - y')^2 + (z + z')^2}.
\end{aligned} \tag{A.2}$$

3. The space with the permittivity and the permeability of the medium ε_1 and μ_1 of the medium and inside the corner region, concluded between two planes and perpendicularly intersecting along the abscissa axis $X0Y$ and $X0Z$ (Fig. A.2).

M. V. Nesterenko et al., *Vibrator and Slot Antenna Arrays for Modern Applications*, Lecture Notes in Electrical Engineering 1244,
https://doi.org/10.1007/978-3-031-68346-6

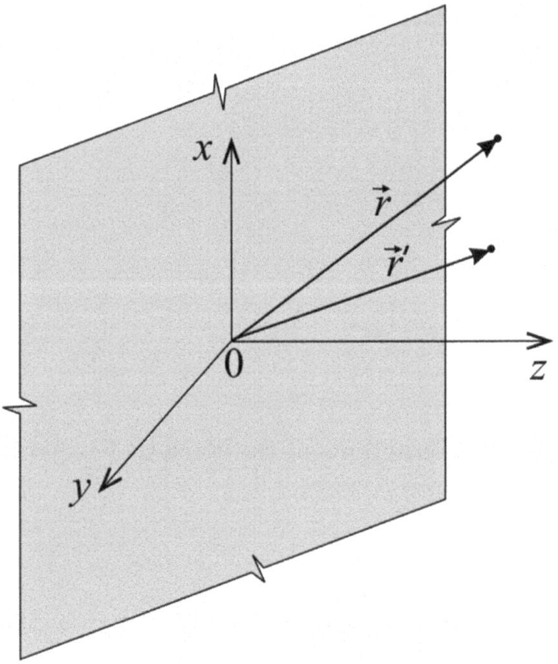

Fig. A.1 The half-space over a perfectly conducting plane

$$G^e_{zz}\left(x, y, z; x', y', z'\right) = \frac{e^{-ik_1 R}}{R} - \frac{e^{-ik_1\sqrt{(x-x')^2+(y+y')^2+(z-z')^2}}}{\sqrt{(x-x')^2+(y+y')^2+(z-z')^2}} + \frac{e^{-ik_1 R_1}}{R_1} - \frac{e^{-ik_1\sqrt{(x-x')^2+(y+y')^2+(z+z')^2}}}{\sqrt{(x-x')^2+(y+y')^2+(z+z')^2}}. \tag{A.3}$$

4. The hollow infinite rectangular waveguide of the cross-section $\{a \times b\}$ with the perfectly conducting walls (Fig. A.3).

$$\hat{G}^e(\vec{r}, \vec{r}') = \frac{2\pi}{ab}\sum_{m=0}^{\infty}\sum_{n=0}^{\infty}\frac{\varepsilon_m\varepsilon_n}{k_z}e^{-k_z|z-z'|}\left[(\vec{e}_x \otimes \vec{e}_{x'})\Phi_x(x, y; x', y') + (\vec{e}_y \otimes \vec{e}_{y'})\Phi_y(x, y; x', y') + (\vec{e}_z \otimes \vec{e}_{z'})\Phi_z(x, y; x', y')\right]. \tag{A.4}$$

5. The hollow half-infinite rectangular waveguide of the cross-section $\{a \times b\}$ with the perfectly conducting walls.

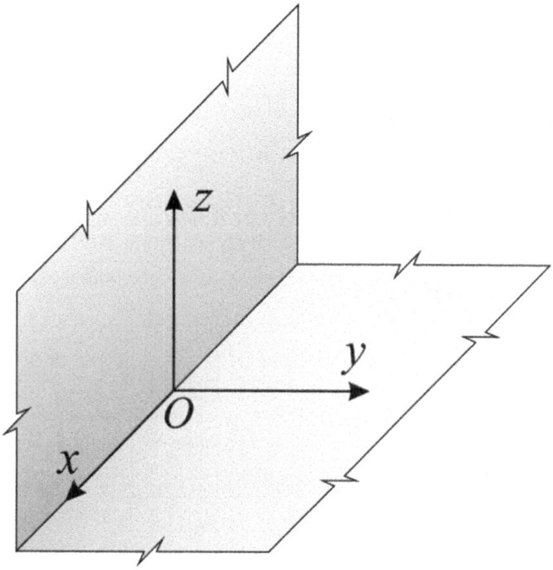

Fig. A.2 The space inside a corner perfectly conducting region

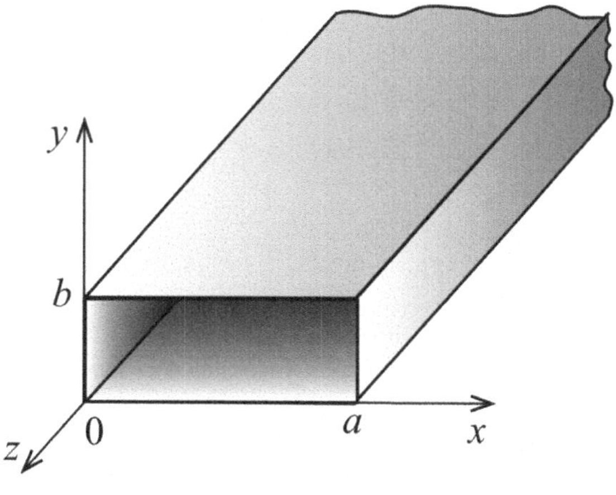

Fig. A.3 The hollow infinite rectangular waveguide

$$\hat{G}^e(\vec{r},\vec{r}') = \frac{2\pi}{ab}\sum_{m=0}^{\infty}\sum_{n=0}^{\infty}\frac{\varepsilon_m\varepsilon_n}{k_z}\Big\{(\vec{e}_x\otimes\vec{e}_{x'})\Phi_x^e(x,y;x',y') \left[e^{-k_z|z-z'|}-e^{-k_z(z+z')}\right] + (\vec{e}_y\otimes\vec{e}_{y'})\Phi_y^e(x,y;x',y')\left[e^{-k_z|z-z'|}-e^{-k_z(z+z')}\right] + (\vec{e}_z\otimes\vec{e}_{z'})\Phi_z^e(x,y;x',y')\left[e^{-k_z|z-z'|}+e^{-k_z(z+z')}\right]\Big\}. \qquad (A.5)$$

6. The hollow rectangular resonator with dimensions $\{a_R \times b_R \times H\}$ and perfectly conducting walls.

$$\hat{G}^e(\vec{r}, \vec{r}') = \frac{2\pi}{a_R b_R} \sum_{m=0}^{\infty} \sum_{n=0}^{\infty} \frac{\varepsilon_m \varepsilon_n}{k_z}$$
$$\times \left\{ (\vec{e}_x \otimes \vec{e}_{x'}) \Phi_x^e(x, y; x', y') \left[\frac{\mathrm{ch} k_z(H - |z - z'|) - \mathrm{ch} k_z(H - |z + z'|)}{\mathrm{sh} k_z H} \right] \right.$$
$$+ (\vec{e}_y \otimes \vec{e}_{y'}) \Phi_y^e(x, y; x', y') \left[\frac{\mathrm{ch} k_z(H - |z - z'|) - \mathrm{ch} k_z(H - |z + z'|)}{\mathrm{sh} k_z H} \right]$$
$$\left. + (\vec{e}_z \otimes \vec{e}_{z'}) \Phi_z^e(x, y; x', y') \left[\frac{\mathrm{ch} k_z(H - |z - z'|) + \mathrm{ch} k_z(H - |z + z'|)}{\mathrm{sh} k_z H} \right] \right\}. \tag{A.6}$$

In the expressions (A.1)–(A.6) the following notations is used:

$$\Phi_x^e(x, y; x', y') = \cos k_x x \cos k_x x' \sin k_y y \sin k_y y',$$
$$\Phi_y^e(x, y; x', y') = \sin k_x x \sin k_x x' \cos k_y y \cos k_y y',$$
$$\Phi_z^e(x, y; x', y') = \sin k_x x \sin k_x x' \sin k_y y \sin k_y y',$$

$\varepsilon_{m,n} = \begin{cases} 1, & m, n = 0 \\ 2, & m, n \neq 0 \end{cases}$, $k_x = \frac{m\pi}{a}$, $k_y = \frac{n\pi}{b}$, $k_z = \sqrt{k_x^2 + k_y^2 - k^2}$, m and n are the integral numbers, $\vec{e}_x$, $\vec{e}_y$ and $\vec{e}_z$ are the unit vectors of the rectangular coordinate system, $\hat{\mathrm{I}} = (\vec{e}_x \otimes \vec{e}_{x'}) + (\vec{e}_y \otimes \vec{e}_{y'}) + (\vec{e}_z \otimes \vec{e}_{z'})$ is the unit dyadic, and "$\otimes$" stands for dyadic product.

7. Space outside the perfectly conducting sphere of the radius $\tilde{R}$ with the permittivity and the permeability of the medium ε_1 and μ_1 (Fig. A.4).

$$\hat{G}^e(\rho, \theta, \varphi; \rho', \theta', \varphi') = \begin{vmatrix} G^e_{\rho\rho'} & 0 & 0 \\ 0 & G^e_{\theta\theta'} & G^e_{\theta\varphi'} \\ 0 & G^e_{\varphi\theta'} & G^e_{\varphi\varphi'} \end{vmatrix}, \tag{A.7}$$

$$G^e_{\rho\rho'}(\rho, \theta, \varphi; \rho', \theta', \varphi') = -\sum_{n=0}^{\infty} \sum_{m=0}^{n} \frac{\varepsilon_m\, h_h^e(\rho, \rho')}{2C_{nm}} P_n^m(\cos\theta)\, P_n^m(\cos\theta') \cos m(\varphi - \varphi'),$$

$$G^e_{\theta\theta'}(\rho, \theta, \varphi; \rho', \theta', \varphi') = -\sum_{n=0}^{\infty} \sum_{m=0}^{n} \frac{\varepsilon_m\, u_n^e(\rho, \rho') \cos m(\varphi - \varphi')}{2n(n+1)\, C_{nm} \sin\theta \sin\theta'}$$

Fig. A.4 The space outside the perfectly conducting sphere

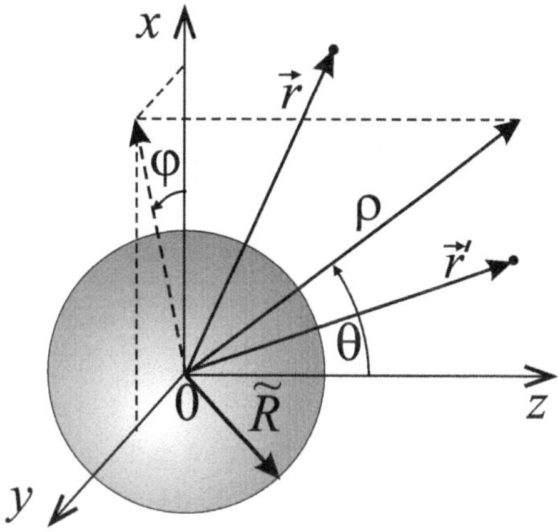

$$\times\left[m^2P_n^m(\cos\theta)P_n^m(\cos\theta')+\sin\theta\sin\theta'\frac{\mathrm{d}P_n^m(\cos\theta)}{\mathrm{d}\theta}\frac{\mathrm{d}P_n^m(\cos\theta')}{\mathrm{d}\theta'}\right],$$

$$G^e_{\theta\varphi'}(\rho,\theta,\varphi;\rho',\theta',\varphi')=\sum_{n=0}^{\infty}\sum_{m=0}^{n}\frac{m\,u_n^e(\rho,\rho')\sin m(\varphi-\varphi')}{n\,(n+1)\,C_{nm}}$$

$$\times\left[\frac{\mathrm{d}P_n^m(\cos\theta)}{\mathrm{d}\theta}\frac{P_n^m(\cos\theta')}{\sin\theta'}+\frac{P_n^m(\cos\theta)}{\sin\theta}\frac{\mathrm{d}P_n^m(\cos\theta')}{\mathrm{d}\theta'}\right],$$

$$G^e_{\varphi\theta'}(\rho,\theta,\varphi;\rho',\theta',\varphi')=-G^e_{\theta\varphi'}(\rho,\theta,\varphi;\rho',\theta',\varphi'),$$
$$G^e_{\varphi\varphi'}(\rho,\theta,\varphi;\rho',\theta',\varphi')=G^e_{\theta\theta'}(\rho,\theta,\varphi;\rho',\theta',\varphi').$$

Here $P_n^m(\cos\theta)$ is the associated Legendre functions of the first sort,

$$C_{nm}=\frac{2\pi\,(n+m)\,!}{(2n+1)\,(n-m)\,!},$$

$$h_n^e(\rho,\rho')=\begin{cases}4\pi k_1 h_n^{(2)}(k_1\rho')\begin{bmatrix}j_n(k_1\rho)\,Q_n\big(y_n(k_1\tilde{R})\big)\\-y_n(k_1\rho)\,Q_n\big(j_n(k_1\tilde{R})\big)\end{bmatrix}, & \tilde{R}\le\rho<\rho',\\ 4\pi k_1 h_n^{(2)}(k_1\rho)\begin{bmatrix}j_n(k_1\rho')\,Q_n\big(y_n(k_1\tilde{R})\big)\\-y_n(k_1\rho')\,Q_n\big(j_n(k_1\tilde{R})\big)\end{bmatrix}, & \rho>\rho',\end{cases}$$

$$Q_n\big(\mathrm{f}_n(k_1\tilde{R})\big)=\frac{n\,\mathrm{f}_n(k_1\tilde{R})-k_1\tilde{R}\,\mathrm{f}_{n+1}(k_1\tilde{R})}{n\,h_n^{(2)}(k_1\tilde{R})-k_1\tilde{R}h_{n+1}^{(2)}(k_1\tilde{R})},$$

$$\mathrm{f}_n(k_1\tilde{R})=y_n(k_1\tilde{R})\big(j_n(k_1\tilde{R})\big),$$
$$\mathrm{f}_{n+1}(k_1\tilde{R})=y_{n+1}(k_1\tilde{R})\big(j_{n+1}(k_1\tilde{R})\big),$$

$$u_n^e(\rho,\rho') = \begin{cases} 4\pi k_1 \dfrac{h_n^{(2)}(k_1\rho')}{h_n^{(2)}(k_1\tilde{R})}\left[j_n(k_1\rho)\,y_n(k_1\tilde{R}) - y_n(k_1\rho)\,j_n(k_1\tilde{R})\right], \\ \qquad \tilde{R} \le \rho < \rho', \\ 4\pi k_1 \dfrac{h_n^{(2)}(k_1\rho)}{h_n^{(2)}(k_1\tilde{R})}\left[j_n(k_1\rho')\,y_n(k_1\tilde{R}) - y_n(k_1\rho')\,j_n(k_1\tilde{R})\right], \\ \qquad \rho > \rho', \end{cases}$$

$h_n^{(2)}(k_1\rho) = j_n(k_1\rho) - iy_n(k_1\rho) = \sqrt{\frac{\pi}{2k_1\rho}} H_{n+1/2}^{(2)}(k_1\rho)$ is the Hankel spherical function of the second sort, $j_n(k_1\rho) = \sqrt{\frac{\pi}{2k_1\rho}}\, J_{n+1/2}(k_1\rho)$ and $y_n(k_1\rho) = \sqrt{\frac{\pi}{2k_1\rho}} N_{n+1/2}(k_1\rho)$ are the Bessel spherical function and the Neumann one, correspondingly, $J_{n+1/2}(k_1\rho)$ is the Bessel function, $N_{n+1/2}(k_1\rho)$ is the Neumann function and $H_{n+1/2}^{(2)}(k_1\rho)$ is the Hankel function of the second sort with the half-integral index [1].

Let us note, that it turns out to be possible to represent the expression for the component of the Green's function $G_{\rho\rho'}^e(\rho, \theta, \varphi; \rho', \theta', \varphi')$ in a more suitable form for numerical realization, having made the transition from the double series to the single one with the help of the summation theorem for the Legendre polynomials:

$$G_{\rho\rho'}^e(\rho, \theta, \varphi; \rho', \theta', \varphi') = -\sum_{n=0}^{\infty} \frac{n+1/2}{2\pi}\, h_n^e(\rho, \rho')$$
$$P_n\big(\cos\theta\cos\theta' + \sin\theta\sin\theta'\cos(\varphi - \varphi')\big).$$

II. **Magnetic Dyadic Green's Functions**

8. The unbounded space with the permittivity and the permeability of the medium ε_1 and μ_1 ($k_1 = k\sqrt{\varepsilon_1\mu_1}$).

$$\hat{G}^m(\vec{r}, \vec{r}') = \hat{\mathrm{I}}\frac{e^{-ik_1|\vec{r}-\vec{r}'|}}{|\vec{r}-\vec{r}'|}. \tag{A.8}$$

9. The half-space over the perfectly conducting plane with the permittivity and the permeability of the medium ε_1 and μ_1 (Fig. A.1).

$$\hat{G}^m(\vec{r}, \vec{r}') = \hat{\mathrm{I}}\frac{e^{-ikR}}{R} + (\vec{e}_x \otimes \vec{e}_{x'})\frac{e^{-ikR_1}}{R_1} + (\vec{e}_y \otimes \vec{e}_{y'})\frac{e^{-ikR_1}}{R_1} - (\vec{e}_z \otimes \vec{e}_{z'})\frac{e^{-ikR_1}}{R_1}, \tag{A.9}$$

$$R = \sqrt{(x-x')^2 + (y-y')^2 + (z-z')^2}\,,$$
$$R_1 = \sqrt{(x-x')^2 + (y-y')^2 + (z+z')^2}\,.$$

10. The space with the permittivity and the permeability of the medium ε_1 and μ_1 of the medium and inside the corner region, concluded between two planes and perpendicularly intersecting along the abscissa axis XOY and XOZ (Fig. A.2).

$$G_{zz}^{m}(x, y, z; x', y', z') = \frac{e^{-ikR}}{R} + \frac{e^{-ik\sqrt{(x-x')^2+(y+y')^2+(z-z')^2}}}{\sqrt{(x-x')^2+(y+y')^2+(z-z')^2}} - \frac{e^{-ikR_1}}{R_1} - \frac{e^{-ik\sqrt{(x-x')^2+(y+y')^2+(z+z')^2}}}{\sqrt{(x-x')^2+(y+y')^2+(z+z')^2}}. \quad (A.10)$$

11. The hollow infinite rectangular waveguide of the cross-section $\{a \times b\}$ with the perfectly conducting walls (Fig. A.3).

$$\hat{G}^{m}(\vec{r}, \vec{r}') = \frac{2\pi}{ab} \sum_{m=0}^{\infty} \sum_{n=0}^{\infty} \frac{\varepsilon_m \varepsilon_n}{k_z} \Big\{ (\vec{e}_x \otimes \vec{e}_{x'}) \Phi_x^m(x, y; x', y')\, e^{-k_z|z-z'|} + (\vec{e}_y \otimes \vec{e}_{y'}) \Phi_y^m(x, y; x', y')\, e^{-k_z|z-z'|} + (\vec{e}_z \otimes \vec{e}_{z'}) \Phi_z^m(x, y; x', y')\, e^{-k_z|z-z'|} \Big\}. \quad (A.11)$$

12. The hollow half-infinite rectangular waveguide of the cross-section $\{a \times b\}$ with the perfectly conducting walls.

$$\hat{G}^{m}(\vec{r}, \vec{r}') = \frac{2\pi}{ab} \sum_{m=0}^{\infty} \sum_{n=0}^{\infty} \frac{\varepsilon_m \varepsilon_n}{k_z} \Big\{ (\vec{e}_x \otimes \vec{e}_{x'}) \Phi_x^m(x, y; x', y') \Big[e^{-k_z|z-z'|} + e^{-k_z(z+z')} \Big] + (\vec{e}_y \otimes \vec{e}_{y'}) \Phi_y^m(x, y; x', y') \Big[e^{-k_z|z-z'|} + e^{-k_z(z+z')} \Big] + (\vec{e}_z \otimes \vec{e}_{z'}) \Phi_z^m(x, y; x', y') \Big[e^{-k_z|z-z'|} - e^{-k_z(z+z')} \Big] \Big\}. \quad (A.12)$$

13. The hollow rectangular resonator with dimensions $\{a_R \times b_R \times H\}$ and perfectly conducting walls.

$$\hat{G}^{m}(\vec{r}, \vec{r}') = \frac{2\pi}{a_R b_R} \sum_{m=0}^{\infty} \sum_{n=0}^{\infty} \frac{\varepsilon_m \varepsilon_n}{k_z} \times \Big\{ (\vec{e}_x \otimes \vec{e}_{x'}) \Phi_x^m(x, y; x', y') \Big[\frac{\mathrm{ch}k_z(H - |z-z'|) + \mathrm{ch}k_z(H - |z+z'|)}{\mathrm{sh}k_z H} \Big] + (\vec{e}_y \otimes \vec{e}_{y'}) \Phi_y^m(x, y; x', y') \Big[\frac{\mathrm{ch}k_z(H - |z-z'|) + \mathrm{ch}k_z(H - |z+z'|)}{\mathrm{sh}k_z H} \Big] + (\vec{e}_z \otimes \vec{e}_{z'}) \Phi_z^m(x, y; x', y') \Big[\frac{\mathrm{ch}k_z(H - |z-z'|) - \mathrm{ch}k_z(H - |z+z'|)}{\mathrm{sh}k_z H} \Big] \Big\}. \quad (A.13)$$

14. The half-infinite rectangular waveguide with an impedance $(\overline{Z}_S)$ end in the case when the sources are on the end surface.

$$\hat{G}^m(\vec{r},\vec{r}') = \frac{2\pi}{ab}\sum_{m=0}^{\infty}\sum_{n=0}^{\infty}\frac{\varepsilon_m\varepsilon_n}{k_z}\{(\vec{e}_x\otimes\vec{e}_{x'})\Phi_x^m(x,y;x',y')f_{\mathrm{II}}(k_z,\overline{Z}_S)2e^{-k_z z}$$
$$+(\vec{e}_y\otimes\vec{e}_{y'})\Phi_y^m(x,y;x',y')f_{\mathrm{II}}(k_z,\overline{Z}_S)2e^{-k_z z}\Big\},$$
$$f_{\mathrm{II}}(k_z,\overline{Z}_S) = \frac{kk_z(1+\overline{Z}_S^2)}{(ik+k_z\overline{Z}_S)(k\overline{Z}_S-ik)}. \tag{A.14}$$

15. The half-infinite rectangular waveguide with an impedance ($\overline{Z}_S$) end in the case longitudinal currents excitation.

$$\hat{G}^m(\vec{r},\vec{r}') = \frac{2\pi}{ab}\sum_{m=0}^{\infty}\sum_{n=0}^{\infty}\frac{\varepsilon_m\varepsilon_n}{k_z}$$
$$\left\{(\vec{e}_z\otimes\vec{e}_{z'})\Phi_z^m(x,y;x',y')\left[e^{-k_z|z-z'|}-f_\perp(k_z,\overline{Z}_S)e^{-k_z(z+z')}\right]\right\},$$
$$f_\perp(k_z,\overline{Z}_S) = \frac{ik-k_z\overline{Z}_S}{ik+k_z\overline{Z}_S}. \tag{A.15}$$

In the expressions (A.8)–(A.15) the following notations is used:

$$\Phi_x^m(x,y;x',y') = \sin k_x x\sin k_x x'\cos k_y y\cos k_y y',$$
$$\Phi_y^m(x,y;x',y') = \cos k_x x\cos k_x x'\sin k_y y\sin k_y y',$$
$$\Phi_z^m(x,y;x',y') = \cos k_x x\cos k_x x'\cos k_y y\cos k_y y'.$$

16. The remaining notations is the same as in (A.1)–(A.6).
Space outside the perfectly conducting sphere of the radius $\tilde{R}$ with the permittivity and the permeability of the medium ε_1 and μ_1 (Fig. A.4).

$$\hat{G}^m(\rho,\theta,\varphi;\rho',\theta',\varphi') = \begin{vmatrix} G^m_{\rho\rho'} & 0 & 0\\ 0 & G^m_{\theta\theta'} & G^m_{\theta\varphi'}\\ 0 & G^m_{\varphi\theta'} & G^m_{\varphi\varphi'}\end{vmatrix}, \tag{A.16}$$

$$G^m_{\rho\rho'}(\rho,\theta,\varphi;\rho',\theta',\varphi') = -\sum_{n=0}^{\infty}\sum_{m=0}^{n}\frac{\varepsilon_m\, h_n^m(\rho,\rho')}{2C_{nm}},$$
$$P_n^m(\cos\theta)P_n^m(\cos\theta')\cos m(\varphi-\varphi')$$

$$G^m_{\theta\theta'}(\rho,\theta,\varphi;\rho',\theta',\varphi') = -\sum_{n=0}^{\infty}\sum_{m=0}^{n}\frac{\varepsilon_m\, u_n^m(\rho,\rho')\cos m(\varphi-\varphi')}{2n(n+1)\,C_{nm}\sin\theta\sin\theta'}$$
$$\times\left[m^2P_n^m(\cos\theta)P_n^m(\cos\theta')+\sin\theta\sin\theta'\frac{\mathrm{d}P_n^m(\cos\theta)}{\mathrm{d}\theta}\frac{\mathrm{d}P_n^m(\cos\theta')}{\mathrm{d}\theta'}\right],$$

$$G^m_{\theta\varphi'}\left(\rho,\theta,\varphi;\rho',\theta',\varphi'\right)=\sum_{n=0}^{\infty}\sum_{m=0}^{n}\frac{m\,u^m_n(\rho,\rho')\sin m(\varphi-\varphi')}{n\,(n+1)\,C_{nm}}$$
$$\times\left[\frac{dP^m_n(\cos\theta)}{d\theta}\frac{P^m_n(\cos\theta')}{\sin\theta'}+\frac{P^m_n(\cos\theta)}{\sin\theta}\frac{dP^m_n(\cos\theta')}{d\theta'}\right],$$

$$G^m_{\varphi\theta'}\left(\rho,\theta,\varphi;\rho',\theta',\varphi'\right)=-G^m_{\theta\varphi'}\left(\rho,\theta,\varphi;\rho',\theta',\varphi'\right),$$
$$G^m_{\varphi\varphi'}\left(\rho,\theta,\varphi;\rho',\theta',\varphi'\right)=G^m_{\theta\theta'}\left(\rho,\theta,\varphi;\rho',\theta',\varphi'\right).$$

Here $h^m_n(\rho,\rho')=u^e_n(\rho,\rho')$, $u^m_n(\rho,\rho')=h^e_n(\rho,\rho')$, as well as other notations from (A.7).

III. Green's Functions of the Free Half-Space Over the Impedance Plane

For a free half-space over an infinite plane (Fig. A.1), $\sqrt{\mu_0/\varepsilon_0}=Z_0=120\pi$ $(\omega\varepsilon_0=k/Z_0,\ \omega\mu_0=Z_0k)$, characterized by distributed constant impedance $\overline{Z}_S=Z_S/Z_0$, provided that the half-space is excited by coupled surface currents $j^m_x=\overline{Z}_S j^e_y$, $j^m_y=-\overline{Z}_S j^e_x$ (based on the requirement of the boundary condition $[\vec{E},\vec{z}_0]=-\overline{Z}_S[[\vec{H},\vec{z}_0],\vec{z}_0]$)

$$G^m_{xx}(x,y,z;x',y',0)=\frac{1}{4\pi^2}\int_{-\infty}^{\infty}\int_{-\infty}^{\infty}C_\chi e^{-ik_x(x-x')-ik_y(y-y')-i\chi z}dk_x dk_y,$$
$$G^e_{yy}(x,y,z;x',y',0)=-\frac{\overline{Z}_S^2}{4\pi^2}\int_{-\infty}^{\infty}\int_{-\infty}^{\infty}C_\chi e^{-ik_x(x-x')-ik_y(y-y')-i\chi z}dk_x dk_y,\qquad\text{(A.17)}$$

where $C_\chi=\frac{4\pi}{i\chi}\frac{\chi k(1+\overline{Z}_S^2)}{\chi k(1+\overline{Z}_S^2)+\overline{Z}_S(k^2+\chi^2)}$, $\chi^2=k^2-k_x^2-k_y^2$.

Reference

1. Abramovits MIA, Stegun IA (1964) Handbook of mathematical functions with formulas, graphs and mathematical tables. Appl Math Ser 55. National Bureau of Standards

Appendix B

Mathematical Aspects of the Asymptotic Averaging Method

A rigorous justification of the asymptotic averaging method is a purely mathematical problem, which has been studied in detail in monographs [1, 2], where the averaging theorems were proved. Let us briefly discuss the principles that will be needed below.

Let a system of ordinary differential equations be written in standard form when the first derivatives $\frac{dx}{ds}$ are proportional to the small parameter

$$\frac{dx}{ds} = \alpha X(s, x), \tag{B.1}$$

where x is the n-dimensional vector, $0 < \alpha < < 1$ is the small parameter. There exist several approaches to reduce the initial integral equations to the form (B.1), but the method of variation of arbitrary constants is most commonly used [2]. If the reduction of the initial equation system to the standard form has already been fulfilled, change of variables can be made in (B.1)

$$|x = \xi + \alpha\tilde{X}(s, \xi), \tag{B.2}$$

where ξ is the new unknowns. Let $\overline{X}(\xi) = \lim\limits_{l\to\infty} \frac{1}{l}\int_0^l X(s, \xi)ds.$, where the bar denotes averaging over the explicit variable s, that is

$$\overline{X}(\xi) = \lim_{l\to\infty} \frac{1}{l}\int_0^l X(s, \xi)ds. \tag{B.3}$$

Then, after a series of transformations, we obtained [1]

$$\frac{d\xi}{ds} = \alpha\overline{X}(\xi) + \alpha^2 \ldots. \tag{B.4}$$

M. V. Nesterenko et al., *Vibrator and Slot Antenna Arrays for Modern Applications*, Lecture Notes in Electrical Engineering 1244,
https://doi.org/10.1007/978-3-031-68346-6

That is, if the variable ξ satisfies equations (B.4), which the right-hand side differs from the right-hand side of the averaged equations

$$\frac{d\xi}{ds} = \alpha\overline{X}(\xi) \tag{B.5}$$

by the terms proportional to α^2, then expression (B.2) is an exact solution of the original equations (B.1). Therefore, the equality $x = \xi$ can be accepted as the first approximation, when ξ is the solution of equations (B.5), satisfying equations (B.1) up to the second order of smallness. Note, that the expression (B.2), satisfying the equation (B.4), is called the improved first approximation.

Thus, the first approximation equations (B.5) can be obtained from the exact equations (B.1) by averaging over the variable s, while ξ is constant. This formal process, consisting in the replacement of exact equations by averaged ones, is called the averaging principle. The essence of this approach was revealed by N. Bogolyubov and Y. Mitropolsky, who showed that the averaging method is associated with the existence of a certain change of variables, which allows to exclude the variable s from the right-hand sides of the equations with any degree of accuracy relative to the small parameter α [1]. Thus, one can construct not only the first approximation system (B.5), but also to find higher-order averaged systems whose solutions approximate solutions (B.1) with an arbitrarily specified accuracy, although only the first approximation can practically be effectively used due to the rapid complication of the higher order formulas. The proof that the error of the first approximation is small was also obtained in [1], where it was found that under very general conditions the difference $x(s) - \xi(s)$ can be made arbitrarily small for a sufficiently small α at arbitrarily large finite interval $0 < s < l$. Thus, the following averaging theorem holds [1, 2].

Theorem *Let the continuous function* X(s, x) *be defined in the domain* Q *(*$S \geq 0$, $x \in D$*) where the following conditions hold:*

1. *the function* $X(s, x) \in \mathrm{Lip}_x(\lambda, Q)$, *i.e., it satisfies the Lipschitz condition over* x *with constants* λ;
2. *the inequality* $\|X(s, x)\| < M$ *holds, i.e.* X(s, x) *is bounded function;*
3. *there exists the limit* (B.3) *for every point* $x \in D$;
4. *the solution* $\xi(s)$ *of the averaged system* (B.5) *is defined for all* $S \geq 0$; *it lies in the domain D within a* ρ *neighborhood.*

Then for any arbitrarily small $\eta > 0$ *and arbitrarily large* $L > 0$, *such* α_0 *can be specified that if* $0 < \alpha < \alpha_0$, *the inequality*

$$\|x(s) - \xi(s)\| < \eta,$$

holds when $0 \leq s \leq L\alpha^{-1}$. *The functions* $x(s)$ *and* $\xi(s)$ *are the solutions of systems* (B.1) *and* (B.5) *coincide when* $s = 0$.

This theorem can be generalized for systems of integral–differential equations in standard form

$$\frac{dx}{ds} = \alpha X\left(s, x, \int_0^s \phi\left(s, s', x\left(s'\right)\right) ds'\right), \tag{B.6}$$

for which the following averaging scheme is possible [1]. First let us compute the integral

$$\psi(s, x) = \int_0^s \phi\left(s, s', x\right) ds' \tag{B.7}$$

over the explicit variable s‘, while s and x are considered to be parameters. Then, along with (B.6), let us consider the system of differential equations

$$\frac{dy}{ds} = \alpha X(s, y, \psi(s, y)), \tag{B.8}$$

which can be averaged if the limit

$$\lim_{l \to \infty} \frac{1}{l} \int_0^l X(s, y, \psi(s, y)) ds = \overline{X}(y) \tag{B.9}$$

exists. Then, the following system of differential equations can be obtained

$$\frac{d\xi}{ds} = \alpha \overline{X}(\xi), \tag{B.10}$$

which is averaged system, corresponding to the system of integral–differential equations (B.6). Thus, the averaging in the systems (B.6) consists in approximating the solutions of this system by that of the differential equation system (B.10). Obviously, this system can be solved much easier than the original system of integral–differential equations. The conditions under which the solutions (B.6), (B.8) and (B.10) are close can be found in the monograph [2]. In contrast to the differential equation system, the systems of integral–differential equations allow various averaging options. In the general case, one system of integral–differential equations can be associated with several different systems of averaged equations. Some of these averaged systems can be the differential equation systems, while the others represent the systems of integral–differential equations. The possibility of choosing the most acceptable averaged system determines the high efficiency of the averaging method for solving applied problems.

Consider the following system of integral–differential equations

$$\frac{dx}{ds} = \alpha X\left(s, x, \frac{dx}{ds} \int_0^s \phi\left(s, s', x\left(s'\right), \frac{dx\left(s'\right)}{ds'}\right) ds'\right) \tag{B.11}$$

which is a standard system not resolved relative to the derivative. In practice, an attempt to resolve a system relative to the derivative $\frac{dx}{ds}$ often encounters a need to perform cumbersome and time-consuming calculations, therefore, appropriate averaging schemes have been developed to avoid this drawback [2]. For example, in the first approximation, instead of (B.11), we can consider a simplified system

$$\frac{dy}{ds} = \alpha X\left(s, y, 0, \int_0^s \phi(s, s', y(s'), 0)ds'\right), \tag{B.12}$$

since the derivatives on the right-hand side of (B.11) begin to manifest itself in the second and subsequent approximations, as well as in the improved first approximation. Under sufficiently general conditions and small α, solutions (B.11) and (B.12) are arbitrarily close on an interval of $L\alpha^{-1}$ order.

The systems of differential and integral–differential equations allow various variants of partial averaging. For example, only some terms or separate equations in the system can be averaged [2]. In particular, if the source system is representable as

$$\frac{dx}{ds} = \alpha X_1(s, x) + \alpha X_2(s, x) \tag{B.13}$$

and the limit

$$\lim_{l\to\infty} \frac{1}{l}\int_0^l X_1(s, x)ds = \overline{X}_1(x), \tag{B.14}$$

exists, then the partially averaged system can be put in correspondence to the system (B.14)

$$\frac{d\xi}{ds} = \alpha\overline{X}_1(\xi) + \alpha X_2(s, \xi). \tag{B.15}$$

The partial averaging schemes are very diverse; they can be also applied to the equation systems (B.6) and (B.11).

References

1. Bogolyubov NN, Mitropolsky YA (1961) Asymptotic methods in the theory of nonlinear fluctuations. Internet Monog Adv Math Phys. Gordon and Breach, New York
2. Philatov AN (1974) Asymptotic methods in the theory of differential and integral-differential equations. PHAN, Tashkent (in Russian)

Appendix C

Surface Impedances of Electrically Thin Vibrators

Formulas determining the distributed surface impedance of electrically thin vibrators (material parameters are ε, μ, σ) have the following form.

M. V. Nesterenko et al., *Vibrator and Slot Antenna Arrays for Modern Applications*, Lecture Notes in Electrical Engineering 1244,
https://doi.org/10.1007/978-3-031-68346-6

No	The vibrator design	Vibrator model	Impedance
1	Solid metal cylinder. The radius satisfy inequality $r >> \Delta^0$, Δ^0 is skin layer thickness.		$\overline{Z}_S = \frac{1+i}{120\pi\sigma\Delta^0}$
2	Metallized dielectric cylinder. Metal layer thickness is $h_R << \Delta^0$		$\overline{Z}_S = \frac{1}{120\pi\sigma h_R + ikr(\varepsilon - 1)/2}$
3	Metal-dielectric cylinder. L_1 is the thickness of a metal disc, L_2 is the thickness of a dielectric disk.		$\overline{Z}_S = -i\frac{L_2}{L_1 + L_2}\frac{2}{kr\varepsilon}$
4	Magnetodielectric metalized cylinder. r_i is the radius of internal conducting cylinder.		$\overline{Z}_S = \frac{1}{120\pi\sigma h_R - i/kr\mu \ln(r/r_i)}$
5	Metal cylinder coated with magnetodielectric layer, which thickness is $r - r_i$, or corrugated cylinder $(L_1 + L_2) << \lambda$, where L_1 is crests thickness where $\overline{Z}_S = 0$, L_2 is the notch width where $\overline{Z}_S \neq 0$	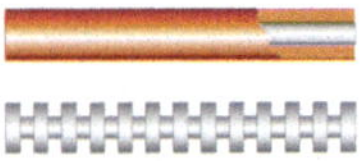	$\overline{Z}_S = ikr\mu \ln(r/r_i)$
		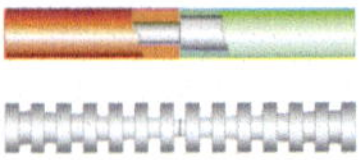	$\overline{Z}_S(s) = \overline{Z}_S \phi(s)$
6	Metal monofilar helix. r is helix radius $kr << 1$, ψ is winding angle		$\overline{Z}_S = (i/2)kr\,\mathrm{ctg}^2\psi$

Formulas for surface impedances of vibrators where derive in the frame of the impedance concept [1] and are valid for thin cylinders $|(k\sqrt{\varepsilon\mu}r)^2 \ln(k\sqrt{\varepsilon\mu}r_i| << 1$ both for finite and infinite cylinders, located in the hollow electrodynamic volume. If vibrators are in a material medium with parameters ε_1 and μ_1, all above formulas must contain the factor $\sqrt{\mu_1/\varepsilon_1}$.

Reference

1. Miller MA, Talanov VI (1961) Using the concept of surface impedance in the theory of sur-face electromagnetic fields (review). Izv Vuz Radiophysics 4(5):795–830 (in Russian)

Appendix D

Electromagnetic Values in CGS and SI Systems of Units

Because we use the CGS system in the book, it is expedient to make brief comparison of main ratios of the electromagnetic theory, represented in the SGS and SI systems.

CGS system of units	SI system of units
Maxwell's equations	
$\mathrm{rot}\,\vec{E} = -\frac{1}{c}\frac{\partial \vec{B}}{\partial t}$,	$\mathrm{rot}\,\vec{E} = -\frac{\partial \vec{B}}{\partial t}$,
$\mathrm{rot}\,\vec{H} = \frac{1}{c}\frac{\partial \vec{D}}{\partial t} + \frac{4\pi}{c}\vec{j}$,	$\mathrm{rot}\,\vec{H} = \frac{\partial \vec{D}}{\partial t} + \vec{j}$,
$\mathrm{div}\vec{D} = 4\pi\rho$,	$\mathrm{div}\vec{D} = \rho$,
$\mathrm{div}\vec{B} = 0$,	$\mathrm{div}\vec{B} = 0$,
where the field sources are:	where the field sources are:
$\vec{j} = ne\vec{v}$, n is the electron density,	$\vec{j}$ is the electric current density in *ampere per square meter* (A/m^2);
$\vec{v}$ is their velocity, e is the charge in CGS units; $\rho = ne$ is the electric charge density,	ρ is the electric charge density in *coulombs per cubic meter* (C/m^3).

$$1 \text{ unit of electric charge in CGS system} = \frac{1}{3\cdot 10^9}\ \textit{coulombs},$$

$$1 \text{ unit of electric current density in CGS system} = \frac{1}{3\cdot 10^9}\ \text{A/m}^2.$$

Constitutive equations:

CGS	SI
$\vec{D} = \varepsilon\vec{E}$,	$\vec{D} = \varepsilon_0\varepsilon\vec{E}$,
$\vec{B} = \mu\vec{H}$,	$\vec{B} = \mu_0\mu\vec{H}$,

where the permittivity and permeability in CGS system do not have dimensions and equal to the relative permittivity and permeability ε and μ in the SI system. ε_0 and μ_0 are the corresponding permittivity and permeability of vacuum here.

M. V. Nesterenko et al., *Vibrator and Slot Antenna Arrays for Modern Applications*, Lecture Notes in Electrical Engineering 1244,
https://doi.org/10.1007/978-3-031-68346-6

The electric and magnetic field intensities in the CGS system have the same dimension $g^{1/2}cm^{-1/2}\,sec^{-1}$, but intensity of the magnetic field is called *oersted* (Oe).

1 Unit of Electric Field Intensity in CGS system = 3.10^4 V/m,

$$1\cdot \mathrm{Oe}=\frac{1}{4\pi}\cdot 10^3\mathrm{A/m},$$

The electric and magnetic inductions ($\vec{D}$ and $\vec{B}$) in CGS system have dimensions of the intensities of corresponding fields. They have different dimensions and different names in SI system.

In SI system the electric induction is measured in *coulombs per square meter* (C/ m^2), at this

1 Unit of the Electric Induction in CGS system = $\frac{1}{12\pi}\cdot 10^{-5}$ C/m^2, and the magnetic induction amount is measured in *webers per square meter* = 1 T (T),

1. *Gauss* (Gs) in CGS system = 10^{-4} T.

The potentials are defined nearly alike in both systems (the potentials of only electric type are represented here).

$$\vec{H}=\frac{1}{\mu}\mathrm{rot}\,\vec{A},\qquad \vec{B}=\mathrm{rot}\,\vec{A},$$
$$\vec{E}=-\mathrm{grad}\varphi-\frac{1}{c}\frac{\partial\vec{A}}{\partial t},\ \vec{E}=-\mathrm{grad}\varphi-\frac{\partial\vec{A}}{\partial t},$$

with the Lorentz's condition.

$$\frac{1}{c}\frac{\partial\varphi}{\partial t}+\mathrm{div}\vec{A}=0,\qquad \frac{1}{v^2}\frac{\partial\varphi}{\partial t}+\mathrm{div}\vec{A}=0,$$

where v is the phase velocity of the wave in the corresponding medium.

The wave equations also differ slightly:

$$\Delta\varphi-\frac{1}{v^2}\frac{\partial^2\varphi}{\partial t^2}=-\frac{4\pi\rho}{\varepsilon},\ \Delta\varphi-\frac{1}{v^2}\frac{\partial^2\varphi}{\partial t^2}=-\frac{\rho}{\varepsilon_0\varepsilon},$$
$$\Delta\vec{A}-\frac{1}{v^2}\frac{\partial^2\vec{A}}{\partial t^2}=-\frac{4\pi\mu}{c}\vec{j},\ \Delta\vec{A}-\frac{1}{v^2}\frac{\partial^2\vec{A}}{\partial t^2}=-\mu_0\mu\vec{j}.$$

and the phase velocity of the plane wave equals

$$v=\frac{c}{\sqrt{\varepsilon\mu}}$$

in the first case and

$$v=\frac{1}{\sqrt{\varepsilon_0\mu_0}}\frac{1}{\sqrt{\varepsilon\mu}}=\frac{3\cdot 10^8[\mathrm{m/sec}]}{\sqrt{\varepsilon\mu}}$$

in the second case because

$$\mu_0 = 4\pi \cdot 10^{-7}\left[\frac{\mathrm{H}}{\mathrm{m}}\right], \quad \varepsilon_0 = \frac{1}{36\pi}10^{-9}\left[\frac{\mathrm{F}}{\mathrm{m}}\right],$$

and *henry·farad* = *sec*^{2}. Thus the finite result is the same.

The wave equations solutions can be proposed in the form of general potentials in both cases, and they are written with the help of the Fourier components of these potentials in the form of:

$$\vec{E}(\vec{r}) = \left(\text{graddiv} + k^2\varepsilon\mu\right)\vec{\pi}(\vec{r}),$$

$$\vec{H}(\vec{r}) = \frac{ik}{w}\text{rot }\vec{\pi}(\vec{r}),$$

where in CGS system

$$k^2 = \frac{\omega^2}{c^2}, \; w = 1, \; \vec{\pi}(\vec{r}) = \frac{1}{i\omega}\int\limits_V \frac{\vec{j}(\vec{r}')}{|\vec{r} - \vec{r}'|}e^{-ik|\vec{r}|\vec{r}'|}d\vec{r}',$$

in SI system.

$k = \omega\sqrt{\varepsilon_0\mu_0}, \cdot w = \sqrt{\frac{\mu_0}{\varepsilon_0}}$ is the wave impedance of free space,

$$\vec{\pi}(\vec{r}) = \frac{1}{4\pi i\omega\varepsilon_0}\int_V \frac{\vec{j}(\vec{r}')}{|\vec{r} - \vec{r}'|}e^{-ik|\vec{r}-r'|}d\vec{r}'.$$

Thus the electromagnetic wave in free space in SGS system is designated as mutually orthogonal triplet of the vectors $\vec{E}$, $\vec{H}$ and $\vec{h}$ where $\vec{E}$ and $\vec{H}$ have equal dimension and value, and this very triplet consists already of the values, which have different names and dimensions—*volt/meter*, *ampere/meter* and the wave vector with dimension 1/*meter* in SI system. But the wave has simple physical meaning—a long line in this case.

The manufacturer's authorised representative in the EU is Springer Nature Customer Service Centre GmbH, Europaplatz 3, 69115 Heidelberg, Germany. If you have any concerns regarding our products, please contact ProductSafety@springernature.com

Printed and bound by CPI Group (UK) Ltd, Croydon, CR0 4YY
01/10/2026
02225970-0013